# modern
# plumbing

by

**E. Keith Blankenbaker**
Associate Professor of Education
Faculty of Industrial Technology, College of Education
The Ohio State University, Columbus, Ohio

**South Holland, Illinois**
# THE GOODHEART-WILLCOX COMPANY, INC.
**Publishers**

Library of Congress Cataloging in Publication Data

Blankenbaker, E. Keith
  Modern plumbing.

  Includes index.
  1.  Plumbing.      I.  Title.
TH6122.B52      696'.1      77—15954.
ISBN 0—87006—325—1

# INTRODUCTION

MODERN PLUMBING provides the basic information about the tools, materials, equipment, processes and career opportunities in the plumbing field.

MODERN PLUMBING is written in a simple language and is profusely illustrated. The hundreds of photographs, drawings and charts will make it easier for you to understand the many technical details. Each illustration is referred to in the copy. This will assist you to link written text and illustrations for a clear understanding.

MODERN PLUMBING includes the latest installation techniques in addition to recent developments in materials, fixtures and appliances. It covers both hand and machine tools, and supplies background knowledge necessary for vocational competence. Each chapter begins by stating objectives which will help you set your learning goals. Test Your Knowledge questions at the end of each chapter will enable you to check your progress.

Apprentices, vocational students, construction trade students and anyone interested in plumbing will find MODERN PLUMBING a valuable aid in learning how modern plumbing systems are designed, installed and maintained. Experienced plumbers who would like to review basic plumbing and/or study the recent developments in the plumbing field will find this book helpful.

E. Keith Blankenbaker

# CONTENTS

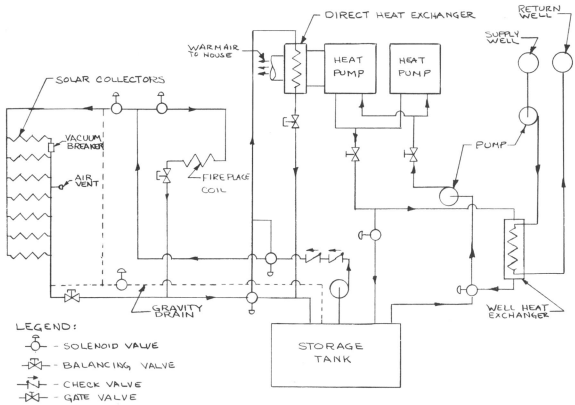

DIRECT HEAT EXCHANGER

WARM AIR TO HOUSE

RETURN WELL

SUPPLY WELL

HEAT PUMP

HEAT PUMP

SOLAR COLLECTORS

VACUUM BREAKER

AIR VENT

FIRE PLACE COIL

PUMP

GRAVITY DRAIN

WELL HEAT EXCHANGER

STORAGE TANK

LEGEND:
- SOLENOID VALVE
- BALANCING VALVE
- CHECK VALVE
- GATE VALVE

TECH HOUSE HEATING SYSTEM SCHEMATIC

Like this experimental solar-energized dwelling built at Langley Research Center, homes of the future will need extensive plumbing to circulate sun-heated water. Above. Rendering of home. Below. Schematic of heating system. Piping from collectors carries heated water to storage tank. When temperature of water drops too low, heat pumps extract heat from water in storage tank or from two wells.   (NASA)

6

# Unit 1
# PLUMBING TOOLS

## Objectives

This unit deals with the function and care of common plumbing tools.

After studying the unit you will be able to:
- Recognize and name each of the tools.
- Explain what each tool is designed to do.
- Select the proper tool in the proper size for the desired task.
- Explain and demonstrate how to keep tools sharp and in good repair.

In plumbing, as in other skilled trades, the plumber's ability and knowledge is closely tied to the tools used. Good tools in the hands of the skillful plumber turn out quality work. Poorly maintained or ill-adapted tools in the hands of the same plumber cannot produce the same quality of work.

## COMMONLY USED TOOLS

Simple plumbing jobs will ordinarily require only a few tools. But, to perform all operations that are part of plumbing work would require a considerable number of various types of tools. This unit will consider the tools most frequently used.

Some of the tools are needed in several sizes. In such cases, you will find guidelines for selecting the proper size for the job.

## MEASURING AND LAYOUT TOOLS

Instruments which measure lengths, heights, diameters, levelness or plumb are classified as measuring tools. Those which are used to produce accurate lines, circles or any other marking are called layout tools. *Plumbing dimensions must be accurate within fractions of an inch and the instruments must be capable of such accuracy over distances of several feet.* Tools the plumber will use include: rules, tapes, squares, levels, transits, plumb bobs, chalk lines, compasses and dividers.

## RULES

The folding wood rule, Fig. 1-1, is equipped with a metal sliding extension. This can be used to take accurate internal measurements. Being rigid, this type of rule can be extended and held above the head to measure heights. Thus, measurements can be taken by one person where a flexible rule would require two persons and a ladder.

Fig. 1-1. Folding wood rule can be carried in a pocket where it is always handy. It is sometimes called a "zigzag" or extension rule.

A plumbers' rule, Fig. 1-2, is a special type of folding rule. It has vertical markings on one side and a 45 deg. scale on the other. It is available in either 6 or 8 ft. lengths. Metric rules are sold in 1 and 2 metre lengths.

*Avoid dropping a folding rule on its end. The stress may loosen the joints enough to cause troublesome inaccuracies.* Even the slightest movement at each joint multiplies into fractions of an inch over several feet.

Dirt and repeated use will make folding the rule difficult. To prevent this problem, a small quantity of lightweight oil or silicone lubricant should be applied to the joints at regular intervals.

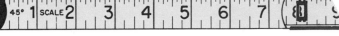

Fig. 1-2. Plumbers' folding rule. The side with vertical inch markings is shown at top. A 45 deg. scale is shown at bottom.
(Stanley Tools Div.)

## TAPES

Some plumbers carry a steel tape measure, Fig. 1-3, for its convenience. Because many of these rules will retract into their case at the push of a button, they can be quickly put

Fig. 1-3. Some steel tapes are retracted by a spring arrangement inside the case.

away with one hand. A hook on the end permits the tape to catch on the end or edge of a piece of stock so that it can be pulled out to make the measurement.

Most steel tapes used for taking measurements during the assembly of pipe and fittings are 6 to 10 ft. long, but units of 16 ft. or more are common. Metric tapes are either 2 or 3 metres long. These sizes are small enough to be carried in a pocket.

Frequently, steel tapes in 25, 50 and 100 ft. lengths, Fig. 1-4, are desirable for locating terminal points for pipe or for measuring the length of pipe required for long runs. Generally, the plumber prefers the 100 ft. size because of its greater capacity. Some steel tapes are marked in both English and metric. Long metric tapes are produced in 10, 15, 20, 25, 30 and 50 metre lengths.

Fig. 1-4. This 100 ft. steel tape is useful for measuring long runs of pipe.

## CARE OF TAPES

*Regardless of the length of the tape, it must be kept clean, dry and free from kinks if it is to provide good service.* Water and mud carried into the case when the tape is rewound can cause rust and damage to the rewinding mechanism.

During the process of winding and unwinding the tape, dirt, sand or other dry abrasive materials tend to wear away the numbers. In time, the tape may become unreadable. Bent tape is difficult to rewind and will not lie straight when extended. These problems can only be prevented if the plumber uses care when the tape is extended and wipes away water and dirt before rewinding. However, the design of the better quality steel tapes permits the replacement of the tape, when the original one becomes damaged. See Fig. 1-5.

Fig. 1-5. Replacement tapes for steel rules can be attached quickly to a metal tang inside the rules case. (Lufkin Div., The Cooper Group)

## SQUARES

Plumbers will find some type of square useful in these situations:
1. When locating the position of fixtures.
2. When marking framing members for cuts which will permit plumbing installation.

The type of square selected will depend on the type of work being done and the preference of the plumber. In any case, a try square, Fig. 1-6, a combination square, Fig. 1-7,

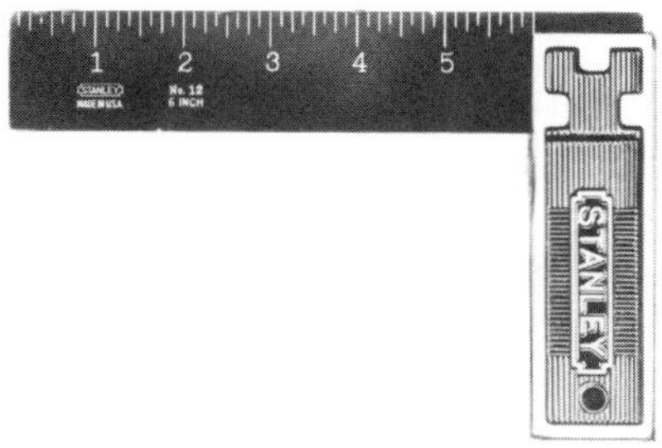

Fig. 1-6. Try square has 6 in. metal blade and metal or wood stock.

and/or framing square, Fig. 1-8, should meet the need.

The try square can be purchased with a 6 in. or 12 in. blade. Combination squares are equipped with a 12 in. blade which can be moved through a head. This head can measure a

Fig. 1-7. Combination square has sliding head and scriber for marking metal. (Stanley Tools Div.)

90 or 45 deg. angle. The framing square has a 24 in. blade and a 16 in. tongue.

*Use care in handling the square. Dropping it or hitting it hard enough could change the angle between the blade and the head or tongue. You will also need to protect it from rusting so that the scales will remain readable.*

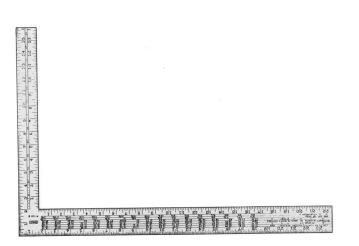

Fig. 1-8. Framing square is used for measuring, squaring and marking cuts to be made on walls and partitions.

## ALIGNMENT TOOLS

When installing pipe and plumbing fixtures, it is frequently necessary to determine if the part is vertical (plumb) or horizontal (level). Several tools are used for these purposes.

The level, Fig. 1-9, is used to check both positions. A good general purpose level has at least three vials. One vial will test

Fig. 1-9. General purpose level should have three vials. Bubble in appropriate vial centers when part being checked is level or plumb.

levelness, Fig. 1-10, when a parallel edge of the level is against the part. A second vial will test levelness when the other parallel edge of the level is against the object. The remaining vials test plumbness of an object regardless of which end of the level is up, Fig. 1-11.

Levels can be purchased in a number of lengths. The most

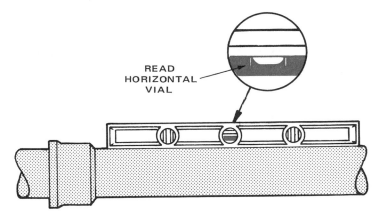

READ HORIZONTAL VIAL

Fig. 1-10. Testing horizontal alignment with level. Reading is taken from vial which is horizontal. (Stanley Tools Div.)

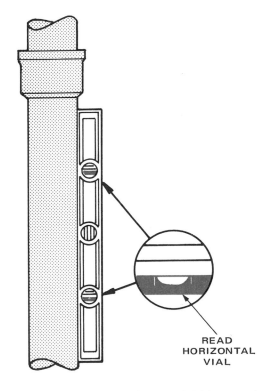

READ HORIZONTAL VIAL

Fig. 1-11. Pipe is plumb when bubble in horizontal vial is centered.

popular are the 2 and 4 ft. models. Generally, an aluminum or magnesium level is recommended for plumbers because it is less likely to be damaged by moisture.

A special plumbers' level, Fig. 1-12, has a movable vial that can be adjusted to measure the slope of a drain line. But many plumbers attach a block to the end of a regular level, see Fig. 1-13. Still others determine slope by reading the bubble in the vial slightly to one side of center.

Another leveling tool used by some plumbers is the line level, Fig. 1-14. By hanging this tool on a string line, it is possible to transfer vertical dimensions over distances without a transit.

*Levels should be handled carefully to prevent the vials from becoming broken.* When not in use, they can be stored where they will not be twisted, bent or forced from their own shape.

Fig. 1-12. This plumbers' level has one vial that can be tilted to measure amount of slope. Each mark on the scale represents 1/16 in. of slope per foot.

BLOCK TAPED
TO LEVEL

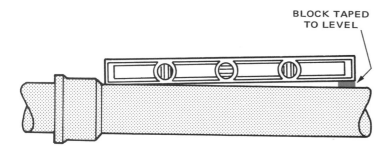

Fig. 1-13. Block placed under level tests slope of pipe. This is not as accurate as using a plumbers' level.

Fig. 1-14. Line level is sometimes used on a tightly stretched line to transfer heights from one distant point to another.

A builders' level, Fig. 1-15, measures elevations (vertical distances) and angles. Unit 11 discusses this tool in more detail.

## PLUMB BOB

With a plumb bob, Fig. 1-16, the plumber can accurately locate the center of vertical runs of pipe and transfer this point from one floor level to another. Though a simple tool, the plumb bob must be made with care if it is to function with accuracy.

In Fig. 1-17 you can see that the string line comes out of the center of the plumb bob, not out of the side. The point of the bob must hang directly below the string in a vertical plane. If canted at an angle the plumb measurement will be inaccurate.

Rounded or bent points on plumb bobs give inaccurate readings. It is desirable, therefore, that this part of the plumb bob be replaceable.

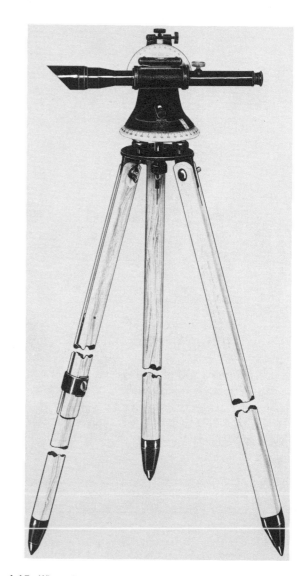

Fig. 1-15. When the leveling job is too big for the level or line level, a builders' level is used for greater accuracy. (The L.S. Starret Co.)

## CHALK LINE

A chalk line is useful in laying out long, straight lines on hard and rather smooth surfaces. The line, coated with chalk,

Fig. 1-16. The plumb bob must be well balanced and its string must be attached at exact top center.

is pulled taut between two points. Then it is carefully snapped against the surface producing a straight line of chalk.

Three precautions must be observed for accurate, clearly visible markings:

1. The line must not be allowed to catch on some object between the two points.
2. It must be stretched tightly.
3. The line must be lifted vertically from the surface on which the chalk mark is to be made and then released.

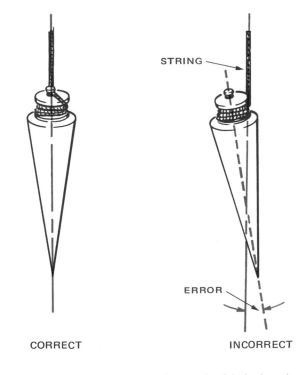

CORRECT          INCORRECT

Fig. 1-17. If string is incorrectly attached to plumb bob, the point will be deflected and reading will be inaccurate.

## CHALK BOX

In an older method of chalking, a piece of chalk is rubbed along the line. A more convenient method is the chalk box, Fig. 1-18. It not only stores the line but chalks it automatically. Powdered chalk or powdered water color pigments are placed inside the box. Chalking takes place as the string is pulled out. Tapping the chalk box as the line is uncoiling, will assure that the line is uniformly covered.

Maintenance includes adding powdered chalk as needed, reattaching the metal clip on the end of the line when the line becomes worn and, on occasion, replacing the string line.

*Because the chalk is water soluble, it is necessary to keep the chalk box dry. Failure to do this results in a clogged chalk box which may be all but impossible to clean.*

Fig. 1-18. Chalk box stores and recoats line between each use. (Stanley Tools Div.)

## COMPASS AND DIVIDERS

Laying out circles and arcs requires a compass or divider, Fig. 1-19. There is a difference in these tools. The compass has a pencil in one leg, whereas, the divider has two metal points.

Each tool has advantages. The pencil mark made by a compass is more easily seen on wood and other light colored materials. But the line scratched in the surface of metals by a divider is more permanent. The divider has the additional advantage of not requiring frequent sharpening.

When it is necessary to sharpen the metal point on the compass, or both points on the divider, the metal should be removed from the outside, Fig. 1-20. *If the compass or divider is to work correctly, the two legs must be the same length.* On the compass this is done by adjusting the pencil each time it wears down. Before sharpening a divider, it is necessary to grind the legs to the same length.

Fig. 1-19. Left. Compass is preferred for marking soft or light-colored surfaces. Right. Dividers have two sharp metal points. It is used for marking hard and smooth surfaces such as metal.

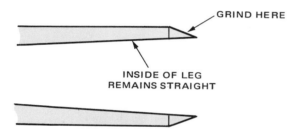

GRIND HERE

INSIDE OF LEG
REMAINS STRAIGHT

Fig. 1-20. To preseve their accuracy, dividers should be sharpened only on the outside of the legs.

## TOOTH-EDGED CUTTING TOOLS

Installation of plumbing requires that the plumber make some alterations to the structure so that pipes can be passed into walls and through roofs and floors. This calls for the use of certain toothed cutting tools.

### SAWS

The SABER SAW, Fig. 1-21, can cut both straight and curved lines in wood and other relatively soft materials including gypsum board. The length of the blade limits the thickness of material it can cut.

Where material thicker than 1 1/2 in. must be cut, the reciprocating saw, Fig. 1-22, is used. With the proper blade, it will also cut pipe.

Sometimes the use of electric power tools is not possible or practical. Then the COMPASS SAW, Fig. 1-23, works well in cutting holes greater than 1 in. in diameter. Because the blade of this saw tapers to a point, it will cut a smaller radius nearer the point. On a large radius, cutting is done with the wider part of the blade nearest the handle.

*The narrow blade can be easily bent if too much pressure is exerted when sawing. Like all saws, it should be stored where the blade will not be damaged or teeth dulled by falling tools or materials.*

Fig. 1-21. Sabre saw will cut openings in materials of less than 1 1/2 in. thickness. (Skil Corp.)

Fig. 1-22. Reciprocating saw, being larger, will cut through thicker material than will the sabre saw.

Fig. 1-23. Compass saw cuts large curves or circles.

HACKSAWS, Fig. 1-24, are the all-purpose tool for cutting metal. Plumbers keep them in their tool box for occasional use in cutting galvanized and black iron pipe. However, this is not

Fig. 1-24. Hacksaw frame has D-handle design. (Stanley Tools Div.)

usually recommended. It is very difficult to produce a square cut and crooked cuts are hard to thread. Should it be necessary to use the hacksaw, install the correct blade. This will improve the quality of the work and lengthen the life of the blade.

Hacksaw blades, Fig. 1-25, are designed and manufactured for different uses. They differ in several respects:

1. Length. Both 10 and 12 in. blades are produced. The smaller is most used.

2. Flexibility. Heat-treating processes can harden all or only a part of a blade. Only the teeth of flexible back blades are hardened. The rest of the blade is soft to withstand

Fig. 1-25. Section of hacksaw blade shows how teeth are pitched forward for better cutting.

bending. It is generally the preferred type for cutting pipe.

3. Set of the teeth. All teeth are angled slightly from the vertical. This widens the cut (kerf) so the back of the blade will not bind and break. See Fig. 1-26. Teeth are set in three different patterns as shown in Fig. 1-27.

4. Coarseness. This refers to the number of teeth per inch. The thinner the material being cut, the finer the blade should be, Fig. 1-28. If teeth are too small they will clog with chips; if too large they catch on the edges of the metal and

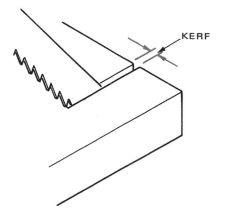

Fig. 1-26. Kerf made by hacksaw blade should be cut from the scrap part of the stock.

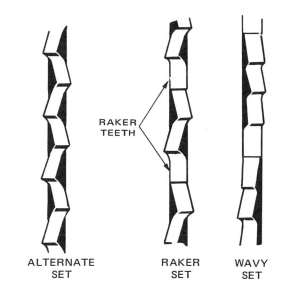

Fig. 1-27. Teeth on all hacksaw blades are bent at a slight angle off vertical. This enables the blade to produce a cut wide enough so that the rest of the blade does not bind or break. In the alternate set, every other tooth is bent at the same angle. In the raker set, every third tooth is left vertical to "rake" out cut material. In the wavy set, teeth are bent at varying angles from left to right so that the line created by the set weaves slowly from left to right.

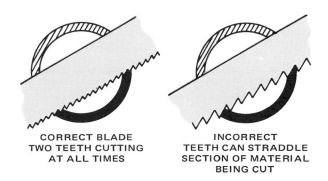

Fig. 1-28. Hacksaw blade should be right for the material being cut. Teeth at right are too coarse for the thin material.

may break off. As a general rule, blades with 32 teeth per inch are suitable for tubing. Blades with 24 teeth per inch work well on galvanized or black iron pipe.

A BACK SAW, mounted in a miter box is a good tool for cutting plastic pipe. See Fig. 1-29. The box insures a square cut that conforms well to the fittings. For smooth cuts use a back saw with 12 to 16 points (teeth) per inch.

## FILES

Another class of tool which has cutting teeth are files, Fig. 1-30. While they have a cutting action, their purpose is to remove small quantities of wood or metal while shaping and smoothing the material. Files have many different shapes, lengths, types of teeth and degrees of coarseness.

Fig. 1-31 shows the most common cross-sectional shapes. The shape selected will depend on the contour the user wishes to produce on the filed surface. For example, flat or convex (bulging) surfaces require a flat or mill file.

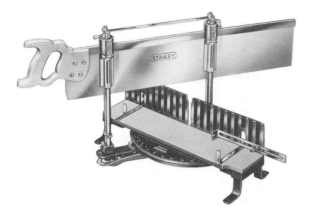

Fig. 1-29. Miter box guides back saw so that cutting is very straight.

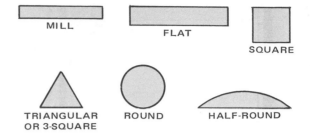

Fig. 1-31. Cross sections show various shapes of files a plumber may use. A contour may be selected to fit the surface being filed.

Teeth on a file are cut in one of several patterns, Fig. 1-32. A single-cut file is generally used for finish work on metal. A double-cut file removes material faster and produces a rougher finish. A rasp-cut file rapidly removes soft material such as wood. The curved-tooth file is preferred when working on soft metals such as aluminum.

Files are also designated by their degree of coarseness:

1. Coarse.
2. Bastard.
3. Second-cut.
4. Smooth.
5. Dead-smooth.

But coarseness is also related to length, Fig. 1-33. With experience, the plumber will learn the full range of teeth sizes for different lengths of files. For general purpose work, a 10 to 12 in. file with bastard or second-cut teeth is recommended.

## SMOOTH-EDGED CUTTING TOOLS

Often, a wood chisel is used, along with a handsaw, Fig. 1-34, to trim openings and make notches for pipe. One with a solid steel shank extending through the handle, as shown in Fig. 1-35, is best.

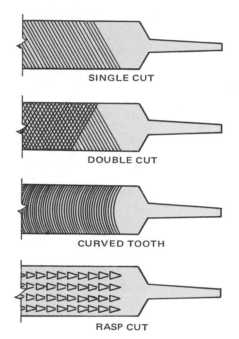

Fig. 1-32. Files can also be identified by the kinds of teeth.

Like all cutting tools, the chisel must be sharp if it is to work well. It will have to be ground occasionally at 25 deg. angle, Fig. 1-36, to remove excess metal. Frequent honing will produce a keen cutting edge. Honing should be done on the beveled side of the chisel, Fig. 1-37. Use a medium and then a

Fig. 1-30. All of the above files are useful in plumbing. Length is measured from heel to point.   (American Saw and Mfg. Co.)

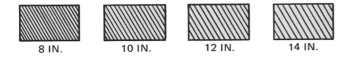

Fig. 1-33. There is a relationship between the length and degree of coarseness of files. The longer the file the coarser the teeth.

## COLD CHISEL

Another useful hand tool is the cold chisel, Fig. 1-38. This term covers several variations such as flat chisel, cape, round nose and diamond point chisel. Most useful to the plumber is the general purpose flat chisel. Cast-iron pipe may be cut this

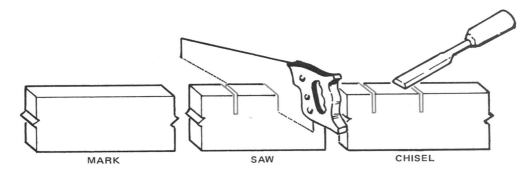

Fig. 1-34. Notches can be cut in studs or joists with wood chisel and hand saw.

Fig. 1-35. Wood chisel has solid steel shank which extends through handle. Metal cap provides striking surface for mallet. Blades range from 1/4 in. to 2 in. wide.

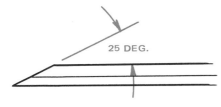

Fig. 1-36. Cutting edge of chisel should be ground to a 25 deg. angle.

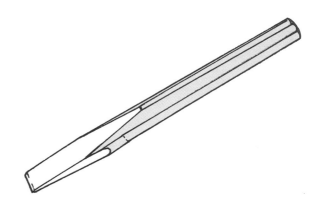

Fig. 1-38. Cold chisel is made of heat-treated steel.

way. The cold chisel is ground to a blunt edge as shown in Fig. 1-39. This cutting edge should be resharpened occasionally. *The head should be ground when necessary to remove the mushrooming caused by hammering.* See Fig. 1-40.

## PIPE CUTTER

A more sophisticated tool is the pipe cutter shown in Fig. 1-41. It has four movable parts, including a cutter wheel, two guide wheels and an adjusting screw. Used properly, it remains serviceable for a long time. However, the cutter wheel will eventually need to be replaced. A dulled cutter tends to crush rather than cut the pipe. A slight amount of lubricating oil applied to the screw and the cutting wheel should be a part of the preventive maintenance.

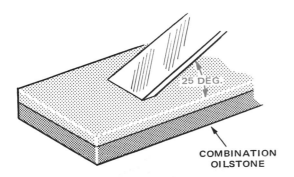

Fig. 1-37. Wood chisel can be honed on an oilstone. The 25 deg. angle used in grinding should be maintained.

fine oilstone to remove the burr raised by the honing. This does not change the angle of the cutting edge.

To produce a polished edge and to remove any remains of the burr which often forms during the honing operation, strop the chisel on leather, coated with buffing compound such as "tripoli."

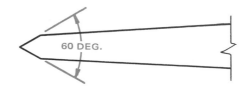

Fig. 1-39. Cutting edge of cold chisel is ground to a 60 deg. angle.

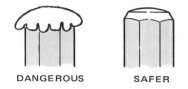

Fig. 1-40. Removing the mushroomed head by grinding reduces danger of injury from flying steel particles.

Fig. 1-43. Ratchet brace can be set to bore holes where the tool cannot be turned full circle. (Stanley Tools Div.)

## SOIL PIPE CUTTER

The soil pipe cutter pictured in Fig. 1-42, is similar to the pipe cutter. It is much faster than the chisel in cutting cast-iron pipe. As the chain is drawn tight and the tool rotated slowly, the cutters are forced into the walls of the pipe forcing it to break cleanly.

moved only a part of a turn. This feature is absolutely essential to a plumber. Square-tanged auger bits are the most useful tool for boring holes with a ratchet brace.

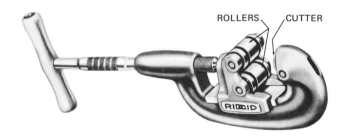

Fig. 1-41. Pipe cutter has threaded handle which moves guide wheels tightly against the pipe being cut. This action helps cutter wheel bite into and cut the metal.

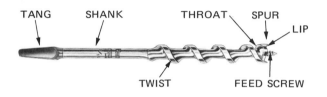

Fig. 1-44. Auger bits are available in diameters from 1/4 in. to 1 in. They are used in the ratchet brace.

## PORTABLE DRILL

The plumber should also have a portable electric drill, Fig. 1-46 with a 1/2 in. chuck. This will be used to bore openings in wood for pipe or to provide a starting point for the compass saw. It can be fitted with multispur bits as shown in Fig. 1-47, a spade bit as in Fig. 1-48, or a hole saw, Fig. 1-49.

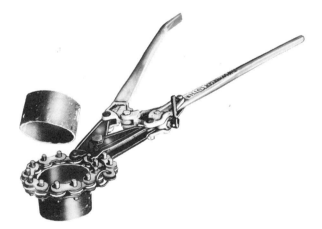

Fig. 1-42. Soil pipe cutter has cutting roller at every link. (Wheeler Mfg. Corp.)

## DRILLING AND BORING TOOLS

Certain boring tools are useful for making holes in wooden structural parts where plumbing is being installed. A ratchet brace, Fig. 1-43 and the auger bit, Fig. 1-44, are useful hand tools.

## RATCHET BRACE

As shown in Fig. 1-45, the ratchet brace allows its operator to drill a hole next to a wall even though the handle may be

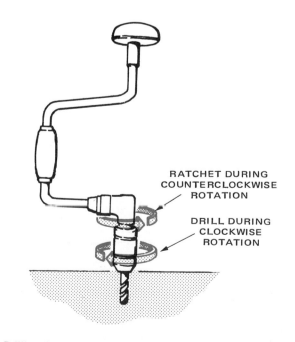

Fig. 1-45. Drilling in corners is possible with ratchet brace when turning radius is limited.

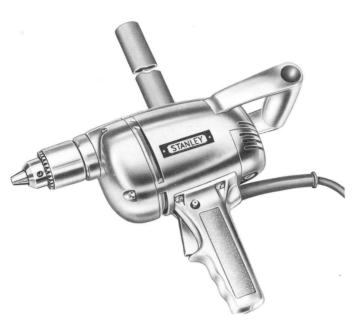

Fig. 1-46. Portable electrical drill speeds up boring of holes.

Fig. 1-47. Multispur bit is generally made with 1/2 in. round shank. It bores large holes in hard and soft wood at any angle. (Milwaukee Electric Tool Corp.)

Fig. 1-48. Spade bit cuts clean, smooth holes. (Stanley Tools Div.)

Fig. 1-49. Hole saw combines drill bit and cylindrical saw blade.

Some plumbers prefer to work with the offset portable electric drill shown in Fig. 1-50. It is simpler to use when drilling holes close to walls or between joists and studs.

## BITS

The multispur bit is a fast-cutting tool which can be used for boring 1/2 in. to 2 in. diameter holes. The spurs must be sharpened with a three-square (triangular) file.

Spade bits produce holes from 3/8 in. to 1 1/2 in. in diameter. They are easily sharpened with a mill file.

## HOLE SAW

The hole saw is designed for cutting holes from 1 to 3 1/2 in. in diameter. There are two different types available. One has a disc-shaped head with concentric circular grooves on one side and a shank on the other side. (Concentric means all circles have a common center.) The shank attaches to the drill. The grooves hold a saw-edged band which does the cutting as the drill turns.

Fig. 1-50. Offset portable electric drill bores hole where space is limited.

To set up this hole saw, the operator inserts the band of the desired diameter in the correct circular groove. It is held in place with lock screws.

This tool, while cheaper, is limited in the depth of cut it will make. The hole saw shown in Fig. 1-51, is made from a single stamped cylinder of steel. It is more expensive but is generally preferred by the plumber. It works better, lasts longer and is easier to use.

## REAMING AND THREADING TOOLS

Reaming the end of a pipe removes the burr formed inside when the pipe is cut. This operation is shown in Fig. 1-52. If not removed, the burr collects deposits that obstruct the flow of water. A pipe reamer, Fig. 1-53, does this job well.

Fig. 1-51. Hole saw with cylindrical blade threaded onto the head works best for large holes.

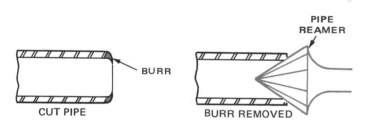

Fig. 1-52. Cutters are on tapered face of the burring reamer. Because of its taper, reamer can be used on any size pipe.

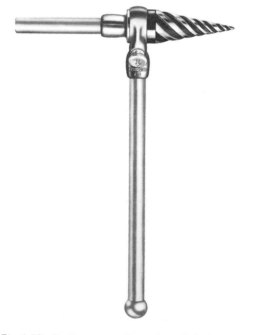

Fig. 1-53. Burring reamer shown has spiraled cutters. (The Rigid Tool Co.)

## DIES

Before galvanized pipe can be assembled in the plumbing system, the ends of the pipe may need to be threaded. For this job, the plumber needs special dies. Pipe dies, Fig. 1-54, cut correctly tapered threads for each of the standard pipe sizes. These sizes range from a 1/8 in. to a 4 in. diameter. The dies must be sharp so that they will cut metal rather than push it around. Dies which push the metal off instead of cutting freely cause threads to break out of the die. *Whenever threads are cut, cutting oil should be applied to reduce friction and heat.*

## DIE STOCKS

Die stocks, Fig. 1-55, are required to turn the dies. The ratchet style die stock is preferred because it permits the worker to use body weight to rotate the die while standing to one side of the pipe.

Fig. 1-54. Pipe die consists of holder and cutters. (Toledo Beaver Tools, Inc.)

## TOOLS FOR ASSEMBLING AND HOLDING

The plumbers' tool box must include a variety of wrenches to turn pipes, fittings and fasteners found in today's plumbing systems. Some pipe wrenches must grip finished surfaces which would be ruined by jaw marks. Others must hold pipes in circumstances where only sharp jaws will do the job.

Still other wrenches with smooth jaws are needed to turn fittings and fasteners such as studs, bolts, spuds, slip nuts and packing nuts. Markings on these parts would deform them or ruin their appearance.

In some situations devices are needed to hold plumbing parts while the plumber performs operations on them. These tools are called vises.

## WRENCHES

The pipe wrench, Fig. 1-56, is used to hold or turn threaded pipe during assembly. At least two pipe wrenches will be

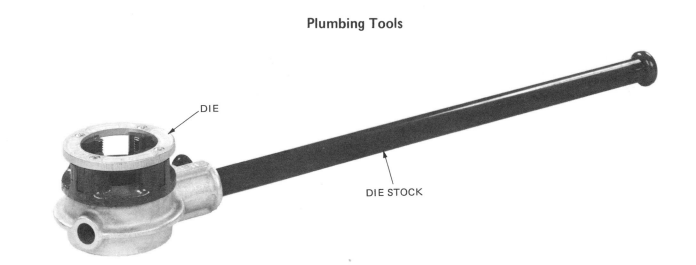

Fig. 1-55. Die stock has a head which holds die securely and a handle to give leverage for turning.

Fig. 1-56. Pipe wrench has heavy toothed jaws for gripping pipe.

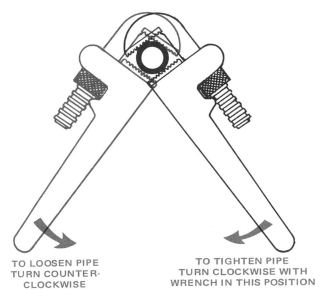

| TO LOOSEN PIPE TURN COUNTER-CLOCKWISE | TO TIGHTEN PIPE TURN CLOCKWISE WITH WRENCH IN THIS POSITION |

Fig. 1-58. Open side of jaws should face in same direction as the force exerted on the handle.

required. Fig. 1-57 indicates the pipe wrench lengths most suitable for various size pipes. Jaws should be adjusted so that the teeth will grip the pipe firmly without crushing it. Fig. 1-58 indicates the correct method of attaching the wrench so that it will grip the pipe or fitting.

*Oil should be applied to the adjustment nut at regular intervals to prevent rusting.* The ability of the pipe wrench to grip pipe is directly related to the condition of the teeth. Cleaning the teeth and sharpening them with a three-square (triangular) file can restore some wrenches to usefulness.

## CHAIN WRENCH

Where diameters of more than 2 in. are involved, it is common practice to use a chain wrench, Fig. 1-59, to hold or rotate the pipe during assembly. *This wrench requires oiling at regular intervals to prevent the chain from becoming stiff and rusted.*

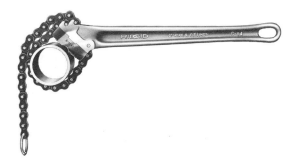

Fig. 1-59. Chain wrench is used on large diameter pipe.

## STRAP WRENCH

A strap wrench, Fig. 1-60, can be used to assemble chrome plated or other finished pipe. Such surfaces would be damaged by the teeth of a pipe wrench.

Often pipe fittings, valves and plumbing fixtures have hex

TABLE OF PIPE WRENCHES
SIZES AND CAPACITIES

| WRENCH LENGTH (INCHES) | PIPE DIAMETER (INCHES) | WRENCH CAPACITY (INCHES) |
|---|---|---|
| 6 | 1/8 - 1/4 | 3/4 |
| 8 | 1/8 - 1/2 | 1 |
| 10 | 1/2 - 3/4 | 1 1/2 |
| 12 | 3/4 - 1 | 2 |
| 14 | 1 - 1 1/2 | 2 |
| 18 | 1 1/2 - 2 | 2 1/2 |

Fig. 1-57. Use this guide for selecting the right wrench for the job.

Fig. 1-60. Strap portion of the strap wrench needs to be coated with rosin to prevent slipping on smooth-surfaced pipe.

(six-sided) or square shoulders which permit the use of a monkey wrench, open end wrench, or adjustable wrench. The monkey wrench, Fig. 1-61, looks like a pipe wrench but is different in two significant ways:
1. The jaws are smooth.
2. There is no provision for the jaws to tighten on the part being turned as pressure is applied to the wrench.

The monkey wrench is only useful for turning or holding objects which have flats. Because of its size and frequent difficulty in adjusting it to hold properly, many people choose another wrench.

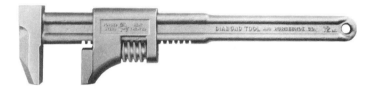

Fig. 1-61. Monkey wrench is like a pipe wrench with smooth jaws. (Diamond Tool and Horseshoe Co.)

## OPEN END WRENCHES

Open end wrenches, Fig. 1-62, are sold individually or in sets. The plumber will find the sizes from 1 in. to 1 3/4 in. useful when assembling pipe and fittings between diameters of 1/2 and 1 in. Metric sizes range from 6 mm to 32 mm in increments of 1 mm. Because the jaws of these wrenches are fixed, they are less likely to slip off a nut.

Fig. 1-62. Open end wrench is less likely to slip off the nut than adjustable jawed tool. (Proto Tool Co.)

## ADJUSTABLE WRENCHES

Adjustable wrenches, Fig. 1-63, are very popular since they can replace several different sizes of open end wrenches. They hold better on nuts and will fit into more places than the monkey wrench. However, they are less satisfactory for most jobs than an open end wrench of the correct size. Fig. 1-64 lists the capacity of the jaws for each of the wrench sizes.

Because of the relative weakness of the movable jaw, it is important that pressure be applied as shown in Fig. 1-65.

Fig. 1-63. Adjustable wrench handles nuts of all sizes up to its capacity.

### SIZE AND CAPACITY OF ADJUSTABLE WRENCHES

| SIZE (INCHES) | CAPACITY (INCHES) |
|---|---|
| 4 | 1/2 |
| 6 | 3/4 |
| 8 | 15/16 |
| 10 | 1 1/8 |
| 12 | 1 5/16 |
| 15 | 1 11/16 |
| 18 | 2 1/16 |
| 24 | 2 1/2 |

Fig. 1-64. Adjustable wrenches will adjust to sizes listed.

*Several drops of oil periodically applied to the adjustment screw will prevent rusting.*

## LOCKING PLIERS

Locking pliers, Fig. 1-66, are a multipurpose tool frequently used when taking apart old plumbing fixtures. Because of the tremendous clamping pressure which can be exerted

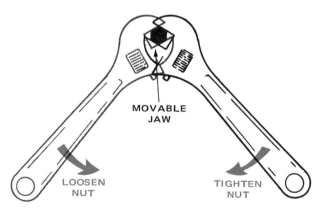

Fig. 1-65. Placing adjustable wrench on nut properly will prevent damage to the jaws.

Fig. 1-66. Locking pliers exert extraordinary clamping pressure.

Fig. 1-67. Basin wrench can fit into small recesses to turn nuts.

with this tool, it is often possible to hold or turn an object on which the square or hex shoulders have worn away. The locking pliers should be used only where the marks made by the hardened jaws will not create an unsightly fixture.

Because of the cramped working space, it is hard to use conventional wrenches on the nuts which secure faucets and some other plumbing equipment. The basin wrench, Fig. 1-67, was developed to solve this difficulty. The offset jaws of this tool permit the plumber to reach into a recess and turn a nut.

## HAMMERS

Two types of hammers should be included in the plumber's list of tools. The carpenters' hammer is used for driving or pulling nails and for tapping a wood chisel. The head is made of forged, hardened and tempered steel. The claw may be straight or curved; the face may be bell shaped or plain.

A ball peen hammer is often used for driving a cold chisel or a punch. Sizes are available in 4, 6, 8 and 12 oz. as well as 1, 1/2 and 2 lb. For heavier work the 12 oz. or 1 1/2 lb. are suitable. Lighter work is best done with a 4 or 6 oz. hammer.

There are other fastening tools occasionally useful to a plumber. Because of their special nature, they will be considered only in the units which relate to their use.

## VISES

The pipe vise, Fig. 1-68, is probably the most frequent used holding device a plumber has. Its hardened jaws permit it to firmly grip the pipe preventing the pipe from turning. Because the jaws tend to leave marks, their use is generally limited to work on pipe which will not be exposed in the finished structure.

A chain type pipe vise, Fig. 1-69, is also available. It serves the same purpose as the conventional pipe vise.

A bench vise equipped with pipe jaws, Fig. 1-70, may be used to secure pipe for cutting and threading. Generally, this type is less satisfactory than a pipe vise.

Among the assembly tools which are necessary for some types of plumbing work are solder pots, propane torches, gasoline torches, and oxyacetylene equipment. Because of the limited application of some of these tools and the space required to discuss them, they will be considered in Unit 10.

Fig. 1-68. Pipe vise holds pipe for other operations being performed.

Fig. 1-69. Chain type vise serves same purpose as regular pipe vise. (The Rigid Tool Co.)

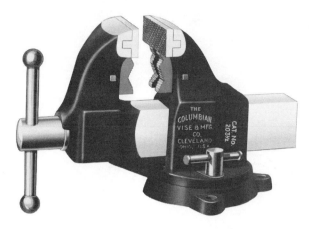

Fig. 1-70. Bench vise with special jaws will also hold pipe while the plumber works on it.

## TEST YOUR KNOWLEDGE — UNIT 1

1. Two types of rules commonly used by plumbers are the _____ _____ rule and the _____ tape.
2. Proper maintenance of a steel tape requires that it be kept _____, _____, and free from _____.
3. The type of square which has a movable head is called a combination square. True or False?
4. The term "plumb" means (circle letter for correct answer):
   a. Horizontal.
   b. 90 deg. angle.
   c. Vertical.
5. What do you call the part of a level which contains the bubble?
6. The tool used for laying out arcs and circles which has a pencil in one leg is called a:
   a. Compass.
   b. Divider.
   c. Protractor.
   d. Scratch awl.
7. For accurate work, it is necessary that both legs of either a compass or a divider be equal in length. True or False?
8. The following statement is concerned with portable electric tools. A _____ saw is designed to make curved cuts through thicker material than the _____ saw can handle.
9. List three tools which can be installed in a portable electric drill to bore holes in wood.
10. If electricity is unavailable, what two tools can be used together to bore holes?
11. As the thickness of the material being cut with a hacksaw increases, the number of teeth per inch on the hacksaw blade should _____.
12. When choosing a file, what characteristics must be considered?
13. Why is is necessary to ream the end of a pipe which has been cut with a pipe cutter?
14. The purpose of applying cutting oil when threading pipe is to reduce _____ and _____.
15. The minimum number of pipe wrenches required to assemble threaded pipe and fittings is three. True or False?
16. The _____ wrench is useful for reaching into recesses such as those found on sinks to turn nuts on such things as faucets.

## SUGGESTED ACTIVITIES

1. Prepare a basic list of tools for a plumber who installs residential plumbing systems. Study supplier catalogs or visit a local hardware store to obtain specific information about tools that could be purchased.
2. Prepare a brief oral or written report on the effect of one of the following on tool quality:
   a. Forging of the tool parts as opposed to casting.
   b. Heat-treating of the metal.

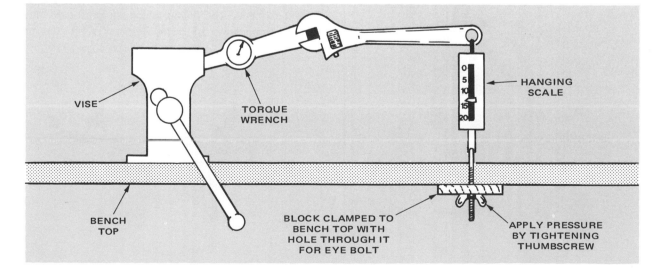

Fig. 1-71. This setup will measure torque developed on wrench handle.

c. High carbon content in steel.

d. High nickel content in steel.

e. Cross-sectional shape of levers (handles) in tools.

3. Compare the amount of torque which can be applied to a pipe or bolt with different length wrenches when specified amounts of pressure (weight) is applied to the wrench. The drawing, Fig. 1-71, illustrates a setup which could be used to conduct the test. Several students could conduct the same test using wrenches of varying lengths. The results can be reported in a chart similar to the one shown in Fig. 1-72.

The results of the separate test could be summarized in a graph, Fig. 1-73, for the purpose of making comparisions regarding the mechanical advantage of tools of different lengths.

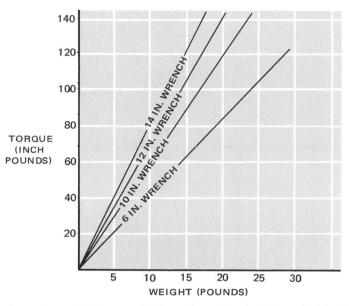

Comparison of the torque developed by varying weights applied to 6 in., 8 in., 10 in., and 12 in. adjustable wrenches.

Fig. 1-73. Sample graph shows comparison of results of several tests.

### TORQUE DEVELOP WITH AN 8 IN. ADJUSTABLE WRENCH

| WEIGHT (POUNDS) | TORQUE (INCH POUNDS) |
|---|---|
|  |  |
|  |  |
|  |  |
|  |  |
|  |  |
|  |  |

Fig. 1-72. Develop a chart like the above for recording torque data.

Builders' level is one of the precision tools the plumber may have to use in constructing building drains. Its use is discussed in Unit 11. (RTP Inc., New York)

23

# Unit 2
# SAFETY

## Objectives

This unit discusses safety attitudes and the practice of safe working habits.

After studying the unit you will be able to:
- Develop a list of general safety rules relating to: clothing, use of ladders, electrical tools and scaffolds, lifting of heavy weights and working with flammable materials.
- Explain why it is necessary to develop safe working habits.

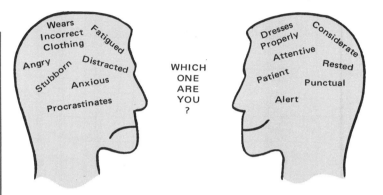

Fig. 2-1. Safety begins with proper attitudes and habits.

Accidents cause discomfort and inconvenience for the person injured. They also result in loss of income and increased medical expenses. Work-related accidents cost the United States an estimated $10 billion annually. For every million hours worked in construction, there are about 14 disabling accidents. This figure does not include the minor injuries which do not cause absence from work. Injuries received from falls or while handling objects account for nearly half the accidents resulting in loss of time from the job. Accidents which injure the hands and eyes, though frequent, are less likely to result in loss of time.

*Your safety and safety of other workers in the shop or on the construction site is directly related to the practice of safe work habits.* In this chapter, general safety practices will be discussed. As you study other units you will learn about specific safety practices that apply to certain tools and tasks.

Safe work habits are the result of attitudes formed by the worker, Fig. 2-1. You must know what is safe and what is not and you must be constantly on the alert in order to anticipate unsafe conditions or practices before they cause an accident. *Accidents do not just happen — they are the result of an unsafe act.*

## CLOTHING

Appropriate dress can reduce the possibility of accident or injury. The following suggestions will help you select your clothing:
1. Avoid trouser legs which drag on the floor or have cuffs that are too wide. They may trip the wearer and cause falls.
2. Check synthetic materials carefully. Some are highly flammable. Welding sparks or other high-temperature heat sources could set them afire.
3. Do not wear neckties and loose or torn clothing. They may catch in revolving machinery parts.
4. Gloves can protect your hands when handling pipe, fittings and fixtures. However, they should not be worn when operating power tools. They often interfere with operation of the controls and have a tendency to catch on revolving equipment.
5. Safety shoes protect your feet from falling objects. Wear them if possible.
6. Wear goggles when using chisels, grinders and other tools which produce chips and/or dust. If you wear corrective glasses, they should be made of safety glass and should be fitted with side shields. If you get foreign matter in your eye do not rub it.
7. Never wear rings and other items of jewelry. They may catch on moving parts and injure your hands and fingers.

## LIFTING

Lifting and carrying various materials incorrectly can cause serious injury to your spinal column. *When lifting objects, keep the back straight and use leg muscles, not your back muscles, to raise the object.* When large pieces of pipe are carried, hand slings make the job easier and safer, Fig. 2-2.

## HOUSEKEEPING

Good housekeeping means keeping the work area as clean and orderly as possible. This is everyone's responsibility. A

Fig. 2-2. Slings make carrying heavy pipe sections easier and safer.

scrap of pipe, an extra fitting or a forgotten tool can cause a bad fall. An additional benefit of good housekeeping is that materials and tools are more easily found. Thus, the job is done more efficiently and with less frustration.

Rubbish should be cleared away and disposed of regularly. If combustible wastes must be stored for any period of time, they should be contained in covered metal receptacles.

Use cleaning materials with care. Oil treated sawdust and other compounds used for cleaning are a fire hazard. Oily mops and rags should be stored in closed metal containers.

Water or oil spilled on the floor should be cleaned up immediately to prevent falls and to reduce the possibility of electrical shocks.

## LADDERS

Ladders are often necessary when running pipe. Certain precautions should be observed:

1. Check the ladder frequently for broken or damaged parts. Look for hidden splits, loose rivets, screws and rungs. Repair damage or discard the ladder.
2. Always face the ladder when ascending or descending.
3. Use a ladder long enough to permit you to reach the work without stretching your body beyond the top of the ladder.
4. Make certain all four legs of a stepladder are resting on a solid base. Never stand on the top platform.
5. Metal ladders are good conductors of electricity. Do not use them near electrical wiring.

6. Straight ladders require firm footing and should never be placed at such a steep angle that they are in danger of falling. If used to reach an upper level, the ladder should project well above the level, Fig. 2-3.
7. Never work higher than the third rung from the top of a straight ladder.
8. Do not use a ladder as a ramp or stairs. Avoid placing ladders at less than a 55 deg. angle. If set at an angle steeper than 75 deg., lash the top to the structure.
9. All ladders should support live loads of at least 200 lb. with a safety factor of 4 on the middle of a rung located midway in its length. A factor of 4 means that it should be able to support up to 800 lb. (4 x 200).
10. To raise or lower a ladder, place its base against the foundation or have someone hold the base at the ground. Then, walk towards the base lifting the ladder by moving the hands along the side rails. To lower, move hands along side rails as you walk away from the base.
11. Ladders should be equipped with safety feet, Fig. 2-4, that will anchor the base to the ground or concrete. This is especially important after a rain when the ground is slippery. Never attempt to place a ladder on loose or muddy ground. Use blocking to provide solid footing.
12. Secure portable ladders against movement. Place blocking strips at the bottom and secure the top by any appropriate means.
13. Never lash two ladders together to increase their length. Do not place them on boxes, barrels or other unstable bases to gain additional height.

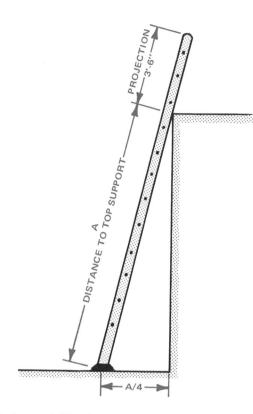

Fig. 2-3. Correct ladder placement prevents the ladder from slipping or tilting backward as you climb.

14. Maximum length of a single portable wooden ladder is 30 ft.; of a portable two-section extension ladder, 60 ft. Stepladders, trestle ladders and platform stepladders may not be longer than 20 ft.

15. Metal ladders may not exceed: 30 ft. for a single straight ladder; 48 ft. for a two-section extension ladder; 60 ft. for greater than two-section ladders. Stepladders, trestle ladders and platform stepladders must not exceed 20 ft. Stepladders must have a metal spreader which holds the front and back sections securely in an open position.

## GROUNDING ELECTRIC TOOLS

Many electrical tools are used by plumbers. They should be properly grounded. Grounded tools are equipped with a

Fig. 2-4. Cleats and special feet provide security against slipping.

three-prong plug, Fig. 2-5. The third prong connects to a ground wire which is attached to the housing of the tool. If the tool develops a short circuit and the housing becomes

Fig. 2-5. Grounded plug has third terminal. For your own protection, do not cut off the ground.

"hot," the electricity flows through the prong into a ground wire. The operator is unharmed, Fig. 2-6.

Plugging an adapter, Fig. 2-7, into a two-pronged receptacle and then plugging in the tool does not ground it. It is necessary to attach the pigtail on the adapter to the grounded conduit or to a specially installed ground wire if conduit is not used.

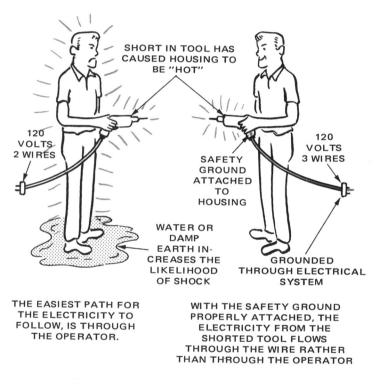

Fig. 2-6. Grounded tools provide safety for the operator.

## OPERATING POWER TOOLS

Use of any type of power tool calls for special precautions:
1. Use the correct tool for the job.
2. Know the proper uses, limitations and the hazards of each power tool on the job.
3. Remove chuck keys and all other adjusting tools before connecting to power.
4. Never use tools with frayed cords or loose and broken switches.
5. Keep area around stationary tools free of clutter.
6. Avoid using power tools in damp areas or near combustible

Fig. 2-7. This adapter provides safety only if the ground wire is correctly attached.

materials.

7. Do not distract anyone using a power tool.
8. Keep tool guards in place and in good working order.

## FLAMMABLE MATERIALS

Plumbers frequently use gasoline, propane, acetylene and other highly flammable materials. These must be handled carefully and should be stored where they will be protected from high temperature. Damage to compressed gas cylinders can cause leaks that result in disastrous explosions. Spilled flammable liquids are an extreme hazard. *Flammable materials should be placed in special containers and stored where they will not be upset.*

## SCAFFOLDS

If scaffolding is used on the job, the following precautions should be observed:

1. Footing or anchoring for all scaffolding should be capable of holding the intended loads without shifting or settling. Barrels, loose bricks, blocks or boxes should not be used to support scaffolds or planks.
2. Where platforms are over 10 ft. above ground or floor, guardrails and toeboards must be placed on all open sides and ends. Scaffolds 4 to 10 ft. high and less than 45 in. wide, must also be guarded with rails.
3. Scaffolds and scaffold parts must be able to support at least four times the maximum intended load. Wire or fiber rope used for scaffold suspension must be capable of supporting at least six times the intended load.
4. Planking or platforms must be overlapped at least 12 in. or be securely fastened against movement.
5. Planking must extend over end supports not less than 6 in. but no more than 18 in. (12 in. at construction sites) and should be fastened to prevent their falling.
6. Never place planks on top of guardrails to gain greater height.
7. Portable or free-standing scaffolding must not be higher than four times their base dimension. Locking devices must be provided on wheels to keep them from moving when in use.
8. Scaffolding should never be moved with workers or materials still on it.
9. Scaffolds must be maintained in a safe condition. Unsafe

scaffolds should be disposed of or repaired.

10. Scaffold uprights must be plumb and rigidly braced to prevent swaying or other movement.
11. Wheels must be secured with locking devices when workers are on the scaffold.
12. At least two of the four casters or wheels on rolling scaffolds must be the swivel type.

## SCAFFOLD PLANKING

On some jobs, metal scaffolds permit the plumber to work safely above the floor level. These scaffolds are easily assembled and are strong enough to support the weight of the plumber and a reasonable number of tools. Metal scaffold components should be inspected for damage as they are assembled.

Generally, the platform which the plumber stands on is made from 2 in. lumber. This lumber must be carefully selected and cared for so that it does not fail under the weight of the worker and the tools. Scaffold planks should be made of structural lumber which is free from knotholes and which has a grain running nearly parallel to the surface of the plank, Fig. 2-8. Because lumber splits more easily along the grain, parallel grain is stronger than grain which angles across the thickness of the plank. To prevent the scaffold planks from rocking and being misaligned, lumber must be selected which is free from bow, twist and other defects which would cause the platform to be unstable.

Because the moisture content of the lumber affects its ability to remain straight, only kiln dried lumber should be used in scaffolds. To prevent rapid changes in the moisture content of scaffold planks, keep them as dry as possible.

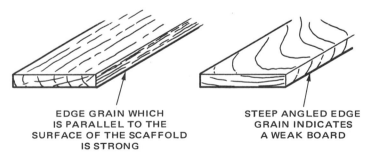

EDGE GRAIN WHICH IS PARALLEL TO THE SURFACE OF THE SCAFFOLD IS STRONG

STEEP ANGLED EDGE GRAIN INDICATES A WEAK BOARD

Fig. 2-8. Careful selection of scaffold lumber can prevent accidents.

OSHA (Occupational Safety and Health Act) regulations require that planking be of scaffold grade for whatever species of wood used. The maximum spans permitted for 2 x 10 in. or wider planks are shown in Fig. 2-9.

## EXCAVATING AND TRENCHING

Where excavations and trenching must be performed by the plumber or the plumber's helpers, underground utilities, if any, should be located and protected. Utility companies and/or other local regulatory agencies must be contacted. Necessary permits and approvals must be obtained before excavating is begun.

Walls and faces of excavations and trenches over 5 ft., which would expose workers to danger, must be guarded with shoring, sloping of the ground or some other equally effective

### MATERIAL

| | FULL THICKNESS UNDRESSED LUMBER | | | NOMINAL THICKNESS LUMBER | |
|---|---|---|---|---|---|
| WORKING LOAD (PSF) | 25 | 50 | 75 | 25 | 50 |
| PERMISSIBLE SPAN (FT.) | 10 | 8 | 6 | 8 | 6 |
| THE MAXIMUM PERMISSIBLE SPAN FOR 1 1/4 x 9 INCH OR WIDER PLANK OF FULL THICKNESS IS FOUR FEET, WITH MEDIUM LOADING OF 50 PSF. | | | | | |

Fig. 2-9. OSHA requirements for scaffold planking.
(National Institute of Occupational Safety and Health)

means. Trenches less than 5 ft. deep may require shoring or sloping if hazardous soil conditions are present, Fig. 2-10. Trench boxes or shields may be used in place of shoring or sloping. Tools, equipment and excavated soil must be kept 2 ft. away from the lip of the trench.

Such trenches and excavations must be inspected daily for safety so that slides and cave-ins are prevented. More frequent inspections are required as work progresses or after rainy weather.

Ladders or steps allowing entrance to trenches must have no more than 25 ft. of lateral travel in trenches 4 ft. or more deep.

Runways and sidewalks must be kept free of debris. If undermined, they must be shored to prevent cave-ins. Barricades and warning signs must be used to prevent falls.

### WELDING AND BRAZING SAFETY

Required safety practices in the use of welding and brazing equipment can be divided into general and specific activities. General safety regulations include:

1. Welding, cutting and brazing of materials, particularly when done in a shop, should be done in an area set aside for this purpose. The area should be open and far away from combustibles. If combustibles cannot be removed, guards should be placed to protect fire hazards from heat and sparks. Fire extinguishing equipment (pails, buckets of sand or portable extinguishers) must be kept on hand.

2. Operators must be trained for safe use of the equipment. Printed rules and safety instructions, covering operation of equipment, should be supplied and strictly followed.

3. No operations shall be performed using drums, barrels, tanks or other containers until they have been cleaned. They should be scrubbed thoroughly to make certain there are no flammable materials present nor any substances such as grease, tar or acid which, when heated, might produce flammable or toxic vapors.

4. The atmosphere in the welding area must be free of flammable gases, liquids and vapors.

5. Suitable eye protection such as goggles, helmets and hand shields must be used during cutting operations.

6. Workers next to welding areas must be protected from ultraviolet rays by noncombustible or flameproof screens, shields or goggles.

7. Protective clothing must be worn by all employees exposed to hazards from welding operations.

### USE OF VENTILATORS AND RESPIRATORS

Some welding and cutting operations will require special safeguards as follows:

1. Mechanical ventilation must be provided when welding or cutting is done where there is less than 10,000 cu. ft. of air per welder, where the ceiling is less than 16 ft. high, or when working in confined quarters.

2. Mechanical ventilation must be able to replace air at the rate of 2000 cu. ft. per minute per welder. This requirement does not apply where hoods or booths are provided with sufficient airflow to maintain a velocity, away from the worker, of at least 100 linear ft. per minute.

3. In the absence of either of the foregoing, NIOSH (National Institute of Occupational Safety and Health) approved supplied-air respirators must be used.

### GAS WELDING SAFETY

Safety regulations for gas welding require that:

1. All cylinders are kept away from radiators and other sources of heat.

2. Cylinders stored inside must be in a well protected, well ventilated, dry location at least 20 ft. from any highly combustible materials. They should not be stored near elevators, stairs or walkways nor in unventilated lockers.

3. Protective caps should be placed over cylinder valves except when the cylinder is in use.

4. Oxygen cylinders must be stored at least 20 ft. from fuel gas cylinders or combustible materials. If this is not possible, a noncombustible barrier at least 5 ft. high with a half hour fire resistance rating will suffice.

5. All cylinder valves must be closed when work is finished. If a special wrench is required, it must be left in position on the stem of the valve while the cylinder is in use. Where cylinders are coupled or attached to a manifold, at least one such wrench shall always be available for immediate use in case of emergency.

6. All cylinders must have a label identifying contents.

7. Cylinders must not be allowed to stand alone without a security provision to keep it from toppling.

8. Do not use acetylene at pressure of more than 15 psig (gauge pressure or 30 psi absolute).

9. No more than 2000 cu. ft. of fuel gas or 300 lb. of liquified petroleum gas may be stored inside.

10. Hoses which show evidence of wear, leaks or burns must be replaced or repaired.

# Safety

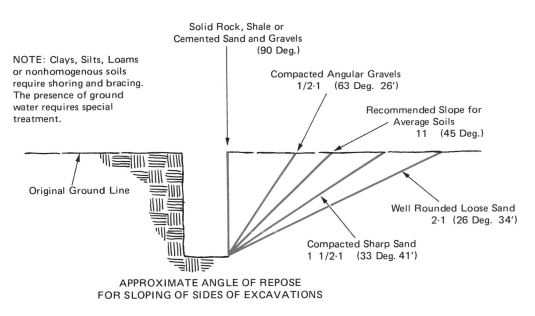

NOTE: Clays, Silts, Loams or nonhomogenous soils require shoring and bracing. The presence of ground water requires special treatment.

Solid Rock, Shale or Cemented Sand and Gravels (90 Deg.)

Compacted Angular Gravels 1/2·1 (63 Deg. 26')

Recommended Slope for Average Soils 1 1 (45 Deg.)

Original Ground Line

Well Rounded Loose Sand 2·1 (26 Deg. 34')

Compacted Sharp Sand 1 1/2·1 (33 Deg. 41')

APPROXIMATE ANGLE OF REPOSE FOR SLOPING OF SIDES OF EXCAVATIONS

Fig. 2-10. Approximate angles for sloping sides of trenches and excavations.

## ELECTRIC ARC WELDING SAFETY

Safety regulations for electric arc welding require that:

1. Welding machines which have become wet must be thoroughly dried and tested before being returned to use.
2. Coiled welding cable must be spread out before use; the ground lead must be securely fastened to the work.
3. Cables must be inspected frequently for damage to conductor and insulation. Repairs must be made immediately before use.
4. Ground and electrode cables may be joined together only with connectors specifically designed for that purpose.
5. Cables spliced within 10 ft. of the operator may not be used.
6. Operator is not permitted to coil cable around his or her body.
7. Welding helmets or hand shields must be worn by the operator. Persons close by must wear eye protection.
8. Workers in the general area must wear shields as protection from arc welding rays.
9. Operators should wear clean, fire resistant gloves and clothing which buttons at the neck and wrists.
10. When not in use, electrode holders must be placed in a safe place.

## HANDLING PLASTIC SOLVENT CEMENTS

Use extreme care when using solvent cements to secure plastic pipe joints. These solvent products are flammable and toxic to varying degrees. Area must be well ventilated, away from heat or flames. Avoid prolonged breathing of fumes and do not allow the material to come in contact with skin or eyes. In case of accidental contact, flush affected area immediately with water. Flush eyes continuously for up to 15 minutes. Carefully read safety precautions on the container before use.

## TEST YOUR KNOWLEDGE — UNIT 2

1. Nearly half of the accidents which result in lost time from work are the result of injuries received from _____ and while _____ objects.
2. Worker traits which contribute to safe work practices include: _____, _____, _____, _____, and _____.
3. Straight ladders should be set up so the distance from the base to the top support is approximately _____ times as great as the horizontal distance from the ladder base to the support.
4. Grounding an electric tool:
   a. Causes it to run faster.
   b. Makes the tool unsafe.
   c. Increases the life of the tool.
   d. Protects the operator if the tool has a short circuit.
5. Guardrails and _____ must be placed on all open sides and ends when scaffold platforms are over _____ ft. above ground.
6. Describe methods of protecting workers against cave-ins from trenching and excavating.
7. During welding, cutting and brazing operations, mechanical ventilation must be provided where there is less than _____ cu. ft. of air per welder or where ceilings are less than _____ ft. high.
8. It is better to coil a welding cable around the body than to risk tripping over it. True or False?

### SUGGESTED ACTIVITIES

1. Check the clothing you wear to see that it meets the standards established in this unit.
2. Using a voltmeter, check all portable electric tools in the shop to see that they are grounded.

# Unit 3
# MATHEMATICS FOR PLUMBERS

## Objectives

This unit reviews all of the basic mathematics likely to be needed by the plumber.

After studying the unit you will be able to:
- Read rule rapidly and accurately to nearest 1/16 inch.
- Add and subtract fractions and whole numbers.

- Compute pipe offsets using the Pythagorean theorem and trigonometric functions.
- Apply the fomrulas for finding area and volume.
- Explain and apply SI metric measure in finding length, area, volume and temperature.
- Convert customary measure to metric measure.

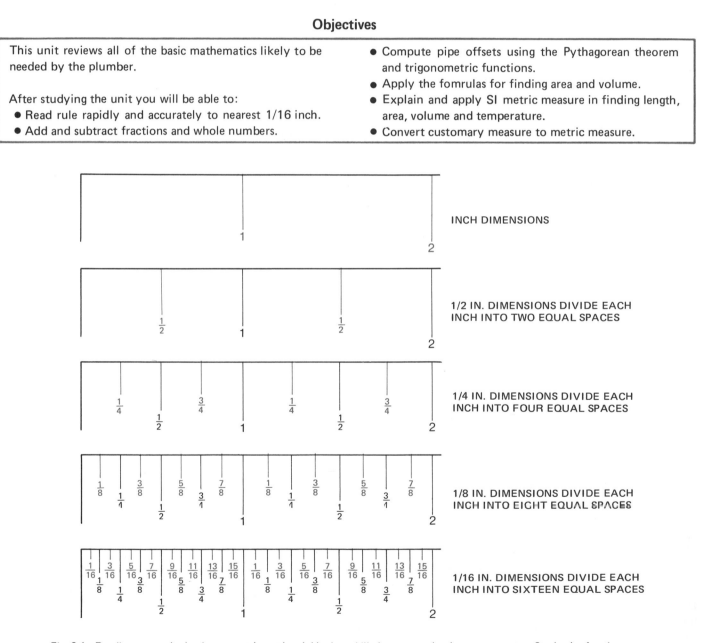

INCH DIMENSIONS

1/2 IN. DIMENSIONS DIVIDE EACH INCH INTO TWO EQUAL SPACES

1/4 IN. DIMENSIONS DIVIDE EACH INCH INTO FOUR EQUAL SPACES

1/8 IN. DIMENSIONS DIVIDE EACH INCH INTO EIGHT EQUAL SPACES

1/16 IN. DIMENSIONS DIVIDE EACH INCH INTO SIXTEEN EQUAL SPACES

Fig. 3-1. Reading a standard rule accurately and quickly is a skill that every plumber must master. Study the fractions above to fix the divisions of an inch firmly in your mind.

Plumbers need to make accurate measurements and calculations. They need to add and subtract dimensions, compute pipe offsets and determine the volumes of tanks. This unit will provide the basic information necessary for the plumber who needs to acquire these skills. In addition, metric measurement is introduced.

*recheck each measurement before cutting materials.* The reduced waste in both time and materials makes this procedure worthwhile.

Reading a scale accurately to the nearest 1/16 in. is easy if the 1/4 in. divisions are used as a starting point and the smaller parts of an inch are added to or subtracted from the larger

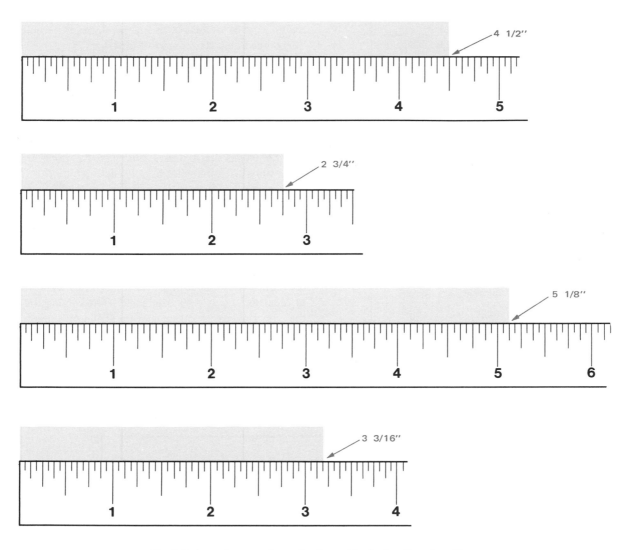

Fig. 3-2. Learn to recognize each rule marking by length and position.

## MEASUREMENT

The basic measuring tool used by the plumber is the folding rule or steel tape. Whichever is used, the scale printed on the tool is the same. Fig. 3-1 illustrates how the basic scale is divided into parts of an inch.

## READING FRACTIONS OF AN INCH

Reading a rule accurately and quickly requires careful attention to the markings on the scale and some practice. Note that the lines marking the scale vary in length. The longest lines are the inch divisions. The shortest marks indicate sixteenths of an inch. Fig. 3-2 gives several examples.

*As a precaution against costly errors, it is a good idea to*

divisions to get the right reading. See Fig. 3-3. This method is much faster and more accuracte than attempting to count the number of spaces.

## ADDING AND SUBTRACTING LENGTHS

Frequently, you will need to add together two lengths which are given in fractions of an inch. Fig. 3-4 illustrates direct addition by putting two scales together. It also shows how to add measurements given in fractions.

Subtracting dimensions given in fractions of an inch is another skill which the plumber will need. The procedure, shown in Fig. 3-5, is nearly identical to addition. Note that the fractions must have common denominators before they can be subtracted.

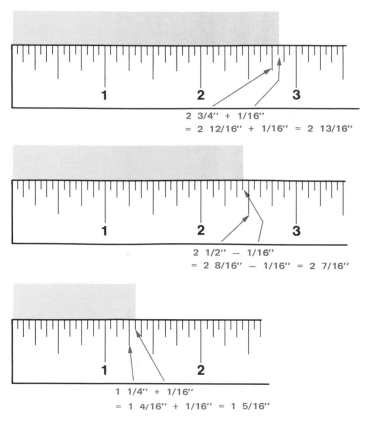

Fig. 3-3. Reading 1/16 in. intervals is easier if you count from the nearest 1/4 in. dimension and add or subtract spaces.

In cases where the denominators of the fractions are not equal (the same number) the procedure shown in Fig. 3-6 must be followed. This procedure requires that the fractions be converted to equal fractions having common (the same) denominators. Each numerator must be increased the same number of times as its denominator to make it equal to the original fraction.

To solve some problems involving the subtraction of fractions, it is necessary to borrow from the whole number. This procedure is described in Fig. 3-7.

## CONVERTING FEET TO INCHES AND INCHES TO FEET

Many times it will be necessary to change dimensions given in inches to equal dimensions in feet and inches. Because there are 12 inches in a foot, this can be done by dividing dimensions given in inches by 12, Fig. 3-8.

Feet can be changed to inches by multiplying the number of feet by 12 (the number of inches in a foot).

## COMPUTING PIPE OFFSETS

The illustration in Fig. 3-9 presents a typical pipe offset problem. Two parallel pipes must be joined by a short length of pipe running at a 45 deg. angle. Determining the length of the short piece of pipe can be difficult unless the right mathematical formulas are used. Two different techniques for finding the length of the diagonal pipe will be discussed.

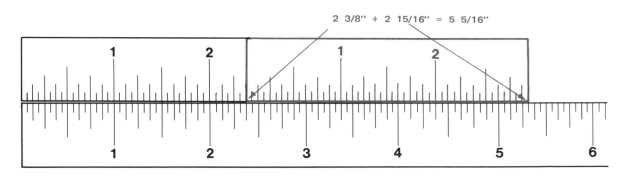

2 3/8″ + 2 15/16″ = _____
DENOMINATOR

FRACTIONS CAN BE ADDED ONLY WHEN THE DENOMINATORS ARE EQUAL. FROM FIG. 3-1, IT CAN BE SEEN THAT 3/8″ = 6/16″. THEREFORE, THE ABOVE PROBLEM CAN BE REWRITTEN AS:

NUMERATORS

2 6/16″ + 2 15/16″ = _____

ADDING FRACTIONS IS ACCOMPLISHED BY ADDING NUMERATORS. THE DENOMINATORS REMAIN UNCHANGED. IN THIS CASE:

NUMERATOR
6/16″ + 15/16″ = 21/16″
DENOMINATOR

BECAUSE THE NUMBERATOR IS LARGER THAN THE DENOMINATOR, THE FRACTION IS GREATER THAN ONE (SIXTEEN 1/16s = ONE). THEREFORE:

21/16″ = 16/16″ + 5/16″ = 1 5/16″

RETURNING TO THE ORIGINAL PROBLEM:

2 3/8″ + 2 15/16″ =
STEP 1 2 6/16″ + 2 15/16″ =
STEP 2 2″ + 2″ + 21/16″ =
STEP 3 2″ + 3″ + 5/16″ = 5 5/16″

Fig. 3-4. You can add two dimensions by laying off the two or more lengths on a scale. But another way, shown above, is to add the fractions together using mathematics.

$5\ 3/4'' - 4\ 1/4'' =$ _____

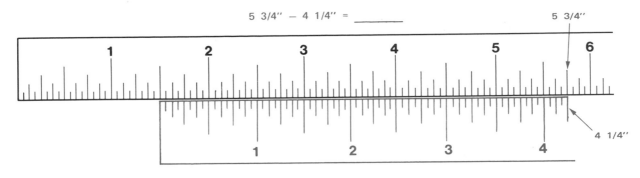

SUBTRACTING THE NUMERATORS OF THE FRACTIONS PRODUCES THE FOLLOWING RESULTS:

$5\ 3/4'' - 4\ 1/4'' =$ \_\_\_$2/4''$

SUBTRACTING THE WHOLE NUMBERS GIVES:

$5\ 3/4'' - 4\ 1/4'' = 1\ 2/4''$

NOW, 2/4'' CAN BE WRITTEN IN A SIMPLER, REDUCED FORM:

$5\ 3/4'' - 4\ 1/4'' = 1\ 1/2''$

Fig. 3-5. Lengths can be subtracted using the same methods shown in Fig. 3-4.

$6\ 1/2'' - 4\ 3/16'' =$ _____

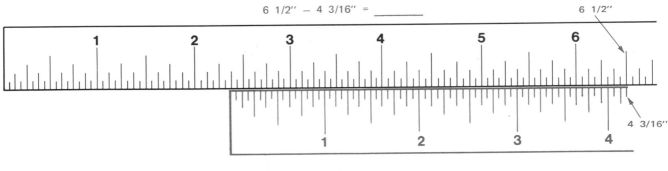

WHEN THE DENOMINATORS OF THE FRACTIONS ARE UNEQUAL, THE FRACTIONS MUST BE CHANGED TO EQUAL FRACTIONS HAVING COMMON (EQUAL) DENOMINATORS. FROM FIG. 3-1, IT CAN BE SEEN THAT 1/2 = 8/16. THEREFORE, THE ABOVE PROBLEM CAN BE WRITTEN AS:

$6\ 8/16'' - 4\ 3/16'' =$ _____

SUBTRACTING THE FRACTIONS PRODUCES THE FOLLOWING RESULTS:

$6\ 8/16'' - 4\ 3/16'' =$ \_\_\_$5/16''$

SUBTRACTING THE WHOLE NUMBERS COMPLETES THE PROBLEM:

$6\ 8/16'' - 4\ 3/16'' = 2\ 5/16''$

Fig. 3-6. Method of subtracting dimensions given in fractions of an inch and having unequal denominators.

$3\ 1/8'' - 1\ 5/16'' =$ _____

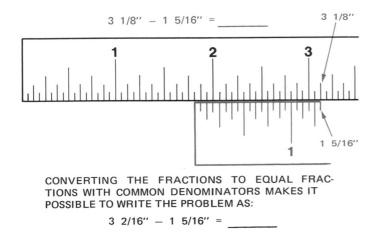

CONVERTING THE FRACTIONS TO EQUAL FRACTIONS WITH COMMON DENOMINATORS MAKES IT POSSIBLE TO WRITE THE PROBLEM AS:

$3\ 2/16'' - 1\ 5/16'' =$ _____

BECAUSE 5/16'' IS GREATER THAN 2/16'' IT IS NOT POSSIBLE TO SUBTRACT. BY BORROWING ONE FROM THE WHOLE NUMBER 3 AND CHANGING THE 1 TO ITS FRACTIONAL EQUIVALENT IN SIXTEENTHS, THE PROBLEM CAN BE WRITTEN AS:

$2 + (16/16 + 2/16)'' - 1\ 5/16'' =$ _____

SIMPLIFIED, THE PROBLEM BECOMES:

$2\ 18/16'' - 1\ 5/16'' =$ _____

SUBTRACTING THE FRACTIONS GIVES:

$2\ 18/16'' - 1\ 5/16'' =$ \_\_\_$13/16''$

SUBTRACTING THE WHOLE NUMBERS COMPLETES THE PROBLEM:

$2\ 18/16'' - 1\ 5/16'' = 1\ 13/16''$

Fig. 3-7. When subtracting dimensions given in fractions of an inch you can borrow from the whole number.

52 INCHES = _____ FEET

BECAUSE 12 INCHES EQUALS ONE FOOT, 12 IS DIVIDED INTO 52:

$$\begin{array}{r} 4 \\ 12\overline{)52} \\ 48 \\ \hline 4 \end{array}$$

THE ANSWER IS WRITTEN:

4'—4''

Fig. 3-8. Method of converting inch dimensions to feet.

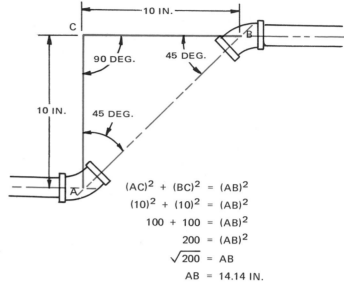

$$(AC)^2 + (BC)^2 = (AB)^2$$
$$(10)^2 + (10)^2 = (AB)^2$$
$$100 + 100 = (AB)^2$$
$$200 = (AB)^2$$
$$\sqrt{200} = AB$$
$$AB = 14.14 \text{ IN.}$$

Fig. 3-10. To find the length of a pipe offset with the Pythagorean theorem it helps to construct an imaginary triangle using the diagonal pipe as one side of the triangle.

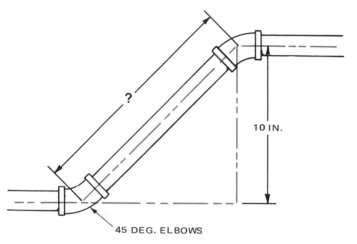

Fig. 3-9. Typical pipe offset problem. Find length of diagonal pipe.

## COMPUTING PIPE OFFSET USING PYTHAGOREAN THEOREM

In the first method we use a formula for finding the length of one side of a right-angle triangle. Known as the PYTHAGOREAN THEOREM, it states that the square of the hypotenuse (third side) of a right-angle triangle is equal to the sum of the squares of the other two sides. Look at Fig. 3-10. Note that the vertical distance between the parallel pipes is 10 in. (line AC). Because 45 deg. elbows are being used, the distance CB is also equal to 10 in. To compute the theoretical length of the diagonal pipe, the Pythagorean theorem is used. Fig. 3-11 illustrates the relationships between the length of the sides of right triangles. *As long as the triangle has one right angle this relationship remains unchanged.*

A difficult task when using the Pythagorean theorem is to compute the square root of a number. To make this less difficult, a table of squares and square roots has been provided under Useful Information, page 255. A second problem is to convert decimal parts of an inch to fractions. A table for this purpose is also provided in the Useful Information section, page 254. When the theoretical length of pipe has been determined, it will be necessary to make allowance for the

actual dimensions of the fittings being used. These dimensions will vary. It depends upon the size and type of pipe and fittings being installed.

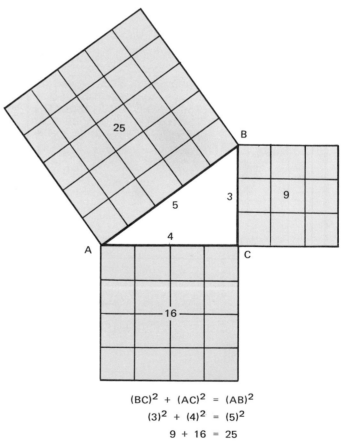

$$(BC)^2 + (AC)^2 = (AB)^2$$
$$(3)^2 + (4)^2 = (5)^2$$
$$9 + 16 = 25$$

Fig. 3-11. The relationship of the length of the sides is shown by the squares constructed along the sides of the triangle.

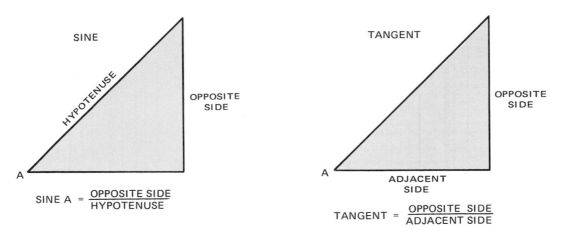

Fig. 3-12. The sine and tangent ratios can be used to compute the length of pipe offset.

## COMPUTING PIPE OFFSET USING TRIGONOMETRIC FUNCTIONS

In many cases, it is easier to use TRIGONOMETRIC FUNCTIONS to compute pipe offsets because of the dimensions which are known. The two functions most likely to be used are the sine and the tangent. These functions give a mathematical relationship or ratio between parts of a triangle. They permit the plumber to find the length of a pipe, if an angle and the length of one side of the triangle is known. Fig. 3-12 shows these ratios.

Let us assume that a 45 deg. elbow and a short diagonal length of pipe are to be installed to connect the two parallel pipes shown in Fig. 3-13. The sine function can be used to

### A BRIEF TABLE OF SINE RATIOS

| ANGLE | SINE |
|---|---|
| 22 1/2 DEG. | .3665 |
| 30 DEG. | .5000 |
| 45 DEG. | .7071 |
| 60 DEG. | .8660 |

Fig. 3-14. This table expresses the relationship of a known angle to the hypotenuse. It says: "If angle A is 45 deg., then the hypotenuse is .7071 times larger than the side opposite the angle."

compute the length of the diagonal pipe. The value for the sine function is taken from the table in Fig. 3-14.

A more complete table is provided in the Useful Informa-

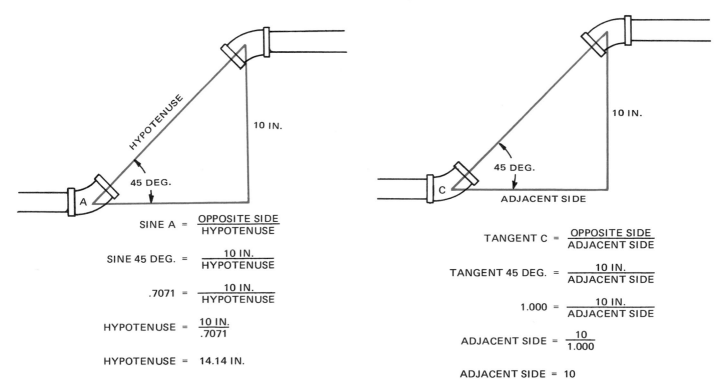

Fig. 3-13. The sine and the tangent function refer to the size relationship of parts of a triangle. Here the function is used to find the length of a pipe.

Fig. 3-15. We compute the horizontal distance between the ends of parallel pipes using the tangent ratio.

tion section, page 255. Again, note that this is a theoretical length. It must be reduced because fittings shorten the distance between the pipes.

The tangent ratio is useful when finding horizontal distances such as length AC shown in Fig. 3-15. On occasion, the plumber may want to compute the distance to assist in the location of pipes. Some values for the tangent ratio are given in the table in Fig. 3-16. A more complete table of tangent ratios is provided. Refer to page 255. The theoretical distance computed must be adjusted to compensate for the actual size of fittings before the pipe is cut.

### BRIEF TABLE OF TANGENT RATIOS

| ANGLE | TANGENT |
|---|---|
| 22 1/2 DEG. | .4142 |
| 30 DEG. | .5774 |
| 45 DEG. | 1.000 |
| 60 DEG. | 1.732 |

Fig. 3-16. In this case, the relationship of the angle's adjacent side and opposite side is 1:1. Thus the two sides are equal.

## COMPUTING AREA AND VOLUME

The surface area of a square or rectangular surface can be computed by multiplying the length times the width, Fig. 3-17. But computing the surface area of circles requires the use of a special formula — AREA = $\pi r^2$. This formula is read: "Area equals Pi times the radius squared."

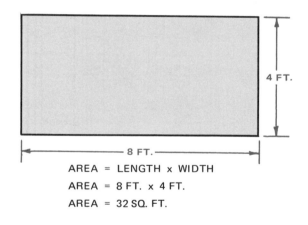

AREA = LENGTH x WIDTH
AREA = 8 FT. x 4 FT.
AREA = 32 SQ. FT.

Fig. 3-17. This formula computes the area of a rectangular surface.

Pi is a mathematical ratio frequently used when making calculations about circles. For most practical purposes, Pi can be assumed to be equal to 3.14. Fig. 3-18 illustrates the use of the area formula for circles.

Volume of a tank is found by multiplying length, width and height as shown in Fig. 3-19. Finding the volume of cylindrical tanks requires the use of the formula: VOLUME = $\pi r^2 h$. This is read: "Volume equals Pi times the radius squared times the height." Fig. 3-20 illustrates the use of this formula.

Having calculated the volume in some convenient units of cubic measure, it may be necessary to convert the cubic units

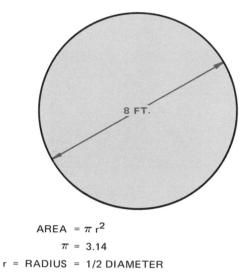

AREA = $\pi r^2$

$\pi$ = 3.14

r = RADIUS = 1/2 DIAMETER

BECAUSE THE DIAMETER IS 8 FT., THE RADIUS IS 4 FT. THEREFORE,

AREA = 3.14 x $(4)^2$

= 3.14 x 16 SQ. FT.

= 50.24 SQ. FT.

Fig. 3-18. Formula for computing the area of a circle.

of measure to a volume measure such as gallons. The most common conversion factors and examples of their use are given in Fig. 3-21. English-Metric conversion factors are provided in the Useful Information section, page 256.

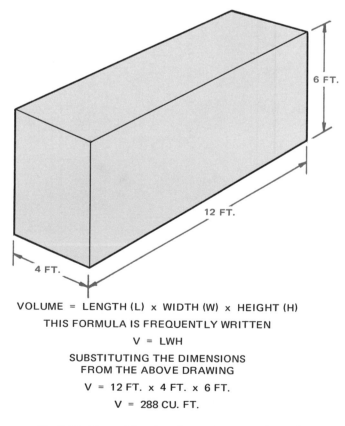

VOLUME = LENGTH (L) x WIDTH (W) x HEIGHT (H)

THIS FORMULA IS FREQUENTLY WRITTEN

V = LWH

SUBSTITUTING THE DIMENSIONS FROM THE ABOVE DRAWING

V = 12 FT. x 4 FT. x 6 FT.

V = 288 CU. FT.

Fig. 3-19. Determining the volume of a rectangular tank.

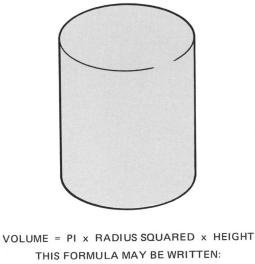

VOLUME = PI x RADIUS SQUARED x HEIGHT

THIS FORMULA MAY BE WRITTEN:

$$V = \pi r^2 h$$

SUBSTITUTING FROM THE ABOVE DRAWING:

$$V = 3.14 \times (2 \text{ FT.})^2 \times 6 \text{ FT.}$$

$$V = 75.36 \text{ CU. FT.}$$

Fig. 3-20. Determining the volume of a cylindrical tank.

## METRIC MEASUREMENT

The United States is expected to follow the lead of Europe and adopt the SI metric system of measurement. The changeover is well under way in some industries. *Because the construction trades, such as plumbing, are not involved in international trade, there is less pressure to change. Even so, it is expected that the change will occur and all plumbers will find it necessary to learn the metric system.*

The base unit of length measurement in the metric system is shown in Fig. 3-22, along with comparable traditional units. One of the chief advantages of using the metric system is the ease of changing from one metric unit to another simply by multiplying or dividing by multiples of 10. This is simpler than attempting to use the various conversion factors required in the traditional inch-pound system.

Unlike English measure, where dimensions are customarily expressed in both feet and inches (2'-3''), the metric system does not mix its units in a single measurement. A distance of 2 metres and 3 centimetres, for example, will be written: 2.03 m or 203 cm, never as 2 m-3 cm.

When the metric system does come to plumbing, the common units will be the following:

1. For long distances, the metre.
2. For short distances, the centimetre.
3. For very small measurements, such as certain pipe diameters and nut sizes, the millimetre.
4. For liquid volume, the litre or the cubic centimetre ($cm^3$).
5. For dry volume, the cubic metre ($m^3$).
6. For pressure, the pascal which is equal to 1 newton per square metre. The newton is the metric unit for force. A newton is the amount of force needed to accelerate a weight of 1 kilogram one metre per second per second.

| TO CONVERT | TO | PROCEDURE | EXAMPLE |
|---|---|---|---|
| CUBIC INCHES | GALLONS | DIVIDE BY 231 | 376 CU. IN. = 376 ÷ 231 = 163 GAL. |
| CUBIC FEET | GALLONS | MULTIPLY BY 7.48 | 6 CU. FT. = 6 x 7.48 = 44.88 GAL. |

Fig. 3-21. How to convert cubic inches and cubic feet to gallons.

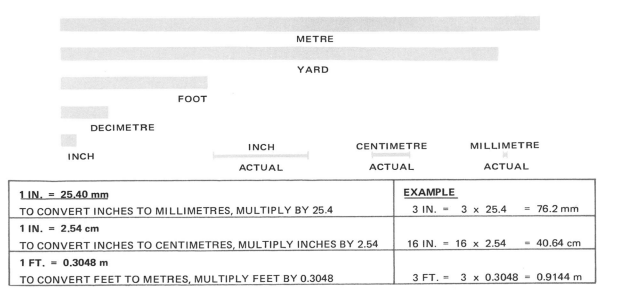

| 1 IN. = 25.40 mm | | EXAMPLE |
|---|---|---|
| TO CONVERT INCHES TO MILLIMETRES, MULTIPLY BY 25.4 | | 3 IN. = 3 x 25.4 = 76.2 mm |
| 1 IN. = 2.54 cm | | |
| TO CONVERT INCHES TO CENTIMETRES, MULTIPLY INCHES BY 2.54 | | 16 IN. = 16 x 2.54 = 40.64 cm |
| 1 FT. = 0.3048 m | | |
| TO CONVERT FEET TO METRES, MULTIPLY FEET BY 0.3048 | | 3 FT. = 3 x 0.3048 = 0.9144 m |

Fig. 3-22. Until the United States is completely metric, we must convert measurements to their equivalent in the metric system. The above illustration compares relative lengths of metric and customary units. (Copyright by Polymetric Services, Inc., Tarzana, CA 91356)

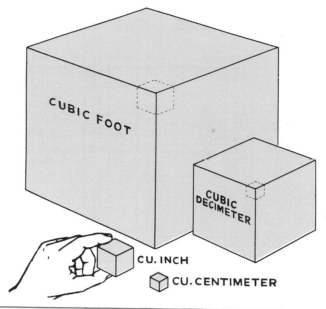

| TO CONVERT FROM CUBIC INCHES TO: | MULTIPLY BY: | EXAMPLE | |
|---|---|---|---|
| GALLONS | 0.004 | 3674 CU. IN. = ____ GAL. | 3674 x .004 = 14.7 GAL. |
| CUBIC FEET | 0.00058 | 4677 CU. IN. = ____ CU. FT. | 4677 x .00058 = 2.712 CU. FT. |
| LITRES | 0.016 | 286 CU. IN. = ____ LITRES | 286 x .016 = 4.576 LITRES |

| TO CONVERT FROM CUBIC FEET TO: | MULTIPLY BY: | EXAMPLE | |
|---|---|---|---|
| GALLONS | 7.48 | 6.5 CU. FT. = ____ GAL. | 6.5 x 7.48 = 48.62 GAL. |
| CUBIC INCHES | 1728.0 | 3 CU. FT. = ____ CU. IN. | 3 x 1728.0 = 5184.0 CU. IN. |
| LITRES | 28.32 | 1.5 CU. FT. = ____ LITRES | 1.5 x 28.32 = 42.48 LITRES |
| CUBIC METRES | 0.028 | 278 CU. FT. = ____ CU. METRES | 278 x .028 = 7.784 CU. M |
| CUBIC YARDS | 0.037 | 5687 CU. FT. = ____ CU. YDS. | 5687 x .037 = 210.42 CU. YD. |

Fig. 3-23. Metric and English dry volume measurements and conversion factors.
(Copyright by Polymetric Services, Inc., Tarzana, CA 91356)

7. For temperatures, the degree Kelvin or the degree Celsius. The degree Celsius is equal to 1 4/5 Fahrenheit degree. Kelvin is in the Celsius scale and measures temperatures in the supercold range (−273 degrees and up).

Fig. 3-23 illustrates the comparative volume measurements and provides factors for converting from the present system. The comparison of liquid measurements in the English and metric systems is shown in Fig. 3-24.

The Kelvin (absolute) or the Celsius scale are easier to understand if seen with the familiar Fahrenheit. Fig. 3-25 compares them and shows how temperatures from one scale can be converted to a like temperature on other scales.

The plumbing industry will see a time when plans, pipe, fittings and fixtures will be dimensioned both in English and metric (dual dimensioning). At some point, standard sizes of all pipe, fittings and fixtures will be changed to standard metric units. This process will require considerable time. For now, the ability to convert from one system to the other and to recognize the size of the various units in both systems will fulfill plumbers' needs.

## TEST YOUR KNOWLEDGE — UNIT 3

1. On a standard 6 ft. folding rule the dimensions are labeled in inches beginning with 1 and ending at 72. Give the equivalent inch dimensions for:
   a. 1'.
   b. 2'.
   c. 3'.
   d. 4'.
   e. 5'.
   f. 6'.

2. Convert the following dimensions given in feet and inches to their equivalent in inches:
   a. 2'-3".
   b. 1'-6".
   c. 4'-6".
   d. 6'-3".
   e. 12'-2".
   f. 3'-9".

3. Convert the following dimensions given in inches to their equivalent in feet and inches:
   a. 17".
   b. 43".
   c. 54 1/2".
   d. 92 3/8".
   e. 35 1/2".
   f. 23".

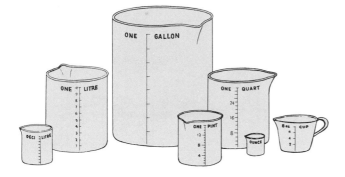

## COMMON LIQUID CONVERSIONS

| GALLON | QUARTS | PINTS | CUPS | OUNCES | LITRES |
|--------|--------|-------|------|--------|--------|
|        |        |       | 1    | 8      | 0.237  |
|        |        | 1     | 2    | 16     | 0.473  |
|        | 1      | 2     | 4    | 32     | 0.946  |
| 1      | 4      | 8     | 16   | 128    | 3.785  |

**EXAMPLES:**

TO CONVERT GALLONS TO LITRES
  MULTIPLY GALLONS BY 3.785

  6 GAL. = _____ LITRES = 6 x 3.78 = 22.68 LITRES

TO CONVERT QUARTS TO LITRES
  MULTIPLY QUARTS BY 0.946

  3 QT. = _____ LITRES = 3 x 0.946 = 2.838 LITRES

Fig. 3-24. Comparing common liquid measures.
(Copyright by Polymetric Services, Inc., Tarzana, CA 91356)

4. Add the following pairs of numbers:
   a. 18 1/2 + 1 1/2.  　　d. 18 3/4 + 1 1/2.
   b. 6 1/4 + 1 1/2.  　　e. 12 1/8 + 3/4.
   c. 10 1/2 + 1 1/4.

5. Subtract the following pairs of numbers:
   a. 10 1/2 − 1 1/2.  　　d. 16 1/8 − 1 1/2.
   b. 8 3/4 − 1 1/2.  　　e. 9 1/4 − 1 3/16.
   c. 12 3/8 − 3/4.

6. Assume that two parallel pipes are to be joined with two 45 deg. elbows and a short length of pipe. The ends are 18 in. apart and one pipe is 18 in. above the other. Using the Pythagorean theorem, find the theoretical length of the diagonal pipe.

7. The problem is the same as No. 6, except that the pipes are spaced 14 in. apart and the sine function is to be used in making the calculation.

8. Compute the volume of a rectangular tank which measures 8 ft. long by 4 ft. wide by 10 ft. tall.

9. How many gallons of water will the above tank hold if it is filled to a depth of 6 ft?

10. What is the capacity, in gallons, of a cylindrical tank which is 8 ft. in diameter at the base and stands 20 ft. tall?

11. A metre is longer or shorter than a yard?

12. A cubic inch is larger or smaller than a cubic centimetre?

13. A quart is smaller or larger than a litre?

14. A metre is _____ times larger than a centimetre?

15. A centimetre is _____ times larger than a millimetre?

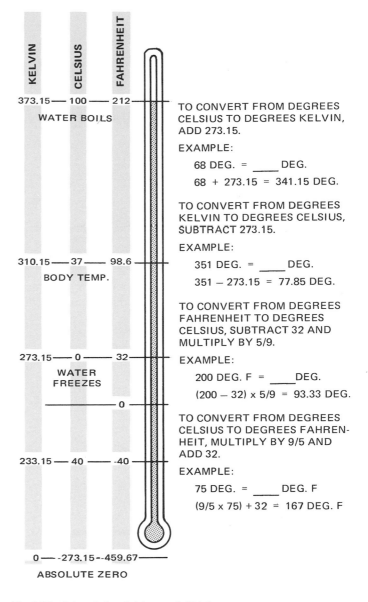

TO CONVERT FROM DEGREES CELSIUS TO DEGREES KELVIN, ADD 273.15.

EXAMPLE:

　68 DEG. = ____ DEG.

　68 + 273.15 = 341.15 DEG.

TO CONVERT FROM DEGREES KELVIN TO DEGREES CELSIUS, SUBTRACT 273.15.

EXAMPLE:

　351 DEG. = ____ DEG.

　351 − 273.15 = 77.85 DEG.

TO CONVERT FROM DEGREES FAHRENHEIT TO DEGREES CELSIUS, SUBTRACT 32 AND MULTIPLY BY 5/9.

EXAMPLE:

　200 DEG. F = ____ DEG.

　(200 − 32) x 5/9 = 93.33 DEG.

TO CONVERT FROM DEGREES CELSIUS TO DEGREES FAHRENHEIT, MULTIPLY BY 9/5 AND ADD 32.

EXAMPLE:

　75 DEG. = ____ DEG. F

　(9/5 x 75) + 32 = 167 DEG. F

Fig. 3-25. Fahrenheit, Celsius and Kelvin temperature measurement and conversion factors.

16. Convert following customary measurements to metric:

| Customary | | Metric |
|-----------|--|--------|
| a. | 18 in. | ____ metre(s). |
| b. | 4 ft. | ____ metre(s). |
| c. | 10 in. | ____ centimetre(s). |
| d. | 453 cu. in. | ____ litre(s). |
| e. | 2.5 cu. ft. | ____ litre(s). |
| f. | 346 cu. ft. | ____ cubic metre(s). |

## SUGGESTED ACTIVITIES

1. Practice measuring the length of various pieces of pipe using both the customary English and the metric scale.

2. Calculate the actual length of pipe required for a variety of typical installations. Include fitting allowances.

3. Compute the volume of a standard tank for water heating or other plumbing installations.

# Unit 4
# PIPING MATERIALS AND FITTINGS

## Objectives

This unit introduces the various pipe and fittings used in residential and light commercial plumbing systems. Emphasis is given to materials, sizes and applications.

After studying this unit you will be able to:
- Name the various materials used in pipe and fittings.
- Suggest appropriate applications for each type of material.
- Recognize and properly name various fittings and pipes.
- List grades and sizes of pipe and fittings.
- Interpret code markings used on plastic pipe.

Pipe and pipe fittings for residential and light commercial plumbing are produced from several different kinds of materials, in different grades and in many sizes. The fittings are of different shapes and designs to meet every need of the modern plumbing system.

Each system will have two sets of piping. One carries away waste water and solid waste. It is called the drainage system or DWV (drainage, waste and venting). The other carries fresh water, under pressure, for drinking, cooking, bathing and laundry. It is called the water supply system. Different pipe and fittings are required by each system. Fig. 4-1 shows a sketch of both systems as installed in a building.

Since the same material may often be used for both supply and drainage, our discussion is organized around the materials. These materials include:
1. Cast iron.
2. Steel.
3. Malleable iron.
4. Copper.
5. Plastic.

## CAST IRON

SOIL PIPE is the term generally used to describe the cast-iron pipe and fittings frequently used in building drainage systems. This piping material is also used in storm drainage systems for roofs, yards and areaways. Use of cast-iron soil pipe is limited to those applications where gravity, not pressure, causes the waste to flow.

Both pipe and fittings are cast from grey iron. This material is both strong and resistant to corrosion. This resistance is due to the formation of large graphite flakes within the material during the casting operation. It serves as an insulation against corrosion. Cast iron pipe will not leak or absorb water.

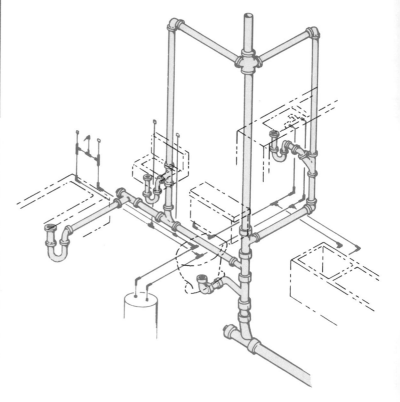

Fig. 4-1. Pictorial drawing shows typical plumbing system as it might be installed in a building. Water supply system is shown in solid color, the drainage system in a lighter color.

## GRADES

The Cast Iron Soil Pipe Institute has standardized specifications for two grades of soil pipe:
1. Service (SV), most frequently used above grade.
2. Extra heavy (XH), used primarily below grade.

See Useful Information, page 257 for detailed dimensions of the basic types of pipe.

## JOINING METHODS

Soil pipe and fittings are made with two types of ends for joining pieces together.
1. The hub and spigot type, Fig. 4-2.
2. The no-hub type, Fig. 4-3.

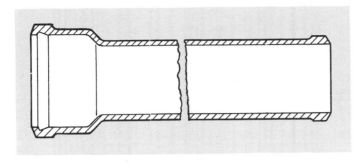

Fig. 4-2. Hub and spigot soil pipe has a "bell" or enlargement at one end.   (Richmond Foundry and Mfg. Co.)

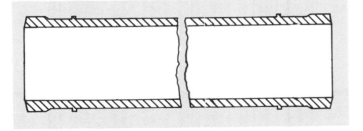

Fig. 4-3. No-hub soil pipe has no bell.

There are two different methods of sealing the joint of the hub and spigot soil pipe:
1. Lead and oakum, Fig. 4-4.
2. Compression joint, Fig. 4-5.

The lead groove on the inside of the hub prevents both the lead and the gasket from coming out. No-hub soil pipe joints are sealed with a neoprene gasket and are held in place with a stainless steel clamp, Fig. 4-6. Detailed information about assembly of pipe is given in Unit 9.

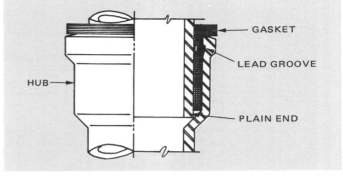

Fig. 4-5. Compression soil pipe joint with gasket.

## SIZES OF SOIL PIPE

Soil pipe and fittings are available in 2, 3, 4, 6, 8, 10, 12 and 15 in. inside diameter (ID). Overall dimensions for hub and spigot soil pipe are given in Fig. 4-7. Soil pipe is sold in 5 and 10 ft. lengths. Double hub pipe can also be purchased. It permits cutting two usable runs of pipe from one length.

## SOIL PIPE FITTINGS

The great variety of fittings available for soil pipe makes nearly any assembly of pipe possible. Only the most frequently used fittings will be discussed.

### Bends

A bend is an angular fitting which permits the pipe system to change direction. The length of the bend affects the length and position of pipe joined with the bend. Different lengths of bends are manufactured in two series.
1. Standard bends have varying lengths depending on the size of the pipe, Fig. 4-8.
2. Long bends are designated with two numbers. For example, 2 x 12 means a 2 in. diameter pipe with a 12 in. bend length, Fig. 4-9.

The chart in Fig. 4-10 indicates the direction change, in degrees, for each of the bends.

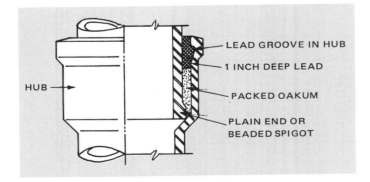

Fig. 4-4. Lead and oakum packed into a soil pipe joint prevent pipe from leaking.   (Cast Iron Pipe Institute)

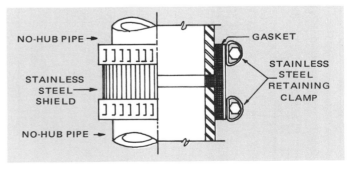

Fig. 4-6. No-hub soil pipe joint is sealed with a rubber-like plastic called neoprene.   (E.I. du Pont de Nemours & Co.)

LAYING LENGTH – 5'0" AND 10'0"

| EXTRA-HEAVY PIPE[1] 'XH' | | | | | | | | |
|---|---|---|---|---|---|---|---|---|
| SIZE (NOM. I.D.) | K MAX. | H MAX. | J | F | Y | E | M | N |
| 2 | 4 1/8 | 3 5/8 | 2 3/8 | 3/4 | 2 1/2 | 2 3/4 | 2 3/4 | 11/16 |
| 3 | 5 3/8 | 4 15/16 | 3 1/2 | 13/16 | 2 3/4 | 3 1/4 | 3 7/8 | 3/4 |
| 4 | 6 3/8 | 5 15/16 | 4 1/2 | 7/8 | 3 | 3 1/2 | 4 7/8 | 13/16 |
| 5 | 7 3/8 | 6 15/16 | 5 1/2 | 7/8 | 3 | 3 1/2 | 5 7/8 | 13/16 |
| 6 | 8 3/8 | 7 15/16 | 6 1/2 | 7/8 | 3 | 3 1/2 | 6 7/8 | 13/16 |
| 8 | 11 1/16 | 10 7/16 | 8 5/8 | 1 3/16 | 3 1/2 | 4 1/8 | 9 | 1 1/8 |
| 10 | 13 5/16 | 12 11/16 | 10 3/4 | 1 3/16 | 3 1/2 | 4 1/8 | 11 1/8 | 1 1/8 |
| 12 | 15 7/16 | 14 13/16 | 12 3/4 | 1 7/16 | 4 1/4 | 5 | 13 1/8 | 1 3/8 |
| 15 | 18 13/16 | 18 3/16 | 15 7/8 | 1 7/16 | 4 1/4 | 5 | 16 1/4 | 1 3/8 |

| SERVICE[1] PIPE 'SV' | | | | | | | | |
|---|---|---|---|---|---|---|---|---|
| SIZE (NOM. I.D.) | K MAX. | H MAX. | J | F | Y | E | M | N |
| 2 | 3 15/16 | 3 3/8 | 2 1/4 | 3/4 | 2 1/2 | 2 3/4 | 2 5/8 | 11/16 |
| 3 | 5 | 4 1/2 | 3 1/4 | 13/16 | 2 3/4 | 3 1/4 | 3 5/8 | 3/4 |
| 4 | 6 | 5 1/2 | 4 1/4 | 7/8 | 3 | 3 1/2 | 4 5/8 | 13/16 |
| 5 | 7 | 6 1/2 | 5 1/4 | 7/8 | 3 | 3 1/2 | 5 5/8 | 13/16 |
| 6 | 8 | 7 1/2 | 6 1/4 | 7/8 | 3 | 3 1/2 | 6 5/8 | 13/16 |
| 8 | 10 1/2 | 9 7/8 | 8 3/8 | 1 3/16 | 3 1/2 | 4 1/8 | 8 3/4 | 1 1/8 |
| 10 | 12 13/16 | 12 3/16 | 10 1/2 | 1 3/16 | 3 1/2 | 4 1/8 | 10 7/8 | 1 1/8 |
| 12 | 14 15/16 | 14 5/16 | 12 1/2 | 1 7/16 | 4 1/4 | 5 | 12 7/8 | 1 3/8 |
| 15 | 18 5/16 | 17 5/8 | 15 5/8 | 1 7/16 | 4 1/4 | 5 | 16 | 1 3/8 |

[1] DIMENSION IN INCHES

Fig. 4-7. Overall dimensions of hub and spigot soil pipe. (Cast Iron Pipe Institute)

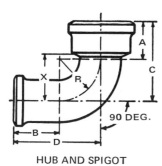

HUB AND SPIGOT

| | DIMENSIONS IN INCHES | | | | | |
|---|---|---|---|---|---|---|
| SIZE (INCHES) | A | B | C | D | R | X |
| 2 | 2 3/4 | 3 | 5 3/4 | 6 | 3 | 3 1/4 |
| 3 | 3 1/4 | 3 1/2 | 6 3/4 | 7 | 3 1/2 | 4 |
| 4 | 3 1/2 | 4 | 7 1/2 | 8 | 4 | 4 1/2 |
| 5 | 3 1/2 | 4 | 8 | 8 1/2 | 4 1/2 | 5 |
| 6 | 3 1/2 | 4 | 8 1/2 | 9 | 5 | 5 1/2 |
| 8 | 4 1/8 | 5 1/2 | 10 1/8 | 11 1/2 | 6 | 6 5/8 |
| 10 | 4 1/8 | 5 1/2 | 11 1/8 | 12 1/2 | 7 | 7 5/8 |
| 12 | 5 | 7 | 13 | 15 | 8 | 8 3/4 |
| 15 | 5 | 7 | 14 1/2 | 16 1/2 | 9 1/2 | 10 1/4 |

NO-HUB

| | DIMENSIONS IN INCHES | | | |
|---|---|---|---|---|
| SIZE (INCHES) | B | D LAYING LENGTH | R | W |
| 1 1/2 | 1 1/2 | 4 1/4 ± 1/8 | 2 3/4 | 1 1/8 |
| 1 | 1 1/2 | 4 1/2 ± 1/8 | 3 | 1 1/8 |
| 3 | 1 1/2 | 5 ± 1/8 | 3 1/2 | 1 1/8 |
| 4 | 1 1/2 | 5 1/2 ± 1/8 | 4 | 1 1/8 |
| 5 | 2 | 6 1/2 ± 1/8 | 4 1/2 | 1 1/2 |
| 6 | 2 | 7 ± 1/8 | 5 | 1 1/2 |

Fig. 4-8. Length of hub and spigot and no-hub 1/4 bends. (Cast Iron Pipe Institute)

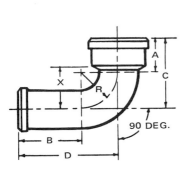

| SIZE (INCHES) | DIMENSIONS IN INCHES | | | | | |
|---|---|---|---|---|---|---|
| | A | B | C | D | R | X |
| 2 by 12 | 2 3/4 | 9 | 5 3/4 | 12 | 3 | 3 1/4 |
| 2 by 18 | 2 3/4 | 15 | 5 3/4 | 18 | 3 | 3 1/4 |
| 2 by 24 | 2 3/4 | 21 | 5 3/4 | 24 | 3 | 3 1/4 |
| 3 by 12 | 3 1/4 | 8 1/2 | 6 3/4 | 12 | 3 1/2 | 4 |
| 3 by 18 | 3 1/4 | 14 1/2 | 6 3/4 | 18 | 3 1/2 | 4 |
| 3 by 24 | 3 1/4 | 20 1/2 | 6 3/4 | 24 | 3 1/2 | 4 |
| 4 by 12 | 3 1/2 | 8 | 7 1/2 | 12 | 4 | 4 1/2 |
| 4 by 18 | 3 1/2 | 14 | 7 1/2 | 18 | 4 | 4 1/2 |
| 4 by 24 | 3 1/2 | 20 | 7 1/2 | 24 | 4 | 4 1/2 |

Fig. 4-9. Length of hub and spigot bends.

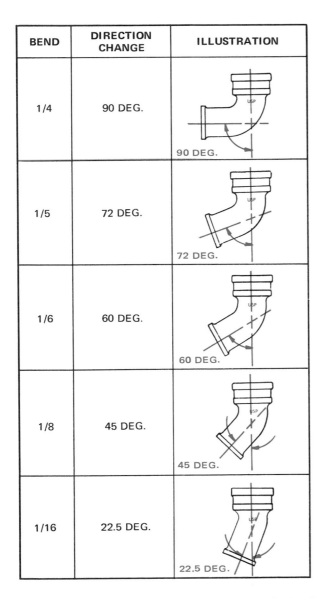

| BEND | DIRECTION CHANGE | ILLUSTRATION |
|---|---|---|
| 1/4 | 90 DEG. | 90 DEG. |
| 1/5 | 72 DEG. | 72 DEG. |
| 1/6 | 60 DEG. | 60 DEG. |
| 1/8 | 45 DEG. | 45 DEG. |
| 1/16 | 22.5 DEG. | 22.5 DEG. |

Fig. 4-10. Degree of direction change produced by five pipe bends.

OFFSET 1/8 BENDS, Fig. 4-11, are useful to install a drain or vent around obstacles. Offsets are specified with two dimensions. A 2 x 10 offset would be 2 in. diameter pipe with an offset of 10 in.

Some bends have a side opening (inlet) where a smaller vent or drain line can be connected. These inlets are designated as right inlet, left inlet or right and left inlet. To identify the type, face the hub with the spigot end down, Fig. 4-12. If the inlet is on the left, the fitting is known as a bend with a left side inlet.

## Branches

The Y BRANCH, Fig. 4-13, or double Y branch, is used where two or three drains converge (join) into one. Note that the branch intersects the main line at a 45 deg. angle. Some branch fittings have threads in either the main or branch hub. These accept cleanout plugs or a threaded pipe from a plumbing fixture drain. An inverted branch, Fig. 4-14, may be required in a vent system.

SANITARY T BRANCHES, Fig. 4-15, are required to make right-angle intersections in the drain piping. Because of the gentle curve of the intersection of the branch line with the main line, solids readily flow through the drainage system.

T branches, Fig. 4-16, may be used in vent lines. Because the flow of air in a vent will not be greatly affected by the abrupt right-angle change in direction, a sanitary T branch is not required.

Some sanitary T branches have side openings so that drain lines may be connected. There are three types:
1. Sanitary T branch with right inlet.
2. Sanitary T branch with left inlet.
3. Sanitary T branch with left and right inlet. Fig. 4-17 shows how to tell which type of T branch it is.

T branches are available with side inlets permitting the connection of vent lines from smaller fixtures. Right and left T branches are determined the same way as sanitary T branches.

## Increasers, double hubs and reducers

INCREASERS, Fig. 4-18, are used in the vent before it goes through the roof. A larger size pipe is required above the heated area of the building to prevent frost buildup from closing the opening.

DOUBLE HUBS and REDUCERS, Fig. 4-19, are available in many different sizes to permit the joining of a wide variety of pipe sizes.

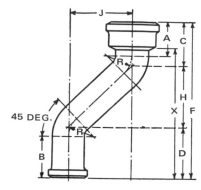

| SIZE (INCHES) | DIMENSIONS IN INCHES | | | | | | | | |
|---|---|---|---|---|---|---|---|---|---|
| | A | B | C | D | F | H | J | R | X |
| 2 by 2 | 2 3/4 | 3 1/2 | 3 1/2 | 4 1/4 | 9 3/4 | 2 | 2 | 2 | 7 1/4 |
| 2 by 4 | 2 3/4 | 3 1/2 | 3 1/2 | 4 1/4 | 11 3/4 | 4 | 4 | 2 | 9 1/4 |
| 2 by 6 | 2 3/4 | 3 1/2 | 3 1/2 | 4 1/4 | 13 3/4 | 6 | 6 | 2 | 11 1/4 |
| 2 by 8 | 2 3/4 | 3 1/2 | 3 1/2 | 4 1/4 | 15 3/4 | 8 | 8 | 2 | 13 1/4 |
| 2 by 10 | 2 3/4 | 3 1/2 | 3 1/2 | 4 1/4 | 17 3/4 | 10 | 10 | 2 | 15 1/4 |
| 2 by 12 | 2 3/4 | 3 1/2 | 3 1/2 | 4 1/4 | 19 3/4 | 12 | 12 | 2 | 17 1/4 |
| 2 by 14 | 2 3/4 | 3 1/2 | 3 1/2 | 4 1/4 | 21 3/4 | 14 | 14 | 2 | 19 1/4 |
| 2 by 16 | 2 3/4 | 3 1/2 | 3 1/2 | 4 1/4 | 23 3/4 | 16 | 16 | 2 | 21 1/4 |
| 2 by 18 | 2 3/4 | 3 1/2 | 3 1/2 | 4 1/4 | 25 3/4 | 18 | 18 | 2 | 23 1/4 |
| 3 by 2 | 3 1/4 | 4 | 4 1/4 | 5 | 11 1/4 | 2 | 2 | 2 1/2 | 8 1/2 |
| 3 by 4 | 3 1/4 | 4 | 4 1/4 | 5 | 13 1/4 | 4 | 4 | 2 1/2 | 10 1/2 |
| 3 by 6 | 3 1/4 | 4 | 4 1/4 | 5 | 15 1/4 | 6 | 6 | 2 1/2 | 12 1/2 |
| 3 by 8 | 3 1/4 | 4 | 4 1/4 | 5 | 17 1/4 | 8 | 8 | 2 1/2 | 14 1/2 |
| 3 by 10 | 3 1/4 | 4 | 4 1/4 | 5 | 19 1/4 | 10 | 10 | 2 1/2 | 16 1/2 |
| 3 by 12 | 3 1/4 | 4 | 4 1/4 | 5 | 21 1/4 | 12 | 12 | 2 1/2 | 18 1/2 |
| 3 by 14 | 3 1/4 | 4 | 4 1/4 | 5 | 23 1/4 | 14 | 14 | 2 1/2 | 20 1/2 |
| 3 by 16 | 3 1/4 | 4 | 4 1/4 | 5 | 25 1/4 | 16 | 16 | 2 1/2 | 22 1/2 |
| 3 by 18 | 3 1/4 | 4 | 4 1/4 | 5 | 27 1/4 | 18 | 18 | 2 1/2 | 24 1/2 |
| 4 by 2 | 3 1/2 | 4 | 4 3/4 | 5 1/4 | 12 | 2 | 2 | 3 | 9 |
| 4 by 4 | 3 1/2 | 4 | 4 3/4 | 5 1/4 | 14 | 4 | 4 | 3 | 11 |
| 4 by 6 | 3 1/2 | 4 | 4 3/4 | 5 1/4 | 16 | 6 | 6 | 3 | 13 |
| 4 by 8 | 3 1/2 | 4 | 4 3/4 | 5 1/4 | 18 | 8 | 8 | 3 | 15 |
| 4 by 10 | 3 1/2 | 4 | 4 3/4 | 5 1/4 | 20 | 10 | 10 | 3 | 17 |
| 4 by 12 | 3 1/2 | 4 | 4 3/4 | 5 1/4 | 22 | 12 | 12 | 3 | 19 |
| 4 by 14 | 3 1/2 | 4 | 4 3/4 | 5 1/4 | 24 | 14 | 14 | 3 | 21 |
| 4 by 16 | 3 1/2 | 4 | 4 3/4 | 5 1/4 | 26 | 16 | 16 | 3 | 23 |
| 4 by 18 | 3 1/2 | 4 | 4 3/4 | 5 1/4 | 28 | 18 | 18 | 3 | 25 |

Fig. 4-11. Offset 1/8 bend is made in many sizes.

The size of increasers or reducers is specified in a particular order to prevent confusion. A 4 x 2 reducer has a 4 in. spigot and a 2 in. hub. In general, even if the fitting is a Y or T branch, the sizes of the connections are given in the following order:

1. Spigot on main.
2. Hub on main.
3. Hub on branch.
4. Hub on branch (if it is a double branch fitting).

For example; a 4 x 2 x 4 Y, means a 4 in. spigot plus a 2 in. hub on the main and a 4 in. hub on the branch of a Y.

TRAPS are installed in the drainage line at fixtures or floor drains. They are meant to stand full of water. This prevents sewer gas from escaping into the building. Fig. 4-20 shows different shapes.

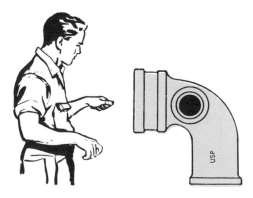

Fig. 4-12. To identify right and left side inlets, face hub toward you. (U.S. Pipe and Foundry)

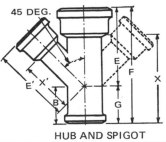

45 DEG.

HUB AND SPIGOT

| SIZE (INCHES) | DIMENSIONS IN INCHES | | | | | | |
|---|---|---|---|---|---|---|---|
| | B MIN. | E | E' | F | G | X | X' |
| 2 | 3 1/2 | 6 1/2 | 6 1/2 | 10 1/2 | 4 | 8 | 4 |
| 3 | 4 | 8 1/4 | 8 1/4 | 13 1/4 | 5 | 10 1/2 | 5 1/2 |
| 4 | 4 | 9 3/4 | 9 3/4 | 15 | 5 1/4 | 12 | 6 3/4 |
| 5 | 4 | 11 | 11 | 16 1/2 | 5 1/2 | 13 1/2 | 8 |
| 6 | 4 | 12 1/4 | 12 1/4 | 18 | 5 3/4 | 15 | 9 1/4 |
| 8 | 5 1/2 | 15 5/16 | 15 5/16 | 23 | 7 11/16 | 19 1/2 | 11 13/16 |
| 10 | 5 1/2 | 18 | 18 | 26 | 8 | 22 1/2 | 14 1/2 |
| 12 | 7 | 21 1/8 | 21 1/8 | 31 1/4 | 10 1/8 | 27 | 16 7/8 |
| 15 | 7 | 25 | 25 | 35 3/4 | 10 3/4 | 31 1/2 | 20 3/4 |
| 3 by 2 | 4 | 7 9/16 | 7 1/2 | 11 3/4 | 4 3/16 | 9 | 5 |
| 4 by 2 | 4 | 8 3/8 | 8 1/4 | 12 | 3 5/8 | 9 | 5 3/4 |
| 4 by 3 | 4 | 9 1/16 | 9 | 13 1/2 | 4 7/16 | 10 1/2 | 6 1/4 |

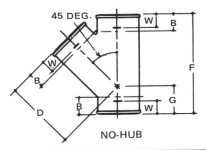

45 DEG.

NO-HUB

| SIZE (INCHES) | DIMENSIONS IN INCHES | | | | |
|---|---|---|---|---|---|
| | B | D | F | G | W |
| 1 1/2 | 1 1/2 | 4 ± 1/8 | 6 ± 1/8 | 2 | 1 1/8 |
| 2 | 1 1/2 | 4 5/8 ± 1/8 | 6 5/8 ± 1/8 | 2 | 1 1/8 |
| 3 | 1 1/2 | 5 3/4 ± 1/8 | 8 ± 1/8 | 2 1/4 | 1 1/8 |
| 4 | 1 1/2 | 7 1/16 ± 1/8 | 9 1/2 ± 1/8 | 2 7/16 | 1 1/8 |
| 5 | 2 | 9 1/2 ± 1/8 | 12 5/8 ± 1/8 | 3 1/8 | 1 1/2 |
| 6 | 2 | 10 3/4 ± 1/8 | 14 1/16 ± 1/8 | 3 5/16 | 1 1/2 |
| 3 x 2 | 1 1/2 | 5 5/16 ± 1/8 | 6 5/8 ± 1/8 | 1 1/2 | 1 1/8 |
| 4 x 2 | 1 1/2 | 6 ± 1/8 | 6 5/8 ± 1/8 | 1 | 1 1/8 |
| 4 x 3 | 1 1/2 | 6 1/2 ± 1/8 | 8 ± 1/8 | 1 11/16 | 1 1/8 |

Fig. 4-13. Y branch and double Y branch fittings are sometimes called "wye" branch. (Cast Iron Soil Pipe Institute)

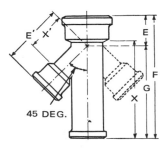

45 DEG.

| SIZE (INCHES) | DIMENSIONS IN INCHES | | | | | |
|---|---|---|---|---|---|---|
| | E | E' | F | G | X | X' |
| 2 | 3 1/4 | 5 7/8 | 12 | 8 3/4 | 9 1/2 | 3 3/8 |
| 3 | 4 | 7 3/8 | 15 1/4 | 11 1/4 | 12 1/2 | 4 5/8 |
| 4 | 4 1/2 | 8 7/8 | 17 | 12 1/2 | 14 | 5 7/8 |
| 3 by 2 | 3 1/4 | 6 5/8 | 13 3/4 | 10 1/2 | 11 | 4 1/8 |
| 4 by 2 | 3 1/16 | 7 3/8 | 14 | 10 15/16 | 11 | 4 7/8 |
| 4 by 3 | 3 3/4 | 8 1/8 | 15 1/2 | 11 3/4 | 12 1/2 | 5 3/8 |

Fig. 4-14. Inverted Y branch, single and double.

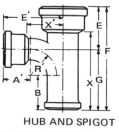

HUB AND SPIGOT

| SIZE (INCHES) | DIMENSIONS IN INCHES | | | | | | | | |
|---|---|---|---|---|---|---|---|---|---|
| | A' | B | E | E' | F | G | R' | X | X' |
| 2 | 2 3/4 | 3 3/4 | 4 1/4 | 5 1/4 | 10 1/2 | 6 1/4 | 2 1/2 | 8 | 2 3/4 |
| 3 | 3 1/4 | 4 | 5 1/4 | 6 3/4 | 12 3/4 | 7 1/2 | 3 1/2 | 10 | 4 |
| 4 | 3 1/2 | 4 | 6 | 7 1/2 | 14 | 8 | 4 | 11 | 4 1/2 |
| 5 | 3 1/2 | 4 | 6 1/2 | 8 | 15 | 8 1/2 | 4 1/2 | 12 | 5 |
| 6 | 3 1/2 | 4 | 7 | 8 1/2 | 16 | 9 | 5 | 13 | 5 1/2 |
| 8 | 4 1/8 | 5 3/4 | 8 3/4 | 10 1/8 | 20 1/2 | 11 3/4 | 6 | 17 | 6 5/8 |
| 10 | 4 1/8 | 5 3/4 | 9 3/4 | 11 1/8 | 22 1/2 | 12 3/4 | 7 | 19 | 7 5/8 |
| 12 | 5 | 7 | 11 3/4 | 13 | 26 3/4 | 15 | 8 | 22 1/2 | 8 3/4 |
| 15 | 5 | 7 | 13 1/4 | 14 1/2 | 29 3/4 | 16 1/2 | 9 1/2 | 25 1/2 | 10 1/4 |
| 3 by 2 | 3 | 4 | 4 3/4 | 6 1/2 | 11 3/4 | 7 | 3 | 9 | 4 |
| 4 by 2 | 3 | 4 | 5 | 7 | 12 | 7 | 3 | 9 | 4 1/2 |
| 4 by 3 | 3 1/4 | 4 | 5 1/2 | 7 1/4 | 13 | 7 1/2 | 3 1/2 | 10 | 4 1/2 |

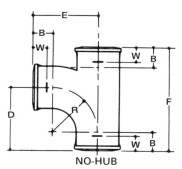

NO-HUB

| SIZE (INCHES) | DIMENSIONS IN INCHES | | | | | |
|---|---|---|---|---|---|---|
| | B | E | F | D | R | W |
| 1 1/2 | 1 1/2 | 4 1/4 ± 1/8 | 6 1/2 ± 1/8 | 4 1/4 | 2 3/4 | 1 1/8 |
| 2 | 1 1/2 | 4 1/2 ± 1/8 | 6 7/8 ± 1/8 | 4 1/2 | 3 | 1 1/8 |
| 3 | 1 1/2 | 5 ± 1/8 | 8 ± 1/8 | 5 | 3 1/2 | 1 1/8 |
| 4 | 1 1/2 | 5 1/2 ± 1/8 | 9 1/8 ± 1/8 | 5 1/2 | 4 | 1 1/8 |
| 3 x 1 1/2 | 1 1/2 | 5 ± 1/8 | 6 1/2 ± 1/8 | 4 1/4 | 2 3/4 | 1 1/8 |
| 3 x 2 | 1 1/2 | 5 ± 1/8 | 6 7/8 ± 1/8 | 4 1/2 | 3 | 1 1/8 |
| 3 x 4 | 1 1/2 | 5 ± 1/8 | 9 ± 1/8 | 5 1/2 | 3 1/2 | 1 1/8 |
| 4 x 2 | 1 1/2 | 5 1/2 ± 1/8 | 6 7/8 ± 1/8 | 4 1/2 | 3 | 1 1/8 |
| 4 x 3 | 1 1/2 | 5 1/2 ± 1/8 | 8 ± 1/8 | 5 | 3 1/2 | 1 1/8 |

Fig. 4-15. Sanitary T branches. (Cast Iron Pipe Institute)

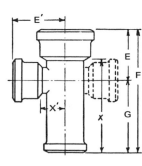

| SIZE (INCHES) | DIMENSIONS IN INCHES | | | | | |
|---|---|---|---|---|---|---|
| | E | E' | F | G | X | X' |
| 2 | 4 1/4 | 4 1/4 | 10 1/2 | 6 1/4 | 8 | 1 3/4 |
| 3 | 5 1/4 | 5 1/4 | 12 3/4 | 7 1/2 | 10 | 2 1/2 |
| 4 | 6 | 6 | 14 | 8 | 11 | 3 |
| 5 | 6 1/2 | 6 1/2 | 15 | 8 1/2 | 12 | 3 1/2 |
| 6 | 7 | 7 | 16 | 9 | 13 | 4 |
| 3 by 2 | 4 3/4 | 5 | 11 3/4 | 7 | 9 | 2 1/2 |
| 4 by 2 | 5 | 5 1/2 | 12 | 7 | 9 | 3 |
| 4 by 3 | 5 1/2 | 5 3/4 | 13 | 7 1/2 | 10 | 3 |

Fig. 4-16. Single and double T branches. (Cast Iron Soil Pipe Institute)

Fig. 4-17. To determine if sanitary T branch inlet is right side or left side, face the hub on the main toward you allowing the branch to point downward. Right-hand inlet is shown. (U.S. Pipe and Foundry)

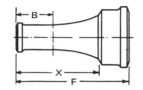

### HUB AND SPIGOT

| SIZE (INCHES) | B<br>INCHES | F<br>INCHES | X<br>INCHES |
|---|---|---|---|
| 2 by 3 | 4 | 11 3/4 | 9 |
| 2 by 4 | 4 | 12 | 9 |
| 2 by 5 | 4 | 12 | 9 |
| 2 by 6 | 4 | 12 | 9 |
| 3 by 4 | 4 | 12 | 9 |
| 3 by 5 | 4 | 12 | 9 |
| 3 by 6 | 4 | 12 | 9 |
| 4 by 5 | 4 | 12 | 9 |
| 4 by 6 | 4 | 12 | 9 |

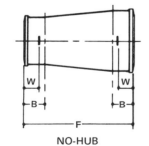

### NO-HUB

| SIZE (INCHES) | DIMENSIONS IN INCHES | | |
|---|---|---|---|
| | B | F | W |
| 2 x 3 | 1 1/2 | 8 ± 1/8 | 1 1/8 |
| 2 x 4 | 1 1/2 | 8 ± 1/8 | 1 1/8 |
| 3 x 4 | 1 1/2 | 8 ± 1/8 | 1 1/8 |

Fig. 4-18. Dimensions of two types of increasers are given for many different sizes. (Cast Iron Soil Pipe Institute)

## Closet fixtures

CLOSET BENDS, Fig. 4-21, connect the water closet to the main drainage line. Both the inlet and outlet ends of this fitting are cast to permit easy cutting. Closet bends are made in a large variety of styles. Some permit the attachment of

### HUB AND SPIGOT DOUBLE HUBS

| SIZE (INCHES) | F<br>INCHES | X<br>INCH |
|---|---|---|
| 2 | 6 | 1 |
| 3 | 6 1/2 | 1 |
| 4 | 7 | 1 |

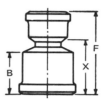

### HUB AND SPIGOT REDUCERS

| SIZE (INCHES) | B<br>INCHES | F<br>INCHES | X<br>INCHES |
|---|---|---|---|
| 3 by 2 | 3 1/4 | 7 1/4 | 4 3/4 |
| 4 by 2 | 4 | 7 1/2 | 5 |
| 4 by 3 | 4 | 7 3/4 | 5 |
| 5 by 2 | 4 | 7 1/2 | 5 |
| 5 by 3 | 4 | 7 3/4 | 5 |
| 5 by 4 | 4 | 8 | 5 |

Fig. 4-19. Hub and spigot double hubs and reducers are available in these sizes and dimensions.

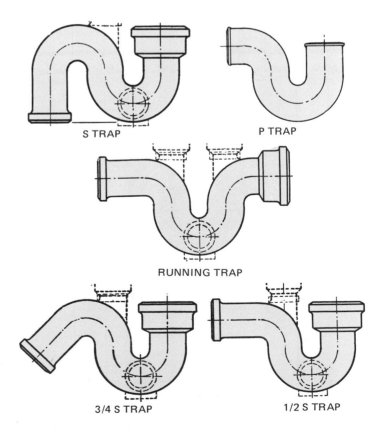

Fig. 4-20. Hub and spigot and no-hub traps have names that usually describe their basic shapes.

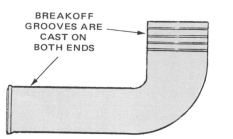

BREAKOFF
GROOVES ARE
CAST ON
BOTH ENDS

Fig. 4-21. Closet bend is adjustable to different length by shortening to different break-off points at either end. (Cast Iron Soil Pipe Institute)

additional drainage lines. Fig. 4-22 will help you determine if the inlet is right or left.

CLOSET FLANGES, Fig. 4-23, connect water closets to the drain system. ROOF DRAINS, Fig. 4-24, permit water to enter the roof storm drainage system and prevent large objects from clogging the pipe or entering storm sewers. FLOOR DRAINS, Fig. 4-25, are installed in concrete floors and permit the escape of water while providing a safe walking surface.

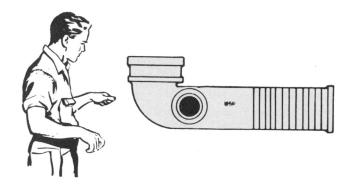

Fig. 4-22. To determine if inlet is right or left-hand, hold closet bend in regular position with hub end nearest you and opening pointing up. Right-hand inlet is shown. (U.S. Pipe and Foundry)

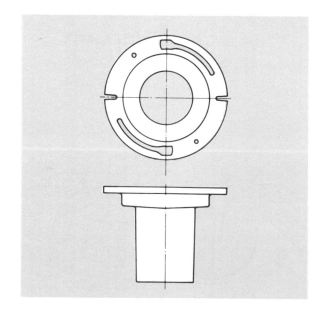

Fig. 4-23. Closet flange attaches water closet to the drainage system. (Cast Iron Soil Pipe Institute)

Fig. 4-24. Roof drain is used on flat roofs to take away water. Its construction prevents loose objects from entering the drain system without completely clogging the opening.

Fig. 4-25. Floor drains are designed to be cemented into concrete floor as the floor is being poured. (V. Andy Smith)

Cast-iron parts for making repairs and changes in systems are also available. REPAIR PLATES, Fig. 4-26, cover leaks in pipes. The HUB Y and TAPPED T, Fig. 4-27, make connection to existing lines much easier.

Fig. 4-26. Repair plates clamp around pipe to seal off leaks.

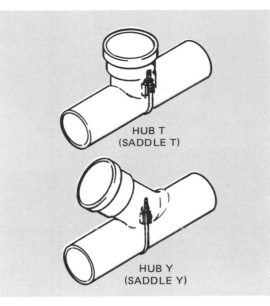

Fig. 4-27. Hub Y and Hub T can be tapped into pipe where new lines are to be added.

nominal sizes as small as 1/8 in. and as large as 2 1/2 in. See Fig. 4-28. There are three grades (strengths) of pipe:

1. Standard.
2. Extra strong.
3. Double extra strong.

Standard weight piping is adequate for most plumbing installations.

## PIPE THREADS

The threads on iron pipe fittings are tapered, Fig. 4-29, so they will form a watertight joint when tightened securely. In Unit 9, the correct allowances are given for each size pipe. This is important so that the threads cut on the pipe will properly join with standard fittings.

## MALLEABLE IRON FITTINGS

Malleable iron fittings are produced by annealing (softening) cast iron. This process produces a fitting which will withstand more bending, pounding and internal pressure than ordinary cast iron.

## STEEL PIPE

Steel pipe is used for:

1. Hot and cold water distribution.
2. Steam and hot water heating systems.
3. Gas and air piping systems.
4. Drainage and vent piping.

Steel pipe is either unfinished (black) or galvanized (zinc coated). *Galvanized pipe is generally required in water supply and drainage piping because it is much less subject to rusting than black pipe.*

The standard length of steel pipe is 21 ft. It is produced in

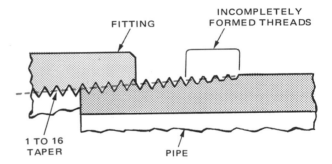

Fig. 4-29. Section view shows how tapered threads go together on pipe sections to form seal against leakage.

|  | NOMINAL SIZE | OUTSIDE DIAMETER | INSIDE DIAMETER | THREADS PER INCH | LENGTH OF PERFECT THREAD | AREA INSIDE (SQ. IN.) | WEIGHT POUND PER FOOT |
|---|---|---|---|---|---|---|---|
| **STANDARD** | 1/8 | 0.405 | 0.269 | 27 | 0.264 | 0.057 | 0.24 |
|  | 1/4 | 0.540 | 0.364 | 18 | 0.402 | 0.104 | 0.42 |
|  | 3/8 | 0.675 | 0.493 | 18 | 0.408 | 0.191 | 0.57 |
|  | 1/2 | 0.840 | 0.622 | 14 | 0.534 | 0.304 | 0.85 |
|  | 3/4 | 1.050 | 0.824 | 14 | 0.546 | 0.533 | 1.13 |
|  | 1 | 1.315 | 1.049 | 11  1/2 | 0.683 | 0.864 | 1.68 |
|  | 1 1/4 | 1.660 | 1.380 | 11  1/2 | 0.707 | 1.495 | 2.27 |
|  | 1 1/2 | 1.900 | 1.610 | 11  1/2 | 0.724 | 2.036 | 2.72 |
|  | 2 | 2.375 | 2.067 | 11  1/2 | 0.757 | 3.355 | 3.65 |
|  | 2 1/2 | 2.875 | 2.469 | 8 | 1.138 | 4.778 | 5.79 |
| **EXTRA STRONG** | 1/8 | 0.405 | 0.215 | 27 | 0.264 | 0.036 | 0.31 |
|  | 1/4 | 0.540 | 0.302 | 18 | 0.402 | 0.072 | 0.54 |
|  | 3/8 | 0.675 | 0.423 | 18 | 0.408 | 0.141 | 0.74 |
|  | 1/2 | 0.840 | 0.546 | 14 | 0.534 | 0.234 | 1.09 |
|  | 3/4 | 1.050 | 0.742 | 14 | 0.546 | 0.433 | 1.47 |
|  | 1 | 1.315 | 0.957 | 11  1/2 | 0.683 | 0.719 | 2.17 |
|  | 1 1/4 | 1.660 | 1.278 | 11  1/2 | 0.707 | 1.283 | 3.00 |
|  | 1 1/2 | 1.900 | 1.500 | 11  1/2 | 0.724 | 1.767 | 3.63 |
|  | 2 | 2.375 | 1.939 | 11  1/2 | 0.757 | 2.953 | 5.02 |
|  | 2 1/2 | 2.875 | 2.323 | 8 | 1.138 | 4.238 | 7.66 |

Fig. 4-28. Dimensions of standard and extra strong steel pipe.

Two types of iron fittings are manufactured:
1. Pressure fittings, Fig. 4-30.
2. Drainage fittings, Fig. 4-31.

Drainage fittings are different from pressure fittings in several ways:
1. The insides of drainage fittings are smooth and shaped for easy flow.
2. Shoulders are recessed so that an unbroken contour is formed when pipes are screwed into the fitting.
3. Drainage fittings are so designed that a horizontal line entering them will have a fall of 1/4 in. per ft.

Only drainage fittings should be used on drainage piping. Either pressure or drainage fittings may be used for installing water, gas or air vent lines.

Both drainage and pressure fittings are produced in many shapes and sizes. Because drainage lines are never less than 1 1/4 in. diameter, drainage fittings are not available in sizes less

90 DEGREE L       45 DEGREE L

90 DEGREE STREET L

DROP-EAR L

90 DEGREE REDUCING L

Fig. 4-32. Elbows are designed for either a 45 deg. or a 90 deg. change of direction. These are the most commonly used. (V. Andy Smith)

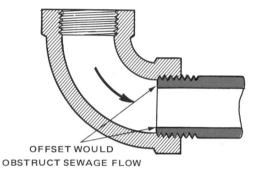

OFFSET WOULD
OBSTRUCT SEWAGE FLOW

Fig. 4-30. Pressure fittings are suitable for air, gas and water lines but not for drainage.

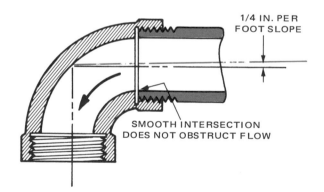

1/4 IN. PER
FOOT SLOPE

SMOOTH INTERSECTION
DOES NOT OBSTRUCT FLOW

Fig. 4-31. This type of fitting is required on all drainage lines.

than 1 1/4 in. However, the basic shapes of the fittings are the same for both types and they will, therefore, be discussed together.

## ELBOWS

Elbows are used to change the direction of a pipeline. They are also called "Ls" or "ells." Several types of elbows are shown in Fig. 4-32.

The DROP elbow or drop ear elbow permits attaching the

pipeline to the building frame. It is frequently used at the last joint before the pipe comes through the wall to be attached to a fixture.

## T FITTINGS

The T, Fig. 4-33, is used to make branches at 90 deg. angles to the main pipe. When all three outlets are the same size, the T is specified as that size. For example, a 1 in. T will connect three 1 in. pipes. When the outlets are not all the same size, the fitting is called a reducing T. It is specified by giving the run (straight through) dimension followed by the side outlet (branch) dimension. For example, a 3/4 x 1/2 T has 3/4 in. openings on the main and a 1/2 in. branch outlet, Fig. 4-34. A 1 x 3/4 x 1/2 T has a 1 in. and a 3/4 in. outlet on the main and a 1/2 in. branch outlet.

## COUPLINGS

COUPLINGS, Fig. 4-35, are short fittings with female threads at both ends. They are used to connect lengths of pipe on straight runs. Reducing couplings are used to connect pipes of different sizes and are specified by the diameter of each opening. For example, a 1/2 x 3/4 coupling has a 1/2 in. and a 3/4 in. opening.

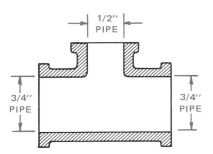

Fig. 4-33. Branch lines are connected to main lines with T fittings like these.

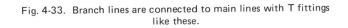

Fig. 4-34. Reducing Ts are specified by the size of the opening on the main followed by the size of the branch opening. T shown is 3/4 x 1/2.

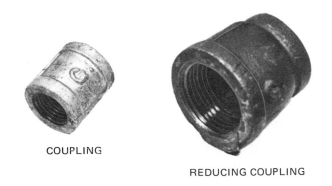

Fig. 4-35. Couplings connect two pipes of equal or unequal sizes. (V. Andy Smith)

## UNIONS

Since all pipes are right-hand threaded, it would be impossible to assemble or disassemble the last length of threaded pipe even in the simplest system without a UNION.

A union can be installed or removed from the system without disturbing other fittings. It consists of three parts:
1. A shoulder, with female threads at one end for attaching to a pipe. The shoulder is shaped to mate with the male part.
2. A collar with female threads for attaching the other two parts over their mating surfaces.
3. A piece with male threads for the collar and mating surface for the shoulder at one end, and female threads at the other end for attaching to a pipe.

In some designs, the shoulder and the male threaded piece, have a machined spherical joint which seals watertight when the collar is securely tightened. There are several different joint shapes but all of them are machined for better fit. Fig. 4-36 shows one type. One or both of the mating surfaces may be of brass to prevent rusting and produce a better seal.

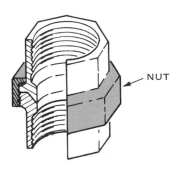

Fig. 4-36. Unions are much alike but may have different shaped joint surfaces. In the spherical type, the end of one part is shaped like a ball with a hole in it. The end of the other part is shaped to fit over it and the two are held together securely by the threaded collar.

## NIPPLES

NIPPLES, are pieces of pipe, 12 in. or less in length threaded on both ends. They are used to join two fittings which are close together. These should be purchased because it is difficult to thread short pieces of pipe with conventional plumbing tools. Do not attempt to make them.

Nipples are specified by diameter and length, Fig. 4-37. A

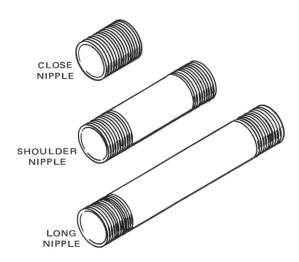

Fig. 4-37. Nipples are short connectors between fittings.

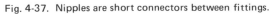

close nipple is threaded along its entire length. Shoulder nipples have a short portion of unthreaded pipe. Length of a pipe nipple is determined by measuring from end to end.

## OTHER MALLEABLE IRON FITTINGS

Pipe PLUGS, Fig. 4-38, have male threads and are used to close openings in other fittings.

Pipe CAPS, Fig. 4-39, have female threads and are used to close the end of a pipe or a pipe nipple.

A BUSHING, Fig. 4-40, has male threads on the outside and female threads inside. It is used to connect a pipe to a larger size fitting.

Fig. 4-40. A bushing takes up the difference in diameter when a smaller pipe must be connected to a larger fitting. (V. Andy Smith)

Fig. 4-38. Plugs close openings in fittings. (V. Andy Smith)

Fig. 4-39. Caps close ends of pipes.

## COPPER PIPE AND FITTINGS

There are four types of copper water pipe, or plumbing tube as it is also often called. In the order of their weight, heaviest to lightest, they are: K, L, M and DWV. The weight refers to thickness of the wall. Heavier pipes have heavier walls.

Pipe size, set by the American Society for Testing and Materials (ASTM), is nominal. (A nominal size is one that is bigger or smaller than actual measurement.) The outside diameter of copper pipe is always 1/8 in. larger than the standard designation. Inside diameter will vary according to wall thickness. Thus, the inside diameter of heavy-walled copper pipe will always be smaller than the inside diameter of thin-walled pipe of the same size.

An important property of copper pipe is its hardness or softness. Fig. 4-41 shows that some types of copper pipe are available in both hard and soft temper while others are manufactured in only one temper. Hard temper pipe is more rigid and better suited for application where it will be exposed. Because it is rigid, it will generally make a more attractive finished project than the easier-to-bend soft temper copper tube.

The primary advantage of soft temper tube is that it comes in long coils which can be bent to nearly any shape.

## SOLDER JOINT FITTINGS

There are two types of solder fittings for copper pipe:
1. Wrought copper fittings.
2. Cast solder fittings.

Wrought copper fittings are made from copper tubing cut and shaped as required. With their thin walls and smooth

| TYPE | COLOR CODE | APPLICATION | STRAIGHT LENGTHS | COILS (SOFT TEMPER ONLY) |
|---|---|---|---|---|
| K | GREEN | UNDERGROUND AND INTERIOR SERVICE | 20 FT. IN DIAMETERS INCLUDING 8 IN. HARD AND SOFT TEMPER | 60 FT. AND 100 FT. FOR DIAMETERS INCLUDING 1 IN. |
| L | BLUE | ABOVE GROUND SERVICE | 20 FT. IN DIAMETERS INCLUDING 10 IN. HARD AND SOFT TEMPER | (SAME AS K) |
| M | RED | ABOVE GROUND WATER SUPPLY DRAINAGE, WASTE, AND VENT | 20 FT. IN ALL DIAMETERS HARD TEMPER ONLY | NOT AVAILABLE |
| DWV | YELLOW | ABOVE GROUND DRAIN, WASTE AND VENT PIPING | 20 FT. IN ALL DIAMETERS 1 1/4 IN. AND GREATER HARD TEMPER ONLY | NOT AVAILABLE |

Fig. 4-41. Grades of copper pipe are given in letters. A color code is also used for instant recognition on the job. Hard temper pipe is also called "drawn." Soft temper pipe is also known as "annealed."

exterior, wrought fittings are easy to recognize. Cast fittings are heavier and have a rougher surface. Cast solder fittings are available in a greater variety of shapes.

Solder joint pressure fittings, Fig. 4-42, perform the same functions as do similar fittings in malleable iron or cast iron. They are designated the same way and fit the various pipe sizes.

Copper drainage fittings, Fig. 4-43, are shaped much like iron fittings and are available with inlets of 1 1/2 in. diameter and greater.

In addition to the basic fittings, there are a group of special fittings for connecting copper to galvanized iron pipe, Fig. 4-44, and soil pipe, Fig. 4-45.

## PLASTIC PIPE AND FITTINGS

Use of plastic pipe for plumbing is increasing rapidly because of the relatively low material costs and ease of installation.

Three grades of plastic pipe and fittings are likely to be

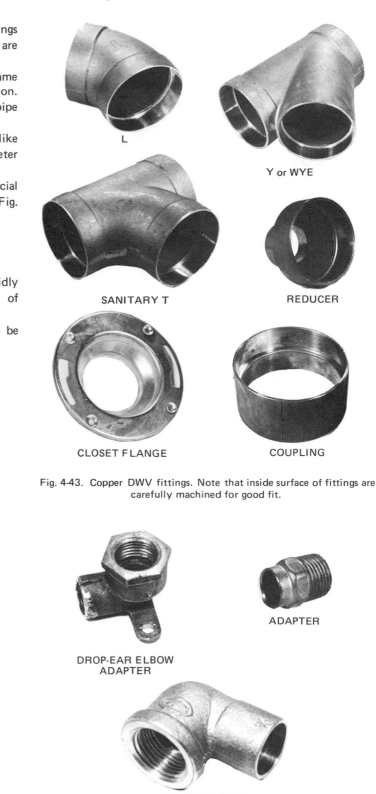

L

Y or WYE

SANITARY T

REDUCER

CLOSET FLANGE

COUPLING

Fig. 4-43. Copper DWV fittings. Note that inside surface of fittings are carefully machined for good fit.

COUPLING  CAP  REDUCER

90 DEGREE L  DROP-EAR L

UNION  BUSHING

45 DEGREE L  T

Fig. 4-42. Copper pipe fittings may be wrought or cast. (V. Andy Smith)

DROP-EAR ELBOW ADAPTER

ADAPTER

ELBOW ADAPTER

Fig. 4-44. Special copper fittings connect copper to galvanized pipe and fittings. (V. Andy Smith)

used in residential or commercial construction. These are:
1. Schedule 40 pipe.
2. Pressure pipe.
3. Pipe used for noncode applications.

Schedule 40 pipe and fittings are manufactured in the same

DWV SOIL PIPE
ADAPTER

DWV COUPLING ADAPTER

Fig. 4-45. Special copper fittings connect copper to cast-iron pipe and fittings. (V. Andy Smith)

standard shapes and sizes as iron pipe and fittings. They are approved by many building codes for DWV and cold water supply. However, other codes will not allow ABS plastic piping in any application except underground waste.

Pressure-rated pipe is designed for water supply. Maximum working pressure for this pipe is given on the label, Fig. 4-46.

Pressure-rated pipe provides an entire system with pipe of uniform strength. Pressure-rated pipe is most useful where a variety of pipe sizes must be installed. Municipal water lines and irrigation systems are two such applications.

In the third group are other types of plastic used in noncode places such as septic tank leach fields and building drains. These pipes are generally lighter and may not be completely watertight.

## PLASTIC FITTINGS

Fig. 4-47 shows standard fittings for plastic pipe. Fig. 4-48 illustrates special fittings for connecting plastic to cast iron, galvanized and copper pipe. Fig. 4-49 will assist you in understanding the usages of the various types of plastic pipe in residential and commercial plumbing.

## TYPES OF PLASTIC

ACRYLONITRILE-BUTADIENE-STYRENE (ABS) plastic pipe and fittings are used primarily for drain, waste and vent

Fig. 4-46. Strength of pressure-rated plastic pipe is marked on the pipe itself. (Celanese Plastics Co.)

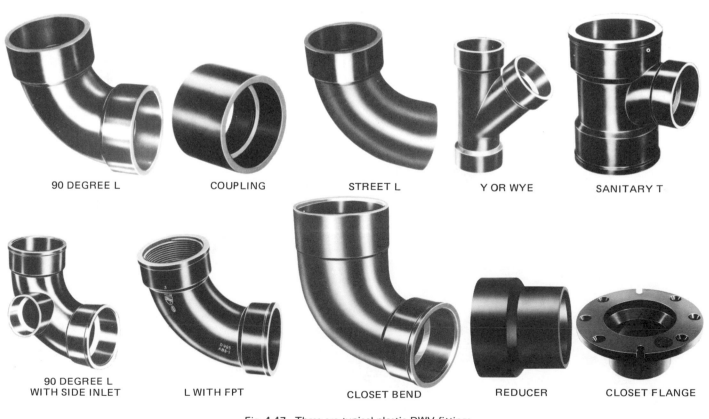

90 DEGREE L     COUPLING     STREET L     Y OR WYE     SANITARY T

90 DEGREE L WITH SIDE INLET     L WITH FPT     CLOSET BEND     REDUCER     CLOSET FLANGE

Fig. 4-47. These are typical plastic DWV fittings.

ADAPTER      ADAPTER      ADAPTER      HUBLESS ADAPTER

LONG TY      Y OR WYE      T WITH INLETS      SANITARY T

Fig. 4-48. Special plastic fittings allow the plumber to connect plastic pipe to copper, galvanized or cast-iron pipe.
(Celanese Plastics Co.)

piping. Two grades of ABS pipe are manufactured, Schedule 40 and Service. Schedule 40 is generally required by building codes for plumbing within a structure. The grade is identified by a stamping on the pipe. See Fig. 4-50.

POLYVINYL-CHLORIDE (PVC) is used primarily in pressure pipe applications such as water distribution, irrigation and natural gas distribution. However, it is also used for DWV piping. PVC provides high impact strength, high tensile strength, excellent weathering characteristics and high corro-

sion resistance to chemicals such as acids and alkalies.

PVC pipe and fittings are available in 3/8 in. to 16 in. diameters. There are several grades which relate to pressure rating. Schedule 40 DWV-PVS is the same as cast-iron DWV.

CHLORINATED POLYVINYL CHLORIDE (CPVC) is a relatively new addition to the plastic pipe field. It is suitable for piping hot water and chemicals. It is normally rated at 180 deg. F and 100 psi (pounds of pressure per sq. in.) rather than the 73 deg. F and 100 psi rating given PVC.

| TYPE PLASTIC | USES | SIZES/GRADES AVAILABLE | JOINTING TECHNIQUES | RIGID/FLEXIBLE |
|---|---|---|---|---|
| ABC | DRAIN/WASTE/VENT | 1 1/4 – 6 SCHEDULE 40 (SCH. 40) | SOLVENT CEMENT | RIGID |
| | SEWER LINES | 3 – 8 STANDARD DIMENSION RATIO (SDR) | | |
| | WATER SUPPLY | 1/2–2 SDR | | |
| PVC | WATER DISTRIBUTION | 1/4 – 12 SCH. 40 & 80 AND SDR | SOLVENT WELDING | RIGID |
| | DWV, SEWERS, & PROCESS PIPING | | SCH. 80 – THREADED & FLANGES | |
| CPVC | HOT & COLD WATER DISTRIBUTION | 1/2–3/4 | SOLVENT WELDING | RIGID |
| | PROCESS PIPING | 1/2–8 SCH. 40 & 80 | THREADING & FLANGES | |
| PE | WATER DISTRIBUTION | 1/2 – 2 IPS | COMPRESSION & FLANGE FITTINGS | FLEXIBLE |
| | NATURAL GAS DISTRIBUTION OIL FIELD PIPING | 1/2 – 6 IPS | | |
| | WATER TRANSMISSION, SEWERS | 3 – 48 SDR | FUSION WELDING | |
| SR | AGRICULTURE FIELD DRAINS STORM DRAINS | 3 – 8 SDR | SOLVENT WELDING | RIGID |

Fig. 4-49. Plastic pipe is designed for certain uses. Chart indicates types, uses, grades, sizes (in inches), jointing techniques, and rigidity.

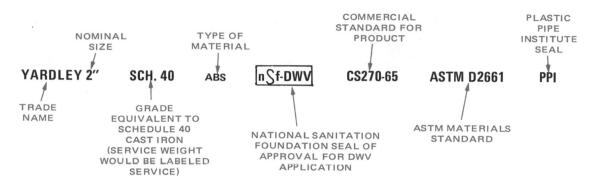

Fig. 4-50. Typical grade stamps used for Schedule 40 and Service weight ABS pipe. Stamps for other types of plastic pipe are almost identical.

POLYETHYLENE (PE) is used most frequently to pipe natural gas but has limited application in water distribution. It is available in low, medium and high density. *Only high density should be used for plumbing.*

STYRENE-RUBBER (SR) is used in septic tanks, drain fields, storm drains, sewers and other noncritical installations.

PVC, ABS, PE and SR make up nearly 90 percent of all plastic pipe and fittings produced commercially. Other kinds are made for a variety of special purposes. Some of them are:

1. Polypropylene. Used for chemical waste piping.
2. Polybutylene. Similar to polyethylene with increased heat resistance.
3. Polyolefin. Used for corrosive waste piping.

Care must be used not to join different types of plastic piping. Their solvents are different and it is impossible to secure a watertight joint.

## LABELING FOR PLASTIC PIPE AND FITTINGS

The labels on plastic pipe and fittings offer some important clues to the quality of the product. Besides indicating the type of material (ABS, PVC, PE or CPVC) the four digit code indicates the type, grade and tensile strength of the pipe to the nearest 100 lb. See Fig. 4-51.

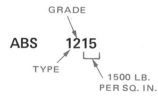

Fig. 4-51. Four-digit code indicates type, grade and tensile strength of pipe and fittings.

To assure users that plastic plumbing products are high quality and suitable for potable water supply, the National Sanitation Foundation (NSF) has a testing program. Under this program a manufacturer may seek the Foundation's seal of approval, Fig. 4-52. It is a sign that the firm's products meet the NSF standards.

To get this approval:

1. Each product line must pass a test for size, strength, durability and toxicity.
2. The manufacturer must permit the NSF to make random inspections of the production steps, inspection techniques and warehousing facilities.
3. The manufacturer must show NSF production records indicating the quality of the materials used.

If, at any time, products are found to be defective, the manufacturer must correct the problem and show that substandard products have been removed from the market. Failure to do so means loss of the NSF seal.

Through the NSF program the quality of a product is known without extensive testing on the job. *Plumbers must understand the coding system and the building code requirements in the area where they work.*

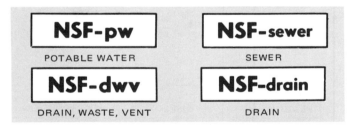

Fig. 4-52. National Sanitation Foundation seals are used on approved plastic pipe.

## VITRIFIED CLAY PIPE

VITRIFIED CLAY PIPE is frequently used for sanitary sewer mains and may be used below grade to connect the house sewer to the sewer main. This kind of pipe is available in a variety of diameters from 4 to 36 in. See Fig. 4-53.

The most recent major change in clay pipe has been the addition of an O-ring compression joint. A rubber compression ring is seated in a polyester casting, Fig. 4-54. With this compression joint, assembly is faster. Some care must be taken in laying clay pipe. Ground supporting it must be well compacted. If soil is loose, additional support must be provided by laying down boards or planking. Backfilling (filling up the excavated trench) can begin as soon as the pipe has been joined.

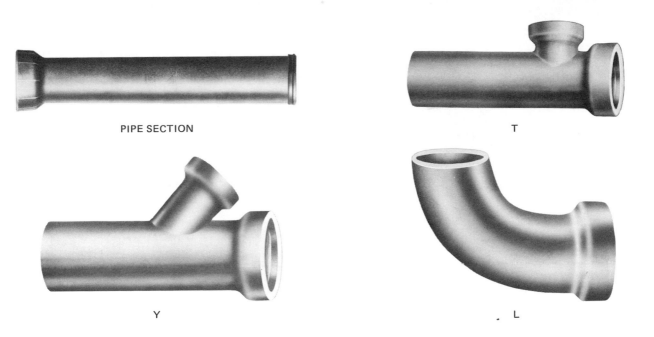

PIPE SECTION

T

Y

. L

Fig. 4-53. Vitrified clay pipe has a standard length of 5 ft. for diameters of 4 in. through 10 in. But most small sizes are also available in 1, 2, 3 and 4 ft. lengths. (The Logan Clay Products Co.)

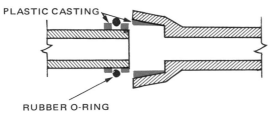

PLASTIC CASTING

RUBBER O-RING

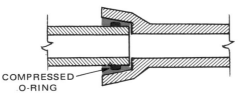

COMPRESSED O-RING

Fig. 4-54. Compression joint in vitrified clay pipe assembles rapidly and needs no additional "setup" time.

## OTHER PIPING MATERIALS

Water supply connection to humidifiers and ice makers are frequently made using copper tubing and brass fittings, Fig. 4-55. Two general types of brass fittings are available:

1. Flared fittings, Fig. 4-56. The end of the tubing must be flared so that the joint seals when the nut is tightened.

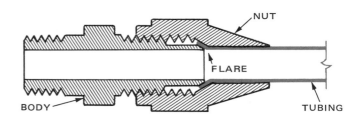

NUT

FLARE

BODY

TUBING

Fig. 4-56. Flared brass fitting depends on flared pipe end for proper seal.

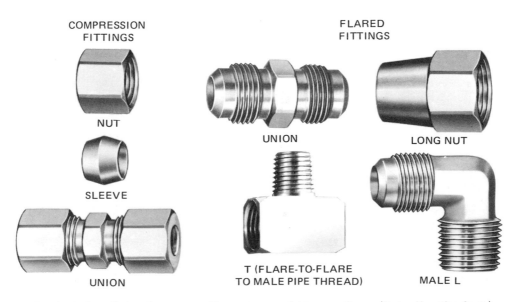

COMPRESSION FITTINGS

NUT

SLEEVE

UNION

FLARED FITTINGS

UNION

LONG NUT

T (FLARE-TO-FLARE TO MALE PIPE THREAD)

MALE L

Fig. 4-55. Brass fittings for copper tubing make watertight connections. (Parker Hannifen Corp.)

2. Compression fittings, Fig. 4-57. A watertight seal is formed when a brass sleeve is compressed between the nut and the fitting.

Glass drain lines, Fig. 4-58, are installed in laboratories where extremely corrosive wastes are likely to be discharged on a regular basis. This type of installation requires special tools and procedures. Stainless steel, aluminum, Monel and several other materials are also used to make corrosion-resistant pipe and fittings. These materials are used almost exclusively on industrial applications.

LEAD PIPE is still in limited use, particularly in repair of older plumbing systems. Because of its flexibility and durability, lead pipe has been used frequently to construct drain and vent piping. The major disadvantages of using lead are the skill required to form and join it, and that it may contaminate the water supply. Lead pipe is available in several grades, Fig. 4-59.

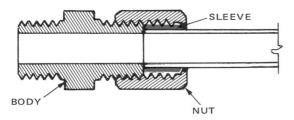

Fig. 4-57. Compression brass fitting presses against body of fitting and pipe surface to form seal.

Fig. 4-58. Glass pipe and fittings are installed where corrosive materials are to be handled by the plumbing. (Owens International Inc.)

## TEST YOUR KNOWLEDGE — UNIT 4

1. The term _____ pipe is frequenty used when referring to cast-iron pipe.
2. The two standardized grades of cast-iron pipe are _____, abbreviated _____, and _____, abbreviated _____.
3. The two types of ends found on soil pipe are (check two):
   - a. Compression.
   - b. Hub and spigot.
   - c. Lead and oakum.
   - d. No-hub.
4. Match the pipe fitting shape shown in the left-hand column of Fig. 4-60 with its correct name in the right-hand column.
5. T branches may be used on drainage piping. True or False?
6. A cast-iron sanitary T specified as a 3 x 2 x 1 1/2 would have a _____ in. spigot on the main, a _____ in. hub on the main and a _____ in. hub on the branch.
7. Write the specifications for a Y which has a 2 in. hub on the branch, a 4 in. spigot on the main, and a 3 in. hub on the main.
8. Steel pipe which is coated to prevent rusting is known as:
   - a. Black pipe.
   - b. Galvanized pipe.
   - c. Heating pipe.
   - d. Rustless pipe.
9. Malleable iron drainage fittings differ from pressure fittings in three ways:
   - a. The inside of the fitting is smooth and _____ for any flow.

| NAME | INSIDE DIAMETER (INCHES) | | | | | | | | | | | |
| | 1/2 | | 3/4 | | 1 | | 1 1/4 | | 1 1/2 | | 2 | |
| | WT.[1] | THK.[2] | WT. | THK. | WT. | THK. | WT. | THK. | WT. | THK. | WT. | THK. |
|---|---|---|---|---|---|---|---|---|---|---|---|---|
| EXTRA LIGHT | 3/4 | .082 | 1 1/2 | .112 | 2 | .115 | 2 1/2 | .118 | 3 1/2 | .135 | 4 | .123 |
| LIGHT | 1 | .106 | 2 | .144 | 2 1/2 | .141 | 3 | .138 | 4 | .153 | 5 | .150 |
| MEDIUM | 1 1/4 | .128 | 2 1/4 | .160 | 3 1/4 | .207 | 3 3/4 | .171 | 5 | .187 | 7 | .204 |
| STRONG | 1 3/4 | .169 | 3 | .201 | 4 | .212 | 4 3/4 | .209 | 6 | .221 | 8 | .231 |
| EXTRA STRONG | 2 1/2 | .223 | 3 1/2 | .230 | 4 3/4 | .246 | 6 | .255 | 7 1/2 | .270 | 9 | .259 |

[1]WEIGHT IN POUNDS PER FOOT
[2]THICKNESS OF WALL IN INCHES

Fig. 4-59. Lead pipe grades and sizes.

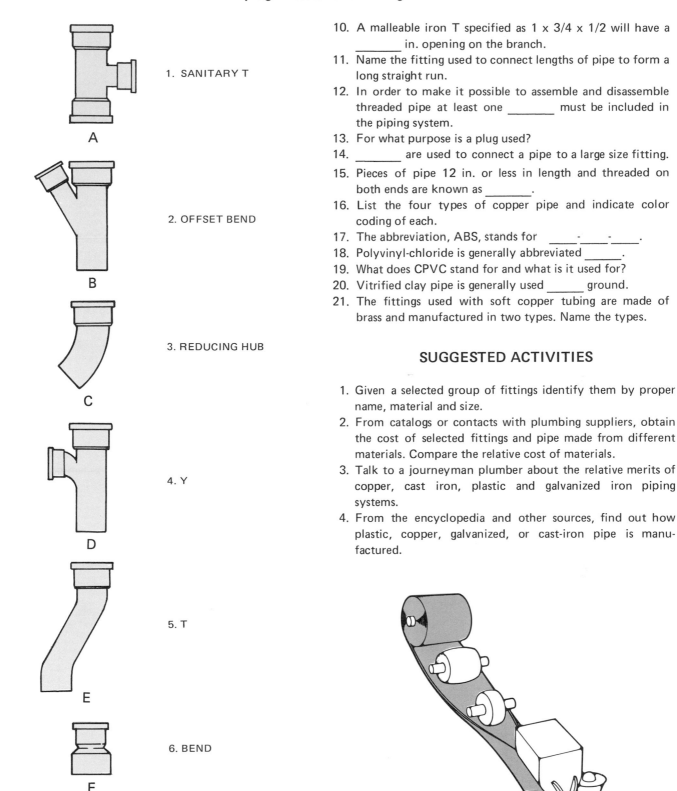

1. SANITARY T

2. OFFSET BEND

3. REDUCING HUB

4. Y

5. T

6. BEND

Fig. 4-60. Match the fitting with the correct name.

10. A malleable iron T specified as 1 x 3/4 x 1/2 will have a _____ in. opening on the branch.
11. Name the fitting used to connect lengths of pipe to form a long straight run.
12. In order to make it possible to assemble and disassemble threaded pipe at least one _____ must be included in the piping system.
13. For what purpose is a plug used?
14. _____ are used to connect a pipe to a large size fitting.
15. Pieces of pipe 12 in. or less in length and threaded on both ends are known as _____.
16. List the four types of copper pipe and indicate color coding of each.
17. The abbreviation, ABS, stands for ____-____-____.
18. Polyvinyl-chloride is generally abbreviated _____.
19. What does CPVC stand for and what is it used for?
20. Vitrified clay pipe is generally used _____ ground.
21. The fittings used with soft copper tubing are made of brass and manufactured in two types. Name the types.

## SUGGESTED ACTIVITIES

1. Given a selected group of fittings identify them by proper name, material and size.
2. From catalogs or contacts with plumbing suppliers, obtain the cost of selected fittings and pipe made from different materials. Compare the relative cost of materials.
3. Talk to a journeyman plumber about the relative merits of copper, cast iron, plastic and galvanized iron piping systems.
4. From the encyclopedia and other sources, find out how plastic, copper, galvanized, or cast-iron pipe is manufactured.

Using electric resistance welding, modern mills form pipe from continuous rolls of sheet metal.

b. Shoulders are _____ so that an unbroken contour is formed when the pipe is installed.
c. Horizontal lines entering drainage fittings will have a slope (fall) of _____ in. per ft.

# Unit 5
# VALVES, FAUCETS AND METERS

## Objectives

Water supply piping systems use several mechanical devices to control, regulate and measure the fluids or gases flowing through them. This unit covers those most common to residential and light commercial plumbing.

After studying the unit you will be able to:
- Recognize and name different types of faucets, valves and meters.
- Tell the application of each type.
- Explain in detail their construction.
- Illustrate how they operate.

*Valves and faucets are hand operated devices which control the flow of liquids and gases such as water, steam, natural gas, oil and air. Meters measure and record the volume of fluid flowing through a pipe.*

## VALVES

Water piping systems employ valves at certain spots so that all or part of the system may be closed down. That is, they simply shut off the water or regulate the rate of flow. Stopping flow becomes necessary when leaks occur. It prevents flooding and allows repairs or alterations to be made on pipes or fixtures.

Valves are made chiefly from water service bronze, brass, malleable iron, cast iron, thermoset plastic and thermoplastic. Other materials with low corrosive qualities are sometimes used. Frequently, two or more different kinds of material are used on a single valve. For example, the valve body may be formed from cast iron, the valve plug from cast bronze, the O-ring seal from rubber and the compression washer from teflon plastic.

Valves are available in standard sizes ranging from 1/4 in. to 12 in. diameter. Overall length, width and height have not been standardized.

When valves are manufactured from metal, they usually have internal threads, Fig. 5-1. Plastic valves, on the other hand, frequently have external threads which require a nut to make a compression joint with the pipe, Fig. 5-2.

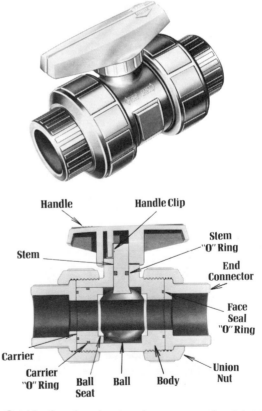

Handle     Handle Clip

Stem

Stem "O" Ring

End Connector

Face Seal "O" Ring

Carrier

Carrier "O" Ring   Ball Seat   Ball   Body   Union Nut

Fig. 5-2. Outside thread and nut make a compression joint for this plastic valve. (Celanese Plastics Co.)

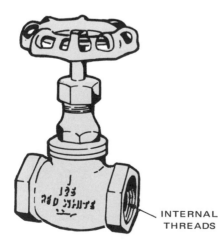

INTERNAL THREADS

Fig 5-1. Valves manufactured from metal are usually equipped with internal threads.

Three common designs or types of valves are:
1. The ground key valve.
2. The compression valve.
3. The gate valve.

## GROUND KEY VALVES

The GROUND KEY or STOP VALVE, Fig. 5-3, generally is used to control flow of liquids or gases between a building and its source of supply or inside the building itself. Its basic design has remained unchanged for nearly 100 years. The ground key offers certain advantages:
1. Simplicity of design.
2. Ability to provide unobstructed flow of liquid.

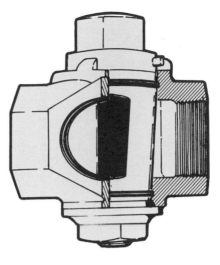

Fig. 5-4. Cutaway of three-way ground key. Note the three ports which control flow of gas or fluid in a T connection. (Hayes Mfg. Co.)

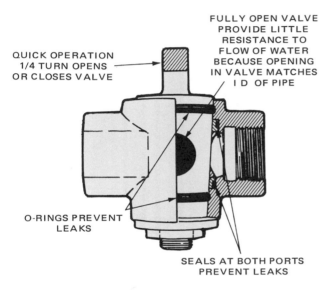

Fig. 5-3. Ground key valve with cutaway to show valve body.

Fig. 5-5. A corporation stop is a valve installed at the water main.

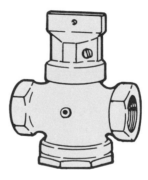

Fig. 5-6. Curb stop permits water supply to be shut off at the curb.

3. Ability to rapidly open or close in only a 1/4 turn of the plug.

Its major disadvantages are:
1. Rapid wear with regular use.
2. Inability to regulate volume of flow.

Several kinds of ground key valves may be found in service. Each has special uses.

THREE-WAY GROUND KEY VALVES control the flow of liquid or gas at points where three pipes connect, Fig. 5-4. They are used in:
1. Gas burner units.
2. Water pumps.
3. Air control on sand blasting equipment.
4. Steam cleaning equipment.
5. Automobile washing equipment.
6. Spraying and dusting equipment.

A CORPORATION VALVE also called a stop or "corp," Fig. 5-5, is a type of ground key valve. Installed in the water main, it controls the water flow to the system after tapping into the city water supply. Corporation stops are available in 1/2 in. through 2 in. diameters.

Other commonly used ground key valves are the CURB STOP, Fig. 5-6, and the METER STOP, Fig. 5-7. Often the

Fig. 5-7. Meter stop valve permits installation of water meter between residential plumbing and the water supply.

stop and waste valve, Fig. 5-8, replaces the meter stop. Curb stops permit the plumber to shut off the water line near the street. See Unit 15 for a more complete explanation of water service installation.

Fig. 5-8. Stop and waste valve can be used in place of meter stop valve.

## COMPRESSION VALVES

COMPRESSION or GLOBE VALVES, Fig. 5-9, are widely used in most piping systems for controlling air, steam and water. The globe shaped body of the valve has a partition in it. This partition closes off the inlet side of the valve from the outlet side, except for a circular opening. The upper side of the opening is ground smooth. A rubber disc or washer attached to the end of the stem presses down against the smooth opening when the handle is turned clockwise. This closes the valve and stops the flow.

The threads on the stem screw into threads in the upper housing of the valve. The top of the housing is hollowed out to receive a packing material. This packing can be replaced if the valve should begin to leak between the packing nut and the valve stem. Major advantages of the globe valve are the following:

1. Critical parts (washer, seat and packing) can be replaced.
2. The valve permits rather accurate control of the flow of water.
3. Valves can be used repeatedly without becoming worn beyond repair.

In spite of its popularity, the globe valve has disadvantages:

1. It partially obstructs flow even when fully open.
2. It is impossible to completely drain the water line. This creates a freezing hazard under certain conditions.

Globe valves are frequently used in water supply lines inside a home, because of their reliability and repairability. Generally, their purpose is to cut off the water supply to parts of the water system. For example, a valve might be installed to cut off the cold water supply to a bathroom during repair of fixtures.

## GATE VALVES

GATE VALVES get their name from the gate-like disc that slides across the path of the flow. They are formed by machining both faces of the metal disc (gate) and both faces of the vertical valve seat, Fig. 5-10. This valve provides an

Fig. 5-10. Gate valve with partial cutaway of valve body. Machined surface of gate slides against machined faces of valve seat. (Red-White Valve Corp.)

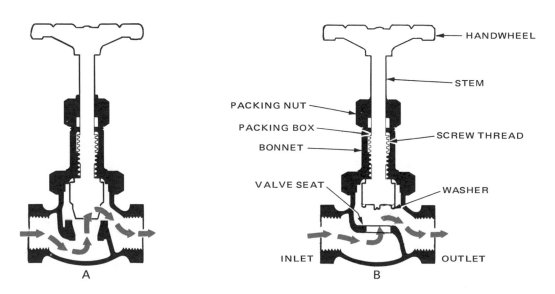

Fig. 5-9. Cutaway of two types of compression or globe valves. A—Plug type. B—Composition (washer) type. (William Powell Co.)

unobstructed waterway when fully open. This feature makes the gate valve useful in large piping installations. It is best suited for main supply lines and pump lines. The gate valve should not be used to regulate flow. It should be fully open or completely closed.

The three basic types of gate valves are:
1. The nonrising stem, Fig. 5-11.
2. The outside screw, Fig. 5-12.
3. The quick closing gate valve, Fig. 5-13.

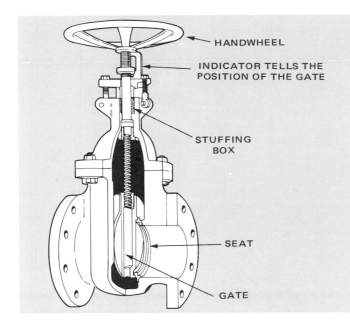

Fig. 5-11. In nonrising stem gate valve outside indicator tells when gate is open or closed.

Fig. 5-12. Outside screw gate valve is similar to nonrising stem gate valve.

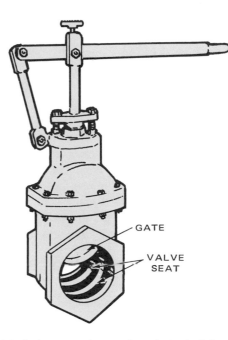

Fig. 5-13. Quick-closing gate valve uses lever instead of threaded stem to raise and lower valve.

## FLUSH VALVES

A water closet or urinal fitted with a FLUSH VALVE has no need for a storage tank. The water flows, under pressure, directly into the fixture. The fixture is flushed with a scouring action. Generally this does a better job of cleaning than gravity flow from a storage tank.

There is an added advantage in direct connection to the water supply piping. The fixture can be flushed again and again without waiting for a storage tank to refill. These two advantages make the flush valve very popular in commercial installations.

There are two types of flush valves:
1. Diaphragm type.
2. Piston type.

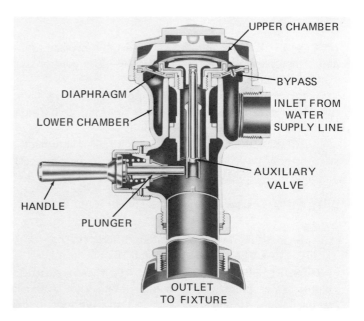

Fig. 5-14. Diaphragm flush valve uses pressure for water from two separate chambers to open and close valve.   (Sloan Valve Co.)

63

The diaphragm type, Fig. 5-14, stops the flow of water when pressure from the water in the upper chamber forces the diaphragm down against the valve seat. Moving the handle slightly in any direction pushes the plunger against the auxiliary valve. Even the slightest tilt causes the diaphragm to leak water out of the upper chamber. As pressure lessens in the upper chamber, the diaphragm raises still further and allows water to flow into the fixture. While the fixture is flushing, a small amount of water flows through the bypass. As it fills the upper chamber, it forces the diaphragm to reseat against the valve seat. This action shuts off the water flow. The piston type flush valve shown in Fig. 5-15, works in a manner much the same.

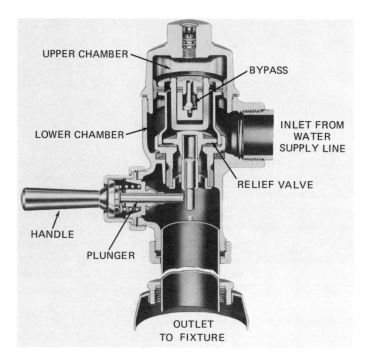

Fig. 5-15. Piston type flush valve.

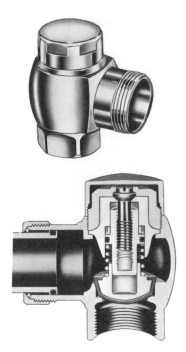

Fig. 5-16. Control stop is installed in water line serving the flush valve. Screw shown in cutaway can be turned in or out to regulate or stop flow of water to the flush valve. (Sloan Valve Co.)

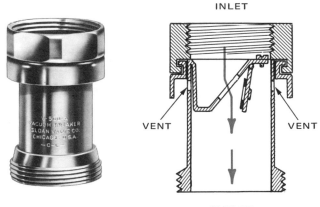

Fig. 5-17. Vacuum breaker has hinged valve which closes if water attempts to flow back through it. Vents underneath flange are always open. They allow free flow of air to prevent vacuum inside pipe.

A CONTROL STOP, Fig. 5-16, is generally installed in the water supply line serving the flush valve. The control stop serves two purposes:
1. It regulates the flow of water entering the flush valve.
2. It can be used to shut off the water supply to each fixture individually when repairs are required.

A VACUUM BREAKER, Fig. 5-17, is installed between the outlet of the flush valve and the fixture to prevent back syphoning of polluted water into water supply piping:

## SPECIAL VALVES

Special application valves are of four different types.
1. CHECK VALVES, as shown in Fig. 5-18.
2. PRESSURE REGULATORS, Fig. 5-19. These are used to reduce water pressure in a building. The valve shown is spring-loaded. It remains open until the pressure in the building reaches a set level. Then it closes and remains closed until the pressure in the building begins to drop.

3. PRESSURE RELIEF VALVES, Fig. 5-20, are a safety device on hot water and steam piping systems. Should the system overheat, the pressure relief safety valve will open permitting hot water or steam to escape. If the pressure were not relieved, the system might explode.
4. FLOAT-CONTROLLED VALVES, Fig. 5-21, maintain a constant level of water in a tank of other containers. The flush tank on a water closet is equipped with such a valve.

## UTILITY FAUCETS

Utility faucets, usually manufactured from extra-grade brass, white metallic alloy and thermoplastic materials, are found in boiler rooms, laundry rooms and on outside walls. Since appearance is not important they are not plated.

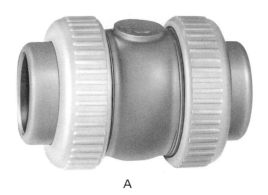

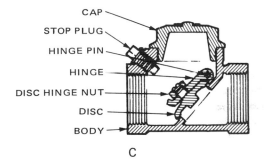

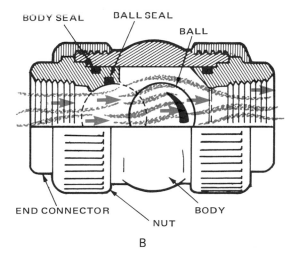

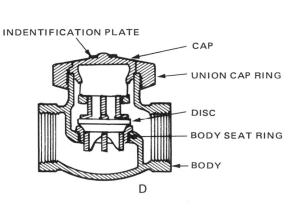

Fig. 5-18. Three types of check valves. A—Outside view of ball check valve. B—Ball check valve allows one-way flow in water supply or drainage lines. (Celanese Plastic Co.) C—Swing check valve has low resistance to flow, making it suited to low-to-moderate pressure of liquid or gas. D—Lift check or horizontal check valve is used for gas, water, steam and air.

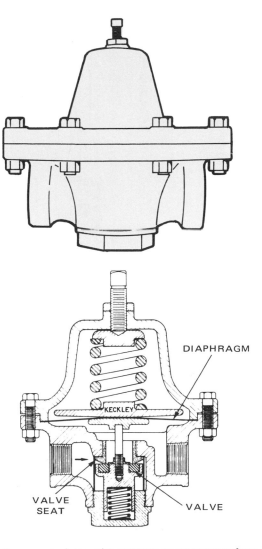

Fig. 5-19. Pressure regulator valve prevents water pressure from building up higher than 80 psi. Spring in dome acts on diaphragm, opening the valve when pressure is low. This pushes down on valve, moving it away from seat. When pressure reaches 80 psi it pushes up on diaphragm and closes the valve.

Fig. 5-20. Pressure relief valve is installed on hot water heaters. Its operation is similar to pressure regulator valve. Excess pressure bleeds off to atmosphere.

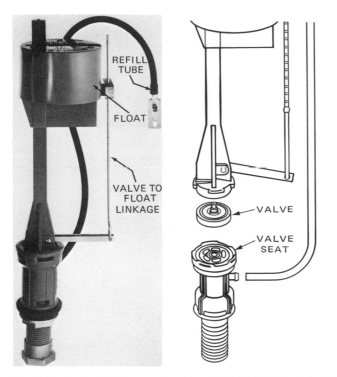

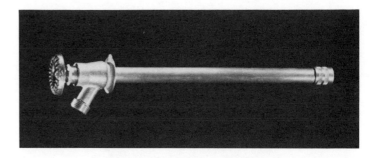

Fig. 5-23. Lawn faucet, often called a freeze proof faucet, has female iron pipe threads at one end. The seat is well inside the building to prevent freezing in cold weather. (Sears Roebuck and Co.)

Fig. 5-21. Float at left controls water supply by opening and closing inlet valve. As water level in toilet tank drops, float pushes linkage down to open valve. Incoming water lifts float which closes valve. See exploded view, right. (Fluidmaster Inc.)

The common utility faucets include:
1. The compression faucet.
2. The lawn faucet.
3. The sediment faucet and boiler drain faucet.

## COMPRESSION FAUCET

The common utility compression faucet, Fig. 5-22, has female iron pipe threads on one end. The water outlet end may be threadless or may have male threads for attaching a garden hose.

Fig. 5-22. Compression faucets are often installed in laundry rooms or outdoors. This one has a flange designed to fit flush against siding or wall covering.

## LAWN FAUCET

The lawn faucet, shown in Fig. 5-23, attaches to the water pipe with female threads. It always has hose threads at the outlet end. Much shorter than other faucets, it hugs the wall for neater appearance.

## SEDIMENT OR BOILER DRAIN FAUCET

A sediment faucet or boiler drain faucet, Fig. 5-24, permits flushing of rust and lime particles from the boiler pipes. In residential applications they are attached to water heaters for the same purpose.

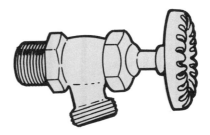

Fig. 5-24. Sediment faucet, or boiler drain faucet — as it is often called — should be installed in drain for hot water tank.

## KITCHEN AND BATHROOM FAUCETS

Faucets and accessories for kitchen sinks, as well as for shower and bath fixtures, are made of many different materials. Among them:
1. Brass.
2. White metallic alloys.
3. Different types of plastic including nylons and acrylics.

Plastic will probably come into greater use as stronger, more workable high-temperature materials are developed.

Several types of faucets are made for the kitchen. There are two basic types:
1. Compression, a design already discussed in relation to valves and utility faucets.
2. Noncompression, which includes washerless, single control and push-button types. These designs seldom have the maintenance problems of the compression type faucet.

## COMPRESSION FAUCETS

Kitchen faucets usually combine two compression valves and a mixer. Water is delivered through a swing spout common to both valves. Typical of this type is the faucet unit shown in Fig. 5-25.

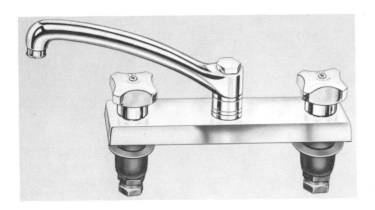

Fig. 5-25. In this kitchen faucet, two compression valves deliver hot and cold water to a mixing spout. (Kohler Co.)

Fig. 5-27. Cutaway of washerless kitchen faucet shows parts which control volume and temperature of water.
(Price Pfister Brass Mfg. Co.)

## NONCOMPRESSION FAUCETS

Washerless faucets need less maintenance than compression faucets simply because there is no repeated compressing of a washer. The water flow is controlled by matching openings in two discs in the valve. One remains stationery while the other rotates with the hand control or lever. See Fig. 5-26.

The term "washerless" is misleading since O-rings may still be used to prevent water leakage around the valve assembly.

## SINGLE CONTROL FAUCETS

Single faucets which control both hot and cold water are now common because of their convenience. The most popular designs are the ROTATING BALL FAUCET, Fig. 5-27, and the ROTATING CYLINDER FAUCET, Fig. 5-28.

A rotating ball replaces the compression washer. The ball, made of metal or plastic, has openings that align with the hot and cold water ports as the ball is rotated by the single lever

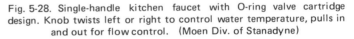

Fig. 5-28. Single-handle kitchen faucet with O-ring valve cartridge design. Knob twists left or right to control water temperature, pulls in and out for flow control. (Moen Div. of Stanadyne)

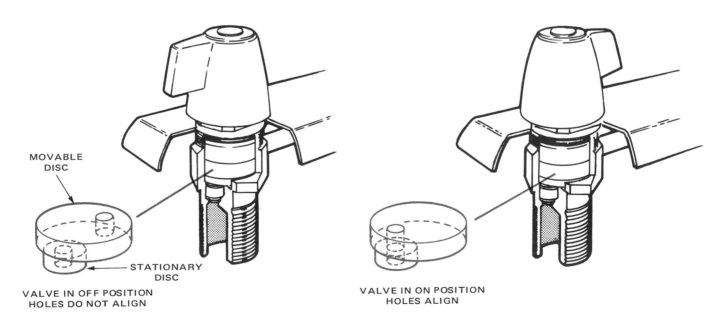

MOVABLE DISC

STATIONARY DISC

VALVE IN OFF POSITION
HOLES DO NOT ALIGN

VALVE IN ON POSITION
HOLES ALIGN

Fig. 5-26. Washerless faucet controls water flow through the alignment of openings in a stationary and movable disc. Two valves are needed. (Delta Faucet Co.)

control, Fig. 5-29. Moving the lever to the right causes cold water to flow. Moving it to the left turns on the hot water. The rate of flow is increased by pushing the lever back.

The rotating cylinder faucet is also called a "cartridge" faucet. It controls the temperature and rate of water flow in a novel way. As the cylinder rotates, both the temperature and rate of flow are set. See Fig. 5-30. This type of valve is simple and easy to maintain.

A third type of single control faucet uses a ceramic disc to control the flow of water, Fig. 5-31. Ceramic is very hard and it is not likely to become eroded. This characteristic makes the faucet long lasting and trouble free.

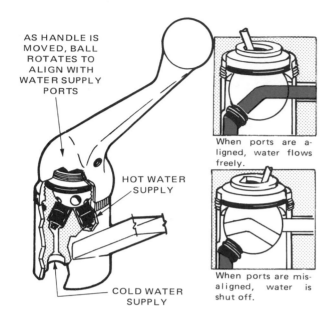

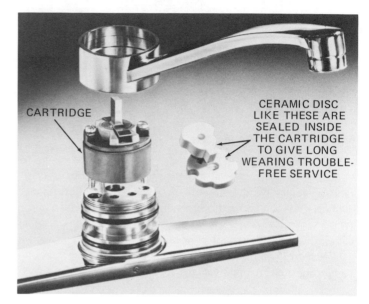

Fig. 5-29. Cutaway view of rotating ball faucet design shows how it controls force and temperature of water. (Delta Faucet Co.)

Fig. 5-31. This single lever faucet uses ceramic discs to control flow of water. (American Standard Inc.)

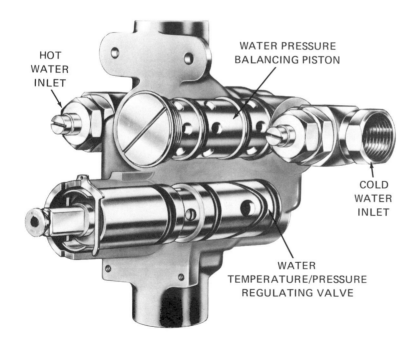

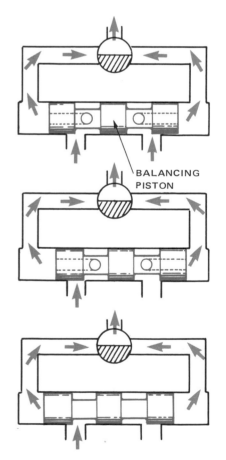

Fig. 5-30. Cutaway view shows assembly of the rotating cylinder faucet. This type is also called a "cartridge" faucet. Water volume is controlled by set screws at inlets on either sides. Balancing valve at center top adjusts when pressure of either hot or cold water changes. (Moen Div. of Stanadyne)

## DISHWASHING UNIT

The special faucet shown in Fig. 5-32, attaches to a kitchen sink. It has a water and detergent mixing tank, hose and brush for washing dishes. It is designed to replace any standard kitchen faucet.

## BATHROOM FAUCETS AND FITTINGS

A combination shower and bath fitting is shown in Fig. 5-33. A diverter in the spout, when lifted, closes the spout and water is sent to the shower head. Flow rate and temperature are adjusted by hot and cold valves. Some units have a single pressure mixing valve to adjust water temperature and pressure. See Fig. 5-34. The telephone shower, Fig. 5-35, is designed for tub and shower installations. The shower head can be hand held or placed in a wall mounted bracket.

Bathroom lavatory supply faucets are usually one of two types. One has separate hot and cold knobs. The other has a single lever control. They are similar in design to kitchen faucets. Fig. 5-36 illustrates a lavatory faucet with separate cold/hot valves. Drain valve is controlled with a plunger located between the supply valves.

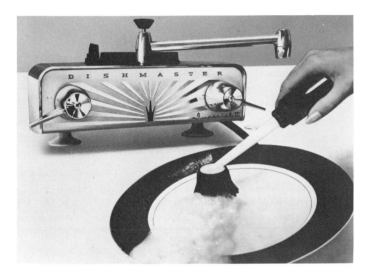

Fig. 5-32. Modern dishwashing unit produces suds automatically when scrubber is turned on. (Manville Mfg. Corp.)

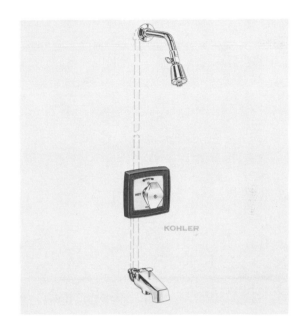

Fig. 5-34. A single-handle mixing valve regulates temperature and volume.

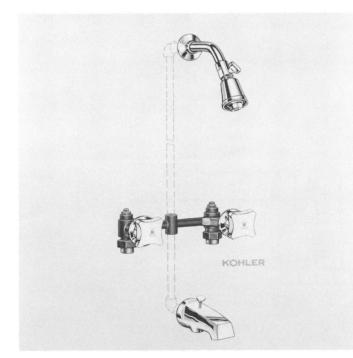

Fig. 5-33. Shower and bath combination fitting with 1/2 in. valves for bathtub installations. (Kohler Co.)

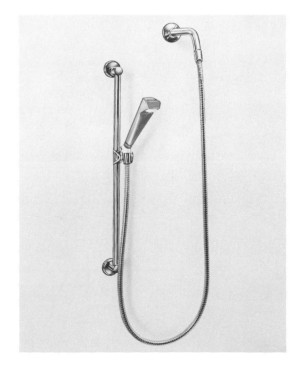

Fig. 5-35. Flexible water line permits this shower head to be raised, lowered or turned in any direction. (Kohler Co.)

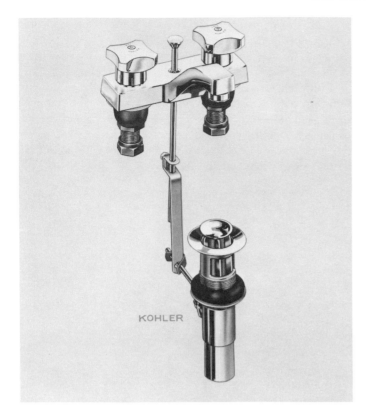

Fig. 5-36. Lavatory supply faucet includes aerator, tailpieces and pop-up drain.

## SINGLE PIPE WATER SYSTEMS

A recent development in home plumbing is the use of single piping to carry both hot and cold water. This system, Fig. 5-37, controls water flow from a central electrically operated valve unit. This valve unit is shown in Fig. 5-38. Electrical devices called solenoids turn on either hot or cold water when a button is pressed at the point where the water is needed. Fig. 5-39, is an exploded view of a solenoid.

Since there is no need for a faucet to turn water on or off, the faucet is replaced with a spout such as the one shown in Fig. 5-40.

To control temperature and rate of water flow, the solenoids for each spout are adjusted as seen in Fig. 5-41. It is possible to get several water temperatures and two rates of flow for each temperature at each spout.

## WATER METERS

Water meters are mechanical devices that measure the amount of water passing through the water service into a building. It is a very delicate instrument requiring careful treatment. It is usually city property which water department employees install and service. Normally, it cannot be removed without permission from the proper municipal authority.

Water meter housings may be made of bronze, cast iron, stainless steel or plastic. Fig. 5-42, shows a plastic unit. Parts of a modern water meter are shown in Fig. 5-43.

Because volume of water used will vary according to the needs of the customer, meters are manufactured with capacities of 20, 30 or 50 gallons per minute. Laying lengths (distance from threaded end to opposite threaded end) are: 7 1/2 in., 9 in. and 10 3/4 in.

Most modern water meters have a magnetic drive, but the inside design varies. The water meter in Fig. 5-44 has a one-piece, cone-shaped disc of hard rubber. The disc rotates a permanent ceramic magnet in a sealed chamber. An opposing magnet in a separate chamber operates the gear train and registers the gallons used.

Meter reading has been made simpler in some cities with the outdoor register units. Portable card punch units, register display units or portable magnetic tape units, will probably become standard in the future.

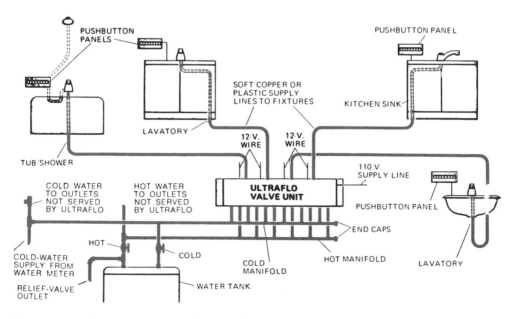

Fig. 5-37. Either hot or cold water travels through single pipe in this system. Valves are located in a centralized control unit located in basement or near water supply source. (Ultraflow Corp.)

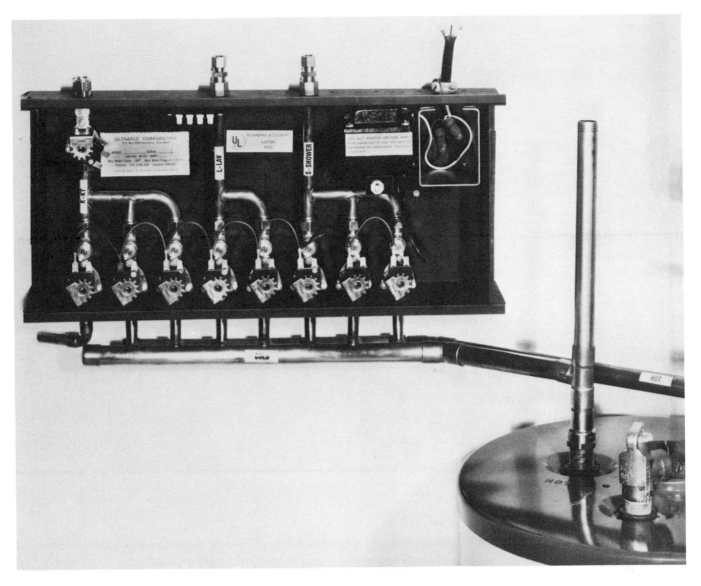

Fig. 5-38. Valve unit of a single-pipe plumbing system is contained in a box. It controls flow of water when button is pressed at the spout. Piping below box, called a manifold, carries both hot and cold water to solenoids located in the box.

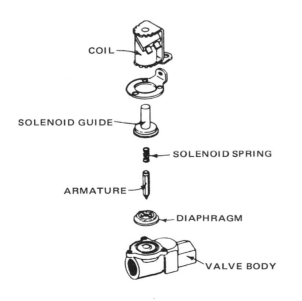

COIL

SOLENOID GUIDE

SOLENOID SPRING

ARMATURE

DIAPHRAGM

VALVE BODY

Fig. 5-39. Exploded view of solenoid. (Ultraflow Corp.)

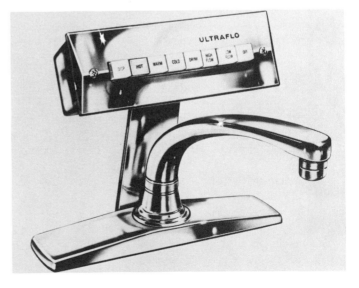

Fig. 5-40. Spout of single pipe system has push button control console. Buttons are switches which control the solenoids.

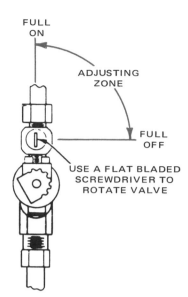

Fig. 5-41. Solenoid unit can be adjusted to control the water temperature and rate of flow.

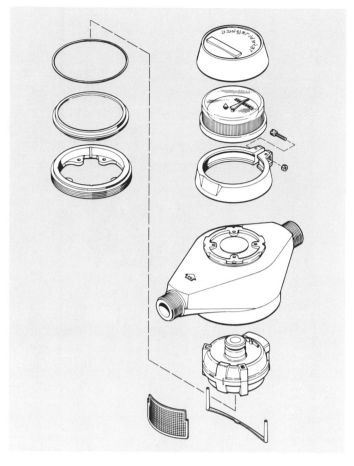

Fig. 5-43. Exploded view of residential meter. Interior parts are made from synthetic polymer, a type of plastic.

Fig. 5-42. This new water meter is designed for residential use. Its housing is of fiber reinforced polycarbonate. (Badger Meter Inc.)

## TEST YOUR KNOWLEDGE — UNIT 5

1. A valve which has a machined metal disc and the advantage of "full-flow" characteristics is known as a _____ valve.
2. A corporation stop, curb stop, and meter stop are examples of:
   a. Compression valves.
   b. Gate valves.
   c. Ground key valves.
3. A _____ valve is installed on water closets and urinals in public buildings.
4. Explain the purpose of check valves.
5. Compression or globe valves have a _____ which is pressed against the seat of the valve.
6. A float-control valve is used on a _____ _____ in a

Fig. 5-44. Cutaway of new meter using magnetic drive mechanism. (Hersey Products Inc.)

bathroom of a home.

7. The compression washer has been eliminated in many bathroom and kitchen faucets by using O-rings or a ball assembly. True or False?

8. Single pipe plumbing systems control the temperature and rate of water flow from a centrally operated _____ _____.

## SUGGESTED ACTIVITIES

1. Study plumbing supplies or catalogs for kitchen and bathroom faucets. Select at least two faucet styles you prefer and secure the list price for these faucets.

2. Examine a ground key, compression valve, and gate valve. Disassemble each valve and identify the various parts. Repair, as required, and reassemble.

3. Using drawings and a written description, explain the function of each water valve and the location of each water control valve in a typical residential installation. Start with the corporation stop at the water main.

As pictured above, many water meters, regardless of size, are made of glass-filled polycarbonate, a type of plastic designed to withstand the corrosive action of minerals and water.

# Unit 6
# HEATING AND COOLING WATER

### Objectives

This unit has to do with the design and operation of the hot water heater/storage tank used in homes and office buildings. It also includes similar information on small units for cooling drinking water.

After studying this unit you will be able to:
- List two basic types of heating/storage tanks used in residences and small commercial buildings.
- Describe the basic differences in the types and explain how they operate.
- Demonstrate, with some technical detail, how their controls and heating elements work.
- Explain the steps for installing a water heater.
- Explain the operation of a water cooler.

Most residential and light commercial structures have centrally located units which heat and store hot water. These units need piping to bring in cold water and to carry heated water to hot water fixtures in the building. Units which use gas as a source of heating energy will also require:
1. Piping to carry the fuel to a burner unit.
2. Venting to carry away smoke and other products of combustion.

Commercial buildings may also require water coolers to supply chilled drinking water. In some cases, these units are also capable of heating small quantities of water.

The plumber is responsible for installing the piping for heaters or coolers. Such workers should, therefore, be familiar with their special plumbing needs. The plumber will also be called upon to place the units in operation or perform minor maintenance on them. Thus, a basic understanding of their operation is necessary and will be supplied in this unit.

## WATER HEATERS AND STORAGE TANKS

The most common water heating equipment in use today combines the heating unit with a storage tank, Fig. 6-1. The basic purpose of such units, is to automatically provide a ready supply of hot water at a preset temperature.

The design of the water tank takes advantage of water's natural properties. Fig. 6-2 describes how a water heater functions. Cold water enters near the bottom of the tank. Hot

Fig. 6-1. Modern automatic water heaters are safe, convenient and attractive.

water is drawn off near the top. As water is heated, it expands. Since expansion makes it lighter, it rises to the top of the tank. Colder, denser water sinks to the bottom of the tank where it is heated. Thus, circulation of water within the tank insures that a reservoir of hot water will be available at all times. This reserve of hot water is necessary because heating large quantities of water rapidly is very expensive. The size of storage tank needed will depend upon the peak demand for water in the building.

## TYPES OF WATER HEATERS

Basically, water heaters are of two types:
1. Those which use a fuel as energy. The fuel is burned by the

74

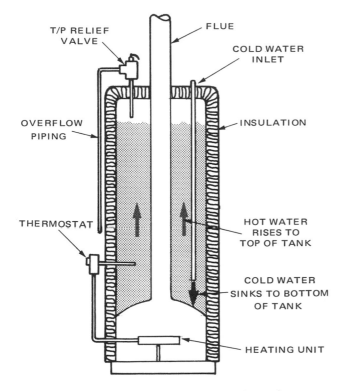

Fig. 6-2. The operation of a water heater requires a heat source, controls, safety devices and an insulated tank.

box located under the tank inside the outer metal jacket. Since the by-products of combustion must be removed, a flue carries it outdoors or to a chimney. The flue is a tubelike passage which usually runs up the center of the heating tank. As the hot burned gases move through it, baffles collect more of their heat and transfer it to the water in the tank.

Heaters using liquefied petroleum are designed and operated the same way as the natural gas heater. However, the fuel and heating orifices are smaller to handle the fuel which is much more concentrated. Although stored in large tanks as a liquid, it vaporizes before it is burned.

Oil hot water heaters are rarely used. They use a pump mechanism to pressurize the oil which is then sprayed into the firepot as a mist. They are similar to burner units of oil-fired furnaces.

Electric water heaters use an insulated heating element to heat water. The element is immersed in the water itself. It needs no flue because it burns no fuel. A 240V electrical service must be provided. Some heaters operate on 120V but do not work efficiently.

The electric heater shown in Fig. 6-4, is a two-stage model. This means it has two heating elements. In normal operation, only the lower coil is working. When larger amounts of hot water are used and hot water must be replaced more rapidly, only the top coil operates. Being larger, it produces more heat.

heater to raise the water temperature. Natural gas is the most common but liquified petroleum and oil are also used.

2. Those which use electricity to heat a resistance coil placed inside the storage tank of the heater.

Natural gas water heaters, Fig. 6-3, burn their fuel in a fire

Fig. 6-3. Cutaway view of gas-fired water heater.

COLD WATER INLET — HOT WATER OUTLET

INSULATION

FLUE WITH BAFFLE

SENSING ELEMENT FOR THERMOSTAT

THERMOSTAT

DRAIN COCK

GAS BURNER

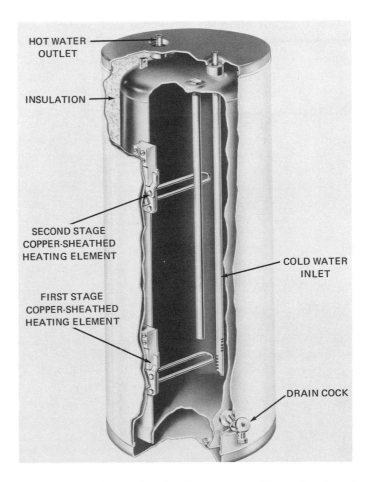

Fig. 6-4. Cutaway view of an electric water heater. The heating element is a long coil of wire which looks like a spring. It becomes red hot as electric current flows through it. (A.O. Smith Corp.)

## FHA MINIMUM DIRECT GAS-FIRED WATER HEATER
### SIZES FOR ONE AND TWO FAMILY UNITS

| NUMBER OF BATHROOMS | 1 – 1 1/2 | | | 2 – 2 1/2 | | | 3 – 3 1/2 | | | |
|---|---|---|---|---|---|---|---|---|---|---|
| NUMBER OF BEDROOMS | 2 | 3 | 4 | 3 | 4 | 5 | 3 | 4 | 5 | 6 |
| STORAGE TANK CAPACITY (GAL.) | 30 | 30 | 40 | 40 | 40 | 50 | 40 | 50 | 50 | 50 |
| INPUT RATING (1000 BTU/HR.) | 30 | 30 | 30 | 33 | 33 | 35 | 33 | 35 | 35 | 35 |

Fig. 6-5. The FHA has set these minimum direct gas-fired water heater sizes for one and two-family units.

## SELECTING WATER HEATERS

Selecting the right type of water heater will depend on several factors. No one unit or type is "right" for every situation. You will need to consider the following:

1. Types of fuel available (or cheapest fuel available). Gas may not be piped to certain areas or gas service may not have been extended to the building. In other instances, electric heat may be too expensive to consider.

2. The capacity of the water heater. If there is to be enough hot water for domestic use the tank must be large enough to keep up with demand. The size of the storage tank and the RECOVERY RATE determine the capacity of a water heater. *The recovery rate is the speed at which cold water can be heated.*

   The table in Fig. 6-5, recommends storage tank size and Btu rating of water heaters.

   When installing an automatic washer, the next larger size storage tank or heater with a higher Btu rating should be selected. This will offset greater hot water requirements.

3. Durability. Ability of the heater to stand up to daily use depends upon the material used in its manufacture. Units are made in galvanized steel, copper and glass lined steel. The last named is usually preferred for both economy and durability. One indicator of the quality of the tank is the guarantee offered by the manufacturer. High quality tanks properly cared for should last 15 to 25 years or more. However, water conditions in some localities will shorten the heater's life.

4. Ability to hold heat. Insulation around the storage tank reduces heat loss and fuel consumption. It is, therefore, wise to select units which are fully insulated.

   Gas water heaters which have met industry standards bear the American Gas Association (AGA) seal of approval. Electric water heaters are tested and approved by the Underwriters Laboratory (UL). They will bear the UL seal.

## CONTROLS AND SAFETY DEVICES

Since water must be neither too hot nor too cold, a temperature control, called a THERMOSTAT, is placed in the water heater. A thermostat is a device which will turn an energy source on and off as needed. The essential part of any thermostat is a sensing element which is moved by the presence of heat. This element can be a tube filled with a liquid or a gas, a strip made up of two metal parts or a spring bellows. The expansion of the sensing element causes the thermostat to open or close a switch or a valve.

When water in the tank reaches a preset temperature, the thermostat turns off the energy supply to the water heater. This stops the heating action.

When cold water enters the tank, the water temperature drops. This moves the sensing element to turn on the energy supply and the water is again heated up to the desired temperature.

A typical control for a gas water heater is shown in Fig. 6-6. When the water heater is operating normally the pilot light heats the thermocouple. This produces a small electric current which affects the solenoid valve. This current creates a magnetic attraction in the solenoid coil which opens the solenoid valve. If the pilot light were to stop burning, the solenoid valve would close and shut off the gas.

The thermostat valve, shown to the right of the solenoid valve, is controlled by the expansion and contraction of the

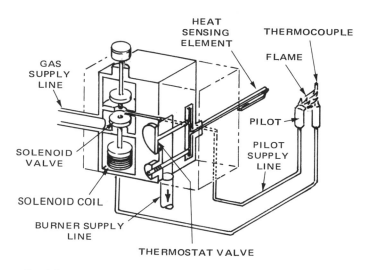

Fig. 6-6. Typical valve control unit for a gas-fired water heater. It has two valves. The solenoid valve, controlled by the pilot flame, remains open as long as the pilot light is burning. The thermostat valve is opened and closed by the heat sensing element as water temperature changes.

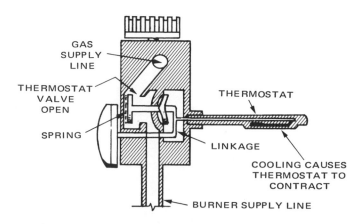

Fig. 6-7. As water cools, the sensing element contracts pulling the thermostat valve open.

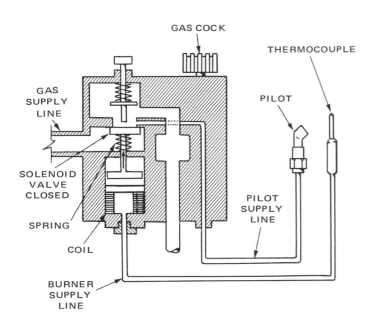

Fig. 6-8. If the pilot light is not burning the thermocouple stops producing electricity. A spring between the valve and the coil then closes the valve.

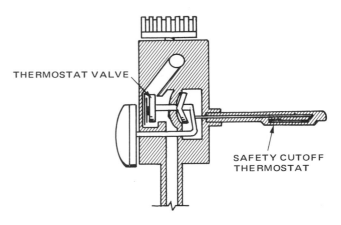

Fig. 6-9. Whenever the water exceeds a safe temperature (200 F or 93 C) the safety cutoff thermostat breaks the circuit between the solenoid valve and the thermocouple. The valve closes. This stops gas flow and prevents water from boiling.

heat sensing unit. As the water temperature decreases, the sensing unit becomes shorter and opens the thermostat valve, Fig. 6-7.

It can be seen from Fig. 6-8 that the pilot light receives a flow of gas through a bypass from the main passage. Thus, it can burn constantly. If the pilot light stops burning, the gas supply is shut off by a spring. However, even if gas continued to flow to the pilot there would be no real danger. This small flow of fuel would escape safely through the flue.

A safety cutoff thermostat, Fig. 6-9, acts as a fail-safe device. It cuts off the current to the solenoid should the water temperature exceed a predetermined safe limit. The safety cutoff thermostat is an electrical device connected to the

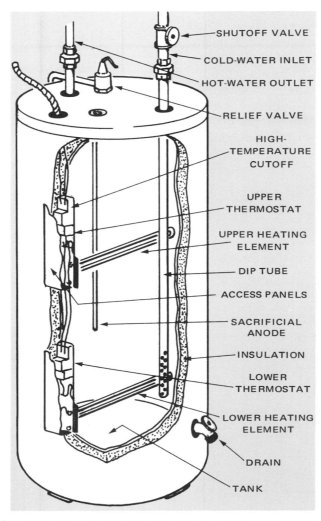

Fig. 6-10. Cutaway shows heating elements and controls for an electric water heater.

circuit containing the solenoid and thermocouple.

Electrically heated water storage tanks (as noted earlier) generally contain two heating elements. The core of the heating element is a high-resistance metal which heats as electricity is passed through it. The metal core is not in contact with its outside metal jacket. Asbestos sheathing separates the two. Each of the heating elements is thermostatically controlled, Fig. 6-10.

Under normal operating conditions, heated water leaves the tank from the top and cold water enters at the bottom. The reduced temperature of the cold water causes the lower heating element to operate until the temperature has reached a predetermined level. When large quantities of hot water are drawn from the tank, the cold water entering the tank may rise to the level where the upper heating element will turn on. When this happens, the lower heating element automatically goes off. This prevents overloading the electrical circuit supplying the water heater. The advantage of this arrangement is that the recovery time of the water heater is reduced.

Fig. 6-11 shows how the electric water heater is wired. Note that only one element can be heating at any one time because of the double-pole switch at the upper element. The high-limit protector is an automatic safety device which shuts off all electrical current to the water heater if the water temperature exceeds 180 F (82 C). Because overheating of the water can only occur when one or both of the thermostats is not functioning properly, it is necessary to repair or replace the defective part before resetting the high-limit protector. (See Unit 19 for additional information on malfunction of water heaters.)

An additional safety device is the temperature/pressure relief valve shown in Fig. 6-12. It is installed in the storage tank or in the line as close to the tank as possible. The purpose of this valve is to prevent the tank from exploding in case the

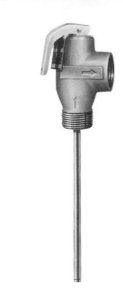

Fig. 6-12. Temperature/pressure relief valves provide protection from explosion in case the water becomes overheated. (Watts Regulator Co.)

thermostat fails to function and the water is overheated. A sensing element will detect the rise in temperature and, together with pressure, open the valve to allow release of water or steam. See Fig. 5-20 for a cutaway view of a T/P relief valve. Release of the excess pressure prevents the explosion of the storage tank. This valve should be tested occasionally by manually lifting the lever. If this does not cause a release of water, the valve should be replaced.

## INSTALLATION OF WATER HEATERS

Gas or oil-fired heaters should be placed near a chimney or flue for proper venting. As a rule, they should be located within 15 ft. of such an outlet. This will insure proper draft. For the same reason, avoid having too many turns in the connecting pipe. Turns tend to block off movement of exhausted gases into the chimney. Flue connectors should be at least as large as the heater flue outlet. Heaters and furnaces may use the same chimney if the heater flue enters the

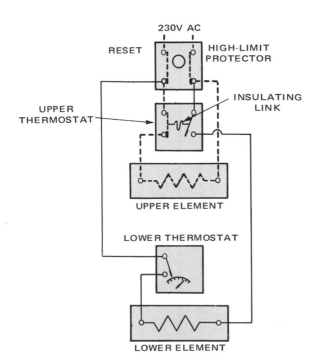

Fig. 6-11. Wiring diagram is typical of circuitry for electric water heater.

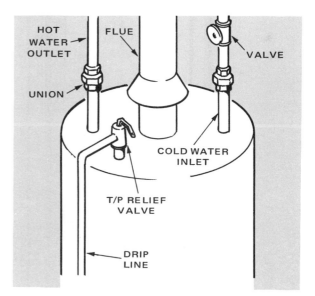

Fig. 6-13. Installation of a hot water heater is made with unions.

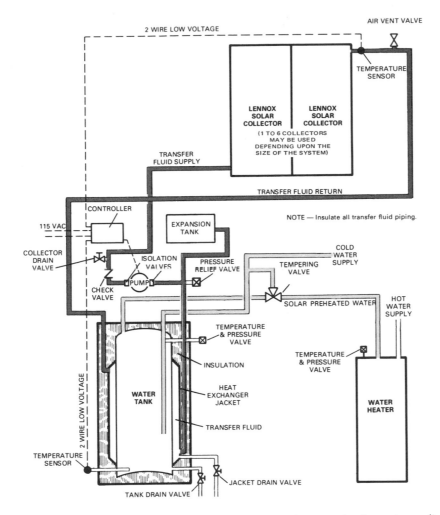

Fig. 6-14. Schematic shows a typical solar energy heat unit for a domestic hot water heating system. (Lennox Industries, Inc.)

chimney above the furnace flue. Horizontal runs of pipe should slope upward 1/4 in. per ft. Connecting to a furnace flue directly is often permitted if a Y connector is used. *Never use a T connector!*

Use black iron piping and fittings to connect the gas or oil supply piping to the heating unit. A shutoff valve should be attached to the piping between the water heater and the fuel supply so that the fuel can be shut off for removal or repair of the water heater.

If local codes allow them, flexible corrugated brass connectors will simplify the hookup of hot and cold water lines. But, should regular pipe connections be required, unions must be installed, Fig. 6-13. A valve on the cold water inlet to the water heater will eliminate the need to turn off the entire cold water supply when draining or replacing the unit.

The T/P relief valve may be installed in a specially designed opening in the top or side of the storage tank or its case. Follow the manufacturer's direction to insure safe operation of the water heater. In addition, a drip line should be attached to the outlet of the T/P relief valve. Extend it to within 6 to 12 in. of a floor drain. Thus, any hot water escaping from the T/P relief valve will be directed safely to the sewer.

## SOLAR HEATING OF DOMESTIC HOT WATER

Increases in the cost of energy and improved efficiency of solar collectors make it practical to install solar domestic water

heating systems in many parts of the United States. Fig. 6-14 illustrates a typical system.

The mechanics of the system are rather simple. First, cold water enters the solar preheat water tank. This water is heated by the transfer fluid (shown in color) which flows through the collector and the heat exchanger jacket. When hot water is used somewhere in the building the water from the solar preheat tank flows into the standard water heater. This conventional gas or electric water heater is required to insure that the hot water temperature will be constant even during extended periods of cloudy weather.

Water pressure from the cold water supply is high enough to cause the water to move from the solar preheat tank to the water heater and into the hot water piping. The tempering valve in the solar preheated water pipe is a safety feature. It prevents high-temperature water above 160 F (71 C) from entering water heater. It is possible for solar-heated water to reach temperatures well above 160 F (71 C) on cloudless days when little hot water is used. The tempering valve mixes cold water with very hot water reducing temperature and preventing user from being burned.

The fluid which circulates through the collector is not the same water which the consumer uses. It is generally an ethylene glycol solution. The solution prevents freezing, protects piping and performs better at elevated temperatures than water. To prevent contamination of domestic (potable) water a triple-walled tank is used in solar preheat water tank.

Temperature sensors at the outlet of the collector and at the base of the solar preheat water tank compare the temperatures at these two points. When the temperature at the collector is higher than the temperature in the tank, the controller turns on the pump. The pump forces the transfer fluid through the collectors. Here it gains heat from the sun. The heated transfer fluid leaves the outlet of the collector and flows through the transfer fluid return to the heat exchanger jacket. While circulating through the heat exchanger jacket, the transfer fluid gives up heat to the water in the solar preheat water tank.

The pressure relief valve near the outlet of the solar preheat tank is a safety device. It allows steam to escape in case the transfer fluid overheats. The expansion tank serves as a holding device and pressure regulator. As the temperature of the transfer liquid changes from well below 0 F (−17 C) during cold winter nights to 220 F (104 C) or higher on hot days, the volume increases greatly. Expansion tank stores excess fluid and helps maintain nearly constant pressure in the system.

The check valve near the pump prevents the transfer fluid from flowing through the collector and heat exchanger jacket when the pump is not running. Without this valve transfer fluid would flow back to the collector at night. Heat from the solar preheat tank could then pass into the atmosphere from the collector.

Air trapped in the system while it is being filled could create vapor locks in the piping. This condition would prevent or seriously restrict the flow of transfer fluid. To keep this from happening, an air vent valve is placed at the outlet to the collector to vent air out of the collector piping when the system is being filled.

The solar preheat water tank illustrated in Fig. 6-15, includes the controls, expansion tank, pump and many of the valves necessary to make the system function. Systems such as this, connected to modular solar collectors, Fig. 6-16, simplify installation and greatly reduce the likelihood of extensive maintenance problems.

## WATER COOLERS

Water coolers, Fig. 6-17, are frequently installed in public buildings to provide readily accessible, cold drinking water. Some water coolers are equipped with a water heating unit which supplies hot water for instant coffee and similar uses. The cutaway drawing, Fig. 6-18, describes the basic components of water coolers.

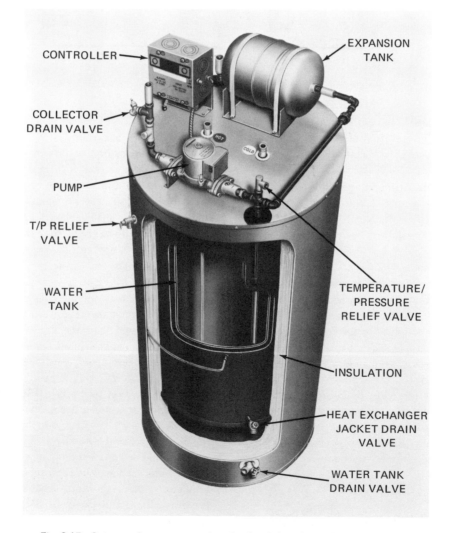

Fig. 6-15. Cutaway shows construction details of the solar preheat storage tank.

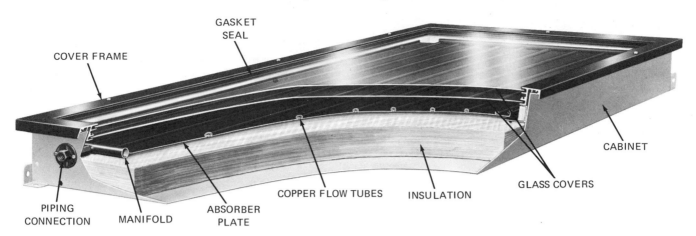

Fig. 6-16. Cutaway of solar collector panel. Absorber plate transfers solar heat to transfer liquid. (Lennox Industries, Inc.)

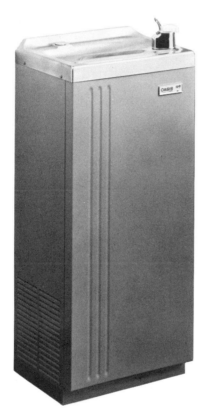

Fig. 6-17. Free-standing water coolers are found in many office buildings. (Ebco Mfg. Co.)

1. Bubbler regulates flow of water.
2. Sanitary top.
3. Cooling tank.
4. Cooling coils bonded to exterior of cooling tank.
5. Insulating jacket reduces transfer of heat.
6. Adjustable thermostat controls water temperature.
7. Precooler in drain precools water entering cooling tank (available on larger models).
8. Hot water spout.
9. Refrigeration system including compressor, motor, fan, and heat exchanger.

Fig. 6-18. Cutaway view shows working parts of a typical water cooler.

## SELECTION OF A WATER COOLER

The first consideration in selecting a water cooler is the number of people to be served by the unit. Manufacturers' literature should be consulted to determine the capacity of each unit. In addition, the desirability of a foot pedal control and a glass filler must be considered.

## INSTALLATION OF WATER COOLERS

A water supply, a drain and an electric outlet are required for the operation of a water cooler. The drain piping is generally 1 1/4 in. The water supply is often connected with 3/8 in. soft copper tubing and compression fittings.

## TEST YOUR KNOWLEDGE — UNIT 6

1. Cold water enters the water heater near the _____ of the tank.
2. As water is heated, it _____ to the _____ of the tank.
3. The temperature of the water in the water heater tank is controlled by a:
   a. Heating element.          c. Thermostat.
   b. Layer of insulation.      d. T/P relief valve.

4. What type valve opens to drain off excess pressure from overheated water?
5. Electric water heaters must be vented to a flue so that fumes can escape to the outside of the buildings. True or False?
6. A three bedroom home with two baths should be equipped with a water heater of the following size:
   a. 30 gallon.
   b. 40 gallon.
   c. 50 gallon.
7. The _____ produces a small electrical current when heated by the pilot light in a gas water heater.
8. Electrically operated water heaters usually have two heating elements. They operate as follows (select correct answer):
   a. If one element burns out the other takes over to heat water.
   b. When rapid hot water recovery is required, both heating elements will operate.
   c. Normally the lower element does all the heating. But when large quantities of water are used, the upper heating element turns on and the lower element shuts off. The result is rapid recovery of hot water supply.
9. A valve should be installed on the cold water inlet to the water heater to:
   a. Control volume of water coming into the heater.
   b. Shut off hot water flow to a fixture needing repair.
   c. Shut off water flow when heater needs repairs or replacement.
10. The drain piping for a water cooler is generally _____ diameter.

## SUGGESTED ACTIVITIES

1. Install a water heater in an existing building or plumbing module.
2. Study manufacturers' literature describing different water heaters and coolers. Make note of how the units differ and what characteristics they have in common.
3. Study cutaway T/P relief valves and thermostats to understand how they function.
4. Study the history of water heaters to learn how water backs, furnace coils and other methods of water heating were used in the past.
5. Study the possibility of heating water with solar energy.

# Unit 7
# BLUEPRINT READING AND SKETCHING

## Objectives

This unit introduces basic skills needed to read drawings and produce piping sketches.

After studying this unit you will be able to:
- Recognize the plumbing symbols and abbreviations used in architectural drawings.
- Recognize and interpret various kinds of plans.
- Take dimensions off drawings in inches and feet.
- Scale drawings using either an architects' scale or a rule.
- Prepare piping sketches in two and three dimensions.

The construction of a building is a complex undertaking. The size and shape of the structure and all its parts are carefully drawn on paper before any building activity begins. Everything should be planned.

To keep the drawings to comparatively small sizes, inches or fractions of inches are made to represent feet. For example, 1/4 in. may represent a foot of actual measure. Such a drawing is called a scale drawing. Use of the scale in preparing drawings will be explained later.

Frequently, the term BLUEPRINTS will be used when referring to drawings. These are actually copies of drawings made from an original set of tracings. Their name comes from their appearance: a white line on a blue background. Many sets of blueprints are made so that building officials, contractors, the owner, suppliers and workers can have them as needed.

Sometimes another process of duplicating is used which produces a blue line on a white background. These copies are called WHITEPRINTS.

## PLANS AND SPECIFICATIONS

A set of drawings for a building will contain the following:
1. Plot plan to describe the location of the structure on the lot. It may also indicate landscaping.
2. Elevation to describe the exterior appearance of each side of the building.
3. Floor plan describing the room arrangement of each floor of the building.
4. Foundation plan describing the size and shape of footings and foundation walls.
5. Plumbing plan locating plumbing fixtures and (sometimes) describing the piping systems which serve the fixtures. (Frequently, this is included on the floor plans for residential structures.)
6. Electrical plan locating service entry, master fuse panel, outlets, switches and light fixtures (this too, may be part of the floor plan).
7. Heating plan describing the placement of the heating plant and duct, piping or electrical resistance wire as may be required for a particular heating system.
8. Details showing the construction of walls, stairs, doors and windows. These are shown in the form of section views.

For a small single family residence, some of the drawings may be combined. For example, the electrical plan may be included on the floor plan. A set of plans, in this instance, may contain no more than three or four 17 x 22 in. sheets.

If the plans are for a large commercial building, however, they may contain more than 100 pages. It is helpful, in either case, to know what kind of information can be located on the drawings.

Ability to interpret drawings and specifications for a building is absolutely necessary if plumbers are to do their work correctly. In plumbing, as in other phases of construction, the old adage, "plan your work and work your plan," is very appropriate.

In most cases, the location of plumbing fixtures and the basic layout of the piping systems will be shown on the drawing for the building. It is the plumber's responsibility to interpret the drawings and install the plumbing system according to the plan.

There are occasions where plans are incorrect or when modifications must be made after the building has been started. Then, it may be necessary for the plumber to sketch changes. Such drawings obviously must be good enough so that architects, building inspectors and other workers can understand them.

## PREPARING SPECIFICATIONS

In addition to describing the shape and size of a building, plans must indicate the type of material and work quality. This information is given in a set of instructions called the SPECIFICATIONS or "specs." The details set forth in the specifications are part of the contract between the plumber and the builder or owner. They are binding on both parties.

Specifications are carefully worded so that there is little chance for misunderstanding. The plumber should read them carefully before preparing the bid or before signing a contract.

In every case, specifications should:

1. Describe materials to be used giving sizes, quality, brand names, style and identification numbers.
2. List all plumbing operations to be performed.
3. Refer all specifications to the detailed working drawings.
4. State quality of workmanship.

Some specifications will be shown on the blueprints in the form of notes. These notes are just as important as any of the symbols, plans or other items written into the specifications.

It is absolutely essential that blueprints and specifications be studied together if the job is to be understood. The partial set of plumbing specifications shown in Fig. 7-1, describes the type and quality of pipe, fittings, valves and fixtures to be installed. When this written information is interpreted as it relates to the blueprints, the plumber has a complete picture of the job.

Fig. 7-2 is the floor plan for a one-story single family residence. The floor plan may be difficult to understand at first. Try to imagine that the roof is removed and that you are looking down into the house from above. From this frame of reference, it is easier to understand the size, shape and location of rooms.

When studying a floor plan, it is desirable to first learn the general layout of the rooms. Locate kitchen, bathrooms and utility rooms. Study each of these areas carefully to determine what plumbing is required in each. Then study the plan for other areas which might require water supply or DWV piping.

In the floor plan shown, the two bathrooms, the kitchen and the utility room are close together. From this floor plan, the plumber can locate each of the major fixtures requiring drains. For example, a drain is needed for the kitchen sink and the three fixtures in each bathroom. Water supply piping will be required for each of the above. In addition, cold water piping will supply the hot water tank and the hose bibs (outside faucets which permit attaching a garden hose).

It is useful to study the foundation plan and the plot plan together. See Figs. 7-3 and 7-4. These plans are helpful in locating sewer and drain piping. Note that the location of each of the plumbing fixtures is indicated on the plot plan. This is important because the house has no basement. Piping, therefore, must be installed in the crawl space. The three brick piers may affect where the drain and supply piping is installed. The plumber will have to work around these structures.

Two other drawings useful to the plumber are the section view, Fig. 7-5 and the front and rear elevations, Fig. 7-6. The section view is what the dwelling would look like if it were sliced in two at the point marked "AA" in Fig. 7-6. This is called the "cutting plane line." The primary value of the drawings is in locating and measuring piping needed for the plumbing stack.

The section view indicates the vertical distances which will be critical to the installation of DWV piping. Using this

## PLUMBING SPECIFICATIONS

1. **GENERAL**
   All provisions of the General Conditions and Supplementary General Conditions sections form a part of this section.
2. **WORK INCLUDED IN THIS SECTION**
   a. DWV and connections to sewer.
   b. Hot, cold water, and gas piping. Hot water heaters and controls.
   c. Trim, valves, traps, drains, cleanouts, access plates and hose bibs.
   d. Roof flashing for vent piping.
   e. Galvanized pipe downspouts, drains and connections to street.
   f. Garbage disposal in kitchen.
   g. Precast shower base.
   h. Cutting, excavation and backfill for plumbing lines.
3. **WORK NOT INCLUDED IN THIS SECTION**
   a. Kitchen equipment except as noted.
   b. Heating equipment.
4. **PERMITS, LICENSES AND INSPECTION**
   The contractor shall pay for all plumbing and sewer permits. All work shall conform to the Plumbing Code of the City _____.
5. **DRAWINGS**
   In general, the drawings provided show the location and type of fixtures required. The contractor will be responsible for preparing any piping drawings required. The contractor shall verify piping drawings with the architect before beginning installation.
6. **SURVEY OF SITE**
   The contractor shall be familiar with the plans and specifications and shall have examined the premises and understood the condition under which she or he will be obliged to operate in performing the contract.
7. **EXCAVATION AND BACKFILL**
   a. Excavate trenches for underground pipes to required depths. After pipelines have been tested and approved, backfill trenches to grade with approved materials, tamped compactly in place as specified under Earthwork Section.
   b. Boring for pipes shall be done by the plumber in such a way as to not weaken the structure.
8. **WATER SUPPLY**
   a. All water piping shall be new copper. Pipe sizes shall conform to the plumbing code of the City of _____.
   b. Provide hot water supply to all fixtures except water closets.
   c. Support piping from the building structure by means of hangers to maintain required slope of lines and to prevent vibration.
9. **GAS SYSTEM**
   All gas lines shall be black iron.
10. **SOIL, WASTE AND VENT LINES**
    a. Comply with the _____ _____ Plumbing Code as to size of pipe and fittings.
    b. All DWV piping shall be new Schedule 80 ABS plastic that is certified by the National Sanitation Foundation.
11. **FIXTURE SCHEDULE**
    a. The tub Kohler No. K-515-F enameled tub, the shower base Swan No. 863-S.
    b. Water closets Kohler No. K-3475-FBA.
    c. Lavatories Kohler No. K2150-C vitreous china.
    d. Water heater A.O. Smith PGX 75 automatic gas fired.
12. **KITCHEN EQUIPMENT**
    a. Furnish and install A.O. Smith No. 55VI sound-shielded disposer.

Fig. 7-1. Plumbing specifications spell out what the plumber is expected to do and describe quality of materials, fixtures and work.

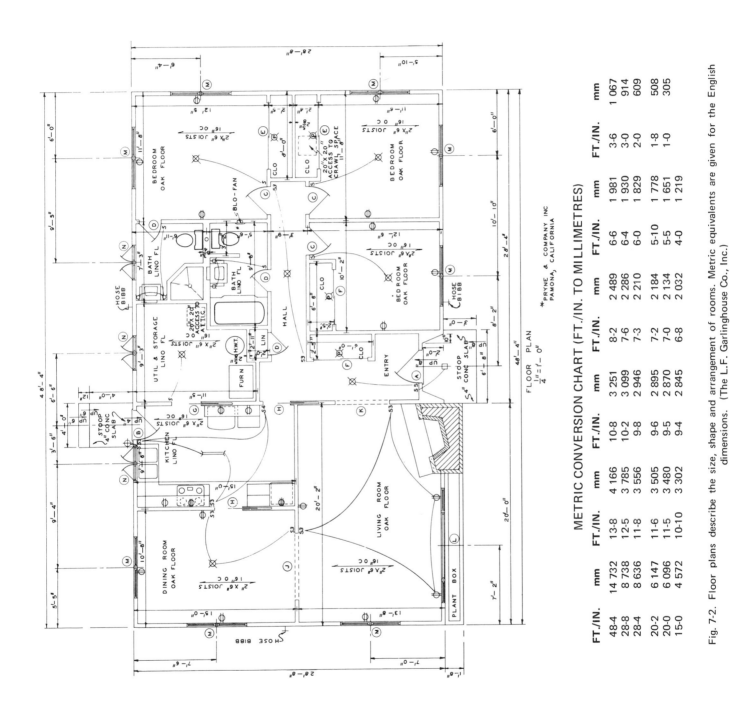

## METRIC CONVERSION CHART (FT./IN. TO MILLIMETRES)

| FT./IN. | mm | FT./IN. | mm | FT./IN. | mm | FT./IN. | mm | FT./IN. | mm |
|---------|------|---------|-------|---------|-------|---------|-------|---------|------|
| 48-4 | 14 732 | 13-8 | 4 166 | 10-8 | 3 251 | 8-2 | 2 489 | 6-6 | 1 981 | 3-6 | 1 067 |
| 28-8 | 8 738 | 12-5 | 3 785 | 10-2 | 3 099 | 7-6 | 2 286 | 6-4 | 1 930 | 3-0 | 914 |
| 28-4 | 8 636 | 11-8 | 3 556 | 9-8 | 2 946 | 7-3 | 2 210 | 6-0 | 1 829 | 2-0 | 609 |
| 20-2 | 6 147 | 11-6 | 3 505 | 9-6 | 2 895 | 7-2 | 2 184 | 5-10 | 1 778 | 1-8 | 508 |
| 20-0 | 6 096 | 11-5 | 3 480 | 9-5 | 2 870 | 7-0 | 2 134 | 5-5 | 1 651 | 1-0 | 305 |
| 15-0 | 4 572 | 10-10 | 3 302 | 9-4 | 2 845 | 6-8 | 2 032 | 4-0 | 1 219 | | |

Fig. 7-2. Floor plans describe the size, shape and arrangement of rooms. Metric equivalents are given for the English dimensions. (The L.F. Garlinghouse Co., Inc.)

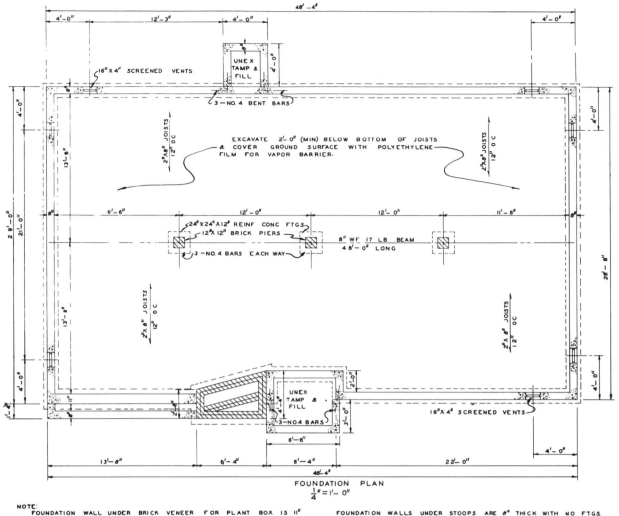

48'-4"

4'-0"  12'-3"  4'-0"  4'-0"

16"X 4" SCREENED VENTS

UNEX
TAMP &
FILL

3—NO. 4 BENT BARS

2"X8" JOISTS
12" OC

EXCAVATE 2'-0" (MIN) BELOW BOTTOM OF JOISTS
& COVER GROUND SURFACE WITH POLYETHYLENE
FILM FOR VAPOR BARRIER.

2"X8" JOISTS
12" OC

11'-6"  12'-0"  12'-0"  11'-6"

24"X24"X12" REINF CONC FTGS.
12"X 12" BRICK PIERS
8" WF 17 LB BEAM
48'-0" LONG
3—NO.4 BARS EACH WAY

2"X 8" JOISTS
12" OC

2"X 8" JOISTS
12" OC

UNEX
TAMP &
FILL

3—NO.4 BARS

6'-8"

16"X 4" SCREENED VENTS

13'-8"  6'-4"  6'-4"  22'-0"  4'-0"

48'-4"

FOUNDATION PLAN
¼" = 1'-0"

NOTE:
FOUNDATION WALL UNDER BRICK VENEER FOR PLANT BOX IS 11"
THICK WITH 22"X 11" REINF CONC FTG.
FOUNDATION WALLS UNDER FRAME WALLS ARE 8" THICK WITH
16" X 8" REINF CONC FTGS.

FOUNDATION WALLS UNDER STOOPS ARE 8" THICK WITH NO FTGS.
FOOTINGS UNDER FIREPLACE IS 12" THICK REINF CONC WITH A
6" PROJECTION AROUND FIREPLACE.

Fig. 7-3. Foundation plan for residence shown in Fig. 7-2.

Fig. 7-4. Plot plan for residence shown in Fig. 7-2. Note that location of various plumbing fixtures is indicated.
(The L.F. Garlinghouse Co., Inc.)

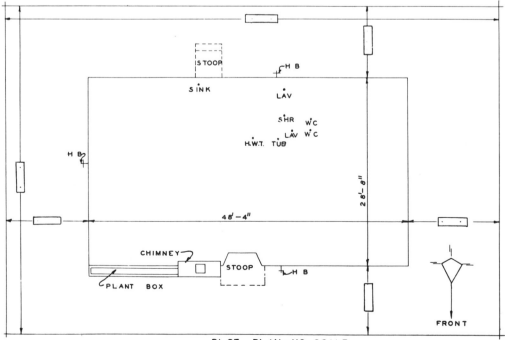

STOOP

SINK

H B

LAV

SHR     WC
LAV     WC
H.W.T.  TUB

H B

28'-8"

48'-4"

CHIMNEY

PLANT BOX

STOOP

H B

FRONT

PLOT PLAN-NO SCALE

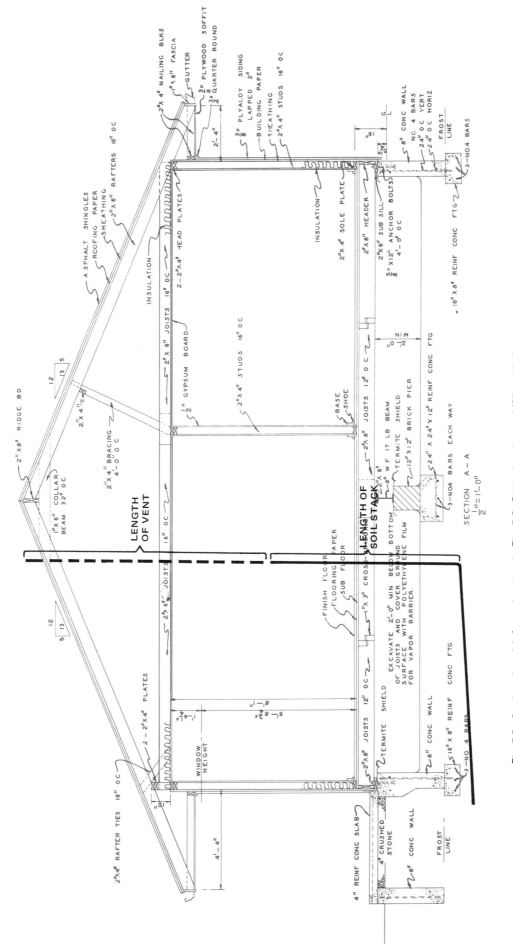

Fig. 7-5. Section view AA from elevation in Fig. 7-6. Original scale was 1/2" = 1'. Reduction of drawing brings it to nearly 1/4 scale.

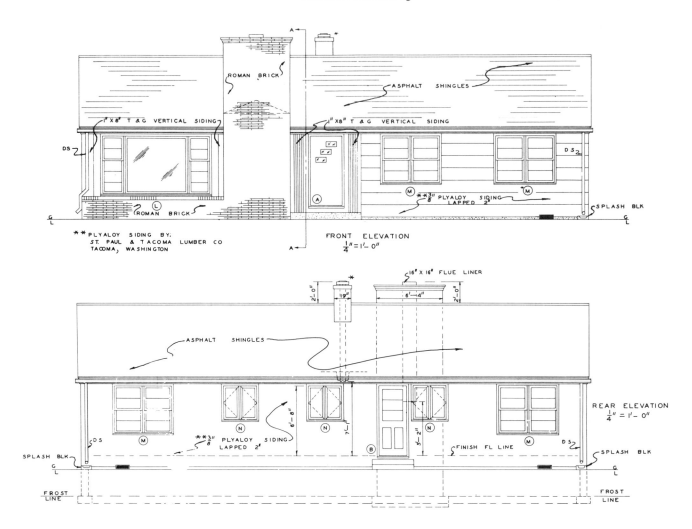

Fig. 7-6. An elevation drawing. Top. Front elevation. Bottom. Rear elevation.

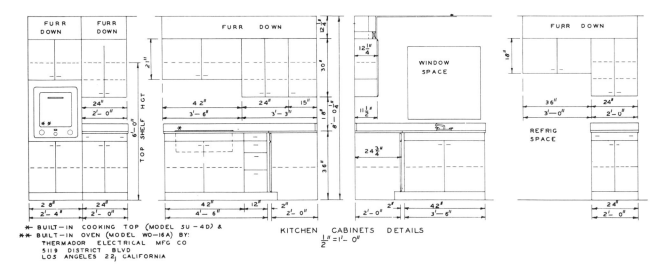

Fig. 7-7. Kitchen cabinet details will help the plumber understand how to install the piping system. (The L.F. Garlinghouse Co., Inc.)

drawing, it is possible to estimate the vertical height of vent stacks and to determine the depth of the building drain and water supply piping.

One detail drawing useful to the plumber is the kitchen cabinet plan, Fig. 7-7. From this drawing, the location and type of sink can be determined. Other information will assist in the location of water supply and waste piping. For example, the window opening over the sink will affect the location of the vent stack.

## DIMENSIONS

Dimensions on architectural drawings are given in feet and inches. For example, a distance of 54 in. would be written 4'-6". The limits of a given distance are shown by extension and dimension lines, Fig. 7-8. Note that the dimension line is unbroken and that it has an arrowhead, dot or diagonal line indicating its limits, Fig. 7-9.

When working from a set of plans, it is frequently necessary

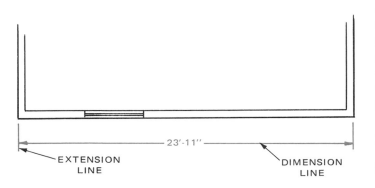

Fig. 7-8. The beginning and end of a given distance are shown by extension and dimension lines.

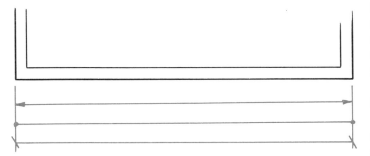

Fig. 7-9. Arrowheads, dots or diagonal lines are used to indicate the limits of a particular dimension.

to add or subtract dimensions to obtain an unknown distance. Note that inches and feet are added separately, Fig. 7-10. When the number of inches reaches or exceeds 12, the answer is simplified by converting the inches to feet. Fig. 7-11, illustrates how dimensions are subtracted. Addition and subtraction of fractions are covered in Unit 3.

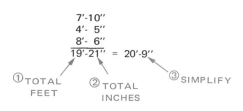

Fig. 7-10. Steps in adding dimensions given in feet and inches.

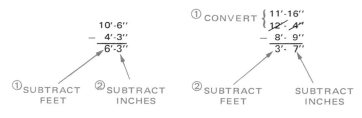

Fig. 7-11. Steps in subtracting dimensions given in feet and inches.

## SYMBOLS

In writing, we use an alphabet of letters to construct words. The architect uses a set of symbols, too, in describing the shape, size and materials used in a building. These symbols are a form of shorthand. They are much easier to draw than an actual picture of the materials. Generally accepted symbols are shown in Figs. 7-12 and 7-13. Study these symbols carefully while looking at the partial set of plans in this unit. It will help you understand how to read plans. For other construction symbols, see Useful Information, pages 259 and 260.

Abbreviations are used on plans to save space. Common plumbing abbreviations are shown in Fig. 7-14.

## SCALING A DRAWING

Where only approximate dimensions are required, the plumber may find it helpful to "scale" the drawing rather than attempt to calculate distances. Because all parts of a structure shown on an architectural drawing are proportionate to the actual size of the building, it is possible to use an architect's scale, Fig. 7-15, to determine an unknown dimension. Always refer to the title block on the drawing, Fig. 7-16, to find out the correct scale to use.

A less accurate method of scaling the drawing is illustrated in Fig. 7-17.

Scaling should not be considered extremely accurate, but it can be a very useful technique when estimating the amount of pipe required for a particular run. Another useful aspect of scaling is that it can be used to check calculations and thus prevent errors. (Drawings presented in this unit have been reduced. Thus, they are not to scale.)

## METRIC DIMENSIONS AND SCALES

At some point, the building construction industry will convert to metric. New sizes will have to be determined for all building components. Brick, plywood, tiles, windows and dimension lumber will receive a new metric measure. These may be close to but not the same as the present sizes.

Let us consider how the changes might affect the size of plumbing materials. For example, a 1 in. diameter pipe may be made to a 24 or 26 mm diameter. Or a 3/4 in. size may become an 18 mm pipe. However, since no standards have been set, it would be difficult to predict what the new metric sizes will be.

Regarding architectural plans, a decision will have to be made on which metric unit to use for drawings. Many

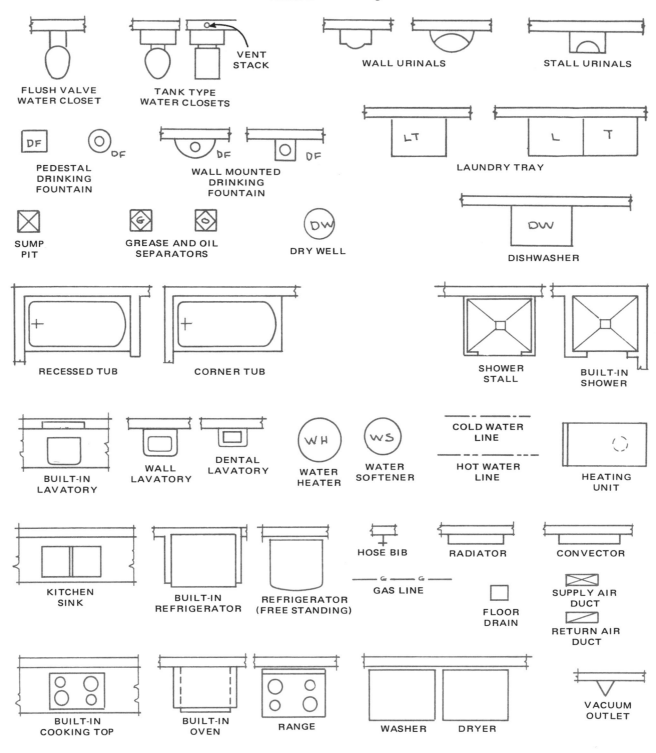

Fig. 7-12. These symbols are used for plumbing fixtures, appliances and mechanical equipment.

countries that have gone metric or are in the process of going metric, are using the millimetre for all drawings except plot plans. In plot plans either the millimetre or metre is used.

Building plans are now customarily drawn to a 1/4 scale. In other words, 1/4 in. on the drawing is equal to 1 ft. of actual size. After conversion to metric, it is possible that the scale will be: 1 mm = 50 mm (a 1:50 scale).

A point to remember is that in SI metrics, different units of measure are never mixed. For example: 1.3 m or 1300 mm is the correct form, not 1 m − 300 mm. This is a departure from

current practice in English measure. For example: a dimension of 3 ft. and 6 in. appears correctly on a drawing as 3'-6", not as 3.5' or 42".

## SKETCHING PIPING INSTALLATIONS

Plans for residential structures do not include pipe drawings. This suggests that plumbers need to develop some skill in sketching piping installations if they are to communicate effectively with other people.

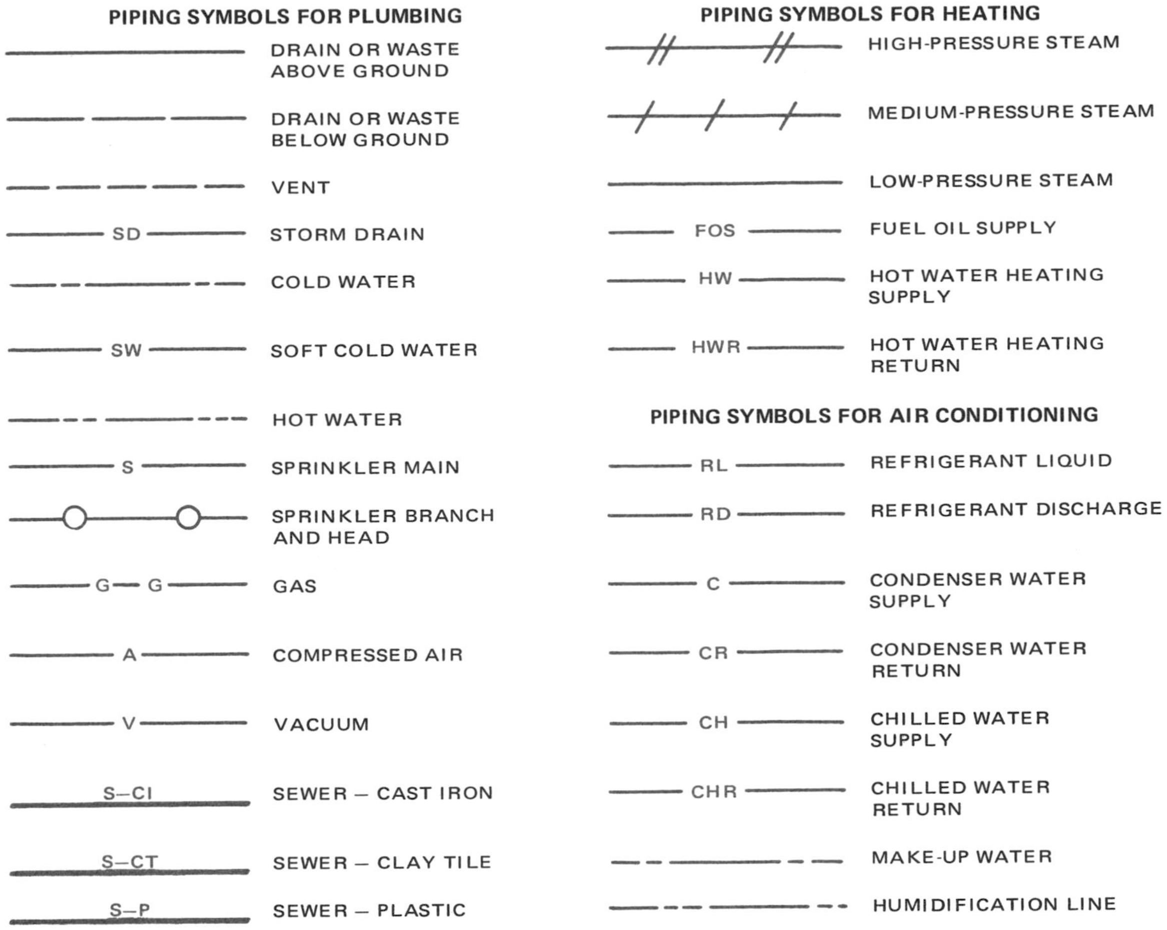

## PIPING SYMBOLS FOR PLUMBING

|  |  |
|---|---|
| ——————— | DRAIN OR WASTE ABOVE GROUND |
| —— — —— — | DRAIN OR WASTE BELOW GROUND |
| — — — — — | VENT |
| —— SD —— | STORM DRAIN |
| — — — — | COLD WATER |
| —— SW —— | SOFT COLD WATER |
| — —— — —— | HOT WATER |
| —— S —— | SPRINKLER MAIN |
| —O——O— | SPRINKLER BRANCH AND HEAD |
| —— G—G —— | GAS |
| —— A —— | COMPRESSED AIR |
| —— V —— | VACUUM |
| S—CI | SEWER — CAST IRON |
| S—CT | SEWER — CLAY TILE |
| S—P | SEWER — PLASTIC |

## PIPING SYMBOLS FOR HEATING

|  |  |
|---|---|
| —//——//— | HIGH-PRESSURE STEAM |
| —/——/——/— | MEDIUM-PRESSURE STEAM |
| ——————— | LOW-PRESSURE STEAM |
| —— FOS —— | FUEL OIL SUPPLY |
| —— HW —— | HOT WATER HEATING SUPPLY |
| —— HWR —— | HOT WATER HEATING RETURN |

## PIPING SYMBOLS FOR AIR CONDITIONING

|  |  |
|---|---|
| —— RL —— | REFRIGERANT LIQUID |
| —— RD —— | REFRIGERANT DISCHARGE |
| —— C —— | CONDENSER WATER SUPPLY |
| —— CR —— | CONDENSER WATER RETURN |
| —— CH —— | CHILLED WATER SUPPLY |
| —— CHR —— | CHILLED WATER RETURN |
| — — — — | MAKE-UP WATER |
| — —— — —— —— | HUMIDIFICATION LINE |

Fig. 7-13. Pipe and fitting symbols.

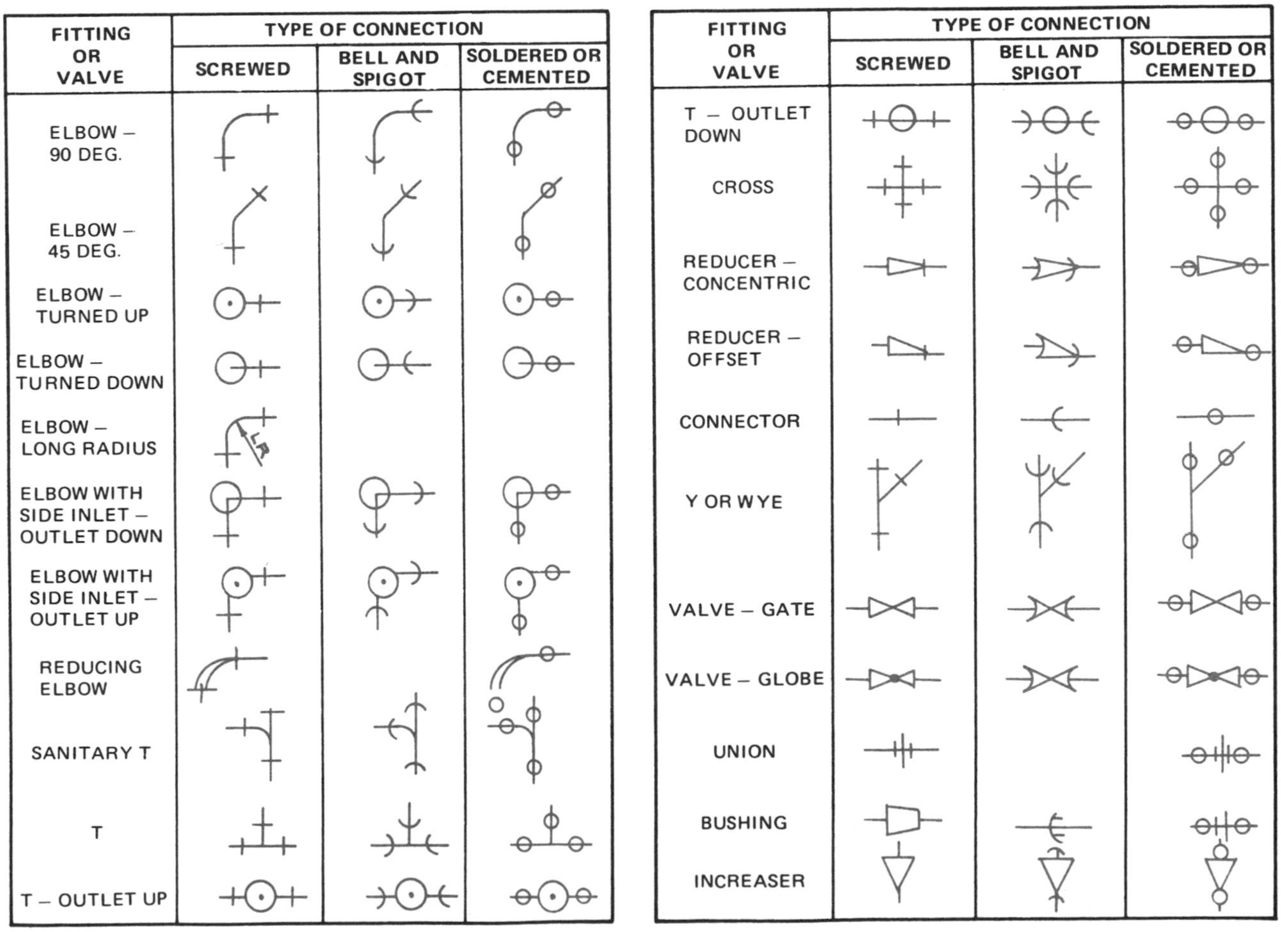

| FITTING OR VALVE | TYPE OF CONNECTION | | |
|---|---|---|---|
| | SCREWED | BELL AND SPIGOT | SOLDERED OR CEMENTED |
| ELBOW — 90 DEG. | | | |
| ELBOW — 45 DEG. | | | |
| ELBOW — TURNED UP | | | |
| ELBOW — TURNED DOWN | | | |
| ELBOW — LONG RADIUS | | | |
| ELBOW WITH SIDE INLET — OUTLET DOWN | | | |
| ELBOW WITH SIDE INLET — OUTLET UP | | | |
| REDUCING ELBOW | | | |
| SANITARY T | | | |
| T | | | |
| T — OUTLET UP | | | |

| FITTING OR VALVE | TYPE OF CONNECTION | | |
|---|---|---|---|
| | SCREWED | BELL AND SPIGOT | SOLDERED OR CEMENTED |
| T — OUTLET DOWN | | | |
| CROSS | | | |
| REDUCER — CONCENTRIC | | | |
| REDUCER — OFFSET | | | |
| CONNECTOR | | | |
| Y OR WYE | | | |
| VALVE — GATE | | | |
| VALVE — GLOBE | | | |
| UNION | | | |
| BUSHING | | | |
| INCREASER | | | |

## PLUMBING ABBREVIATIONS

| ITEM | ABBR. | ITEM | ABBR. |
|------|-------|------|-------|
| CAST IRON | CI | HOT WATER | HW |
| CENTERLINE | CL | LAUNDRY TRAY | LT |
| CLEANOUT | CO | LAVATORY | LAV. |
| COLD WATER | CW | MEDICINE CABINET | MC |
| COPPER | COP. | PLASTIC | PLAS. |
| DISHWASHER | DW | PLUMBING | PLBG. |
| FLOOR DRAIN | FD | WATER CLOSET | WC |
| GALVANIZED IRON | GAL. I | WATER HEATER | WH |
| HOSE BIB | HB | WATER SOFTENER | WS |

Fig. 7-14. Plumbing abbreviations are used on plans.

Fig. 7-15. Part of architects' scale. Each division represents 1 ft. Fine markings to left of "0" represents inches and parts of inches.

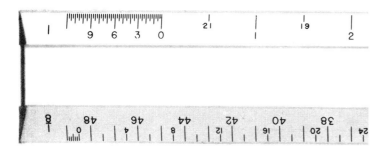

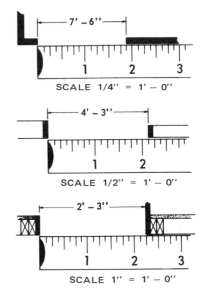

Fig. 7-17. Folding rule can be used to scale a plan. Top. On 1/4 scale, each full inch equals 4 ft. and each 1/8 in. equals 6 in. Middle. If scaled to 1/2 in., each inch of rule equals 2 ft. Each 1/8 in. equals 3 in. Bottom. On the 1 in. scale, an inch equals 1 ft. Each 1/4 in. equals 3 in.

RESIDENCE for  B.E. BIRKHEAD

11658 MARK TWAIN DRIVE

BRIDGETON, MISSOURI

SCALE   1/4" = 1'

Page 4 of 5

Fig. 7-16. In scaling a drawing, be sure to use the same scale in which it was prepared. This information can be found on the title block.

Basically, there are three types of piping sketches.
1. The riser diagram, Fig. 7-18, which is two-dimensional.
2. The plan view sketch, Fig. 7-19.
3. The isometric sketch, Fig. 7-20, which gives the illusion of three-dimensions.

The riser diagram and the plan view sketch are simpler to make. The riser diagram can be used successfully to illustrate pipe runs that are essentially in one plane.

The steps in making a plan view sketch are illustrated in Fig. 7-21. This type of sketching is frequently done on the job site. The sketch may be on the back of a set of plans, on a scrap of wood or even on the ground. The object is to communicate. Therefore, it is only necessary that the appropriate symbols and reasonable porportion be used in making the sketch.

Isometric sketches are helpful in illustrating more complex piping systems. The basic element in isometric sketching is the isometric axis, Fig. 7-22. While learning to make isometric

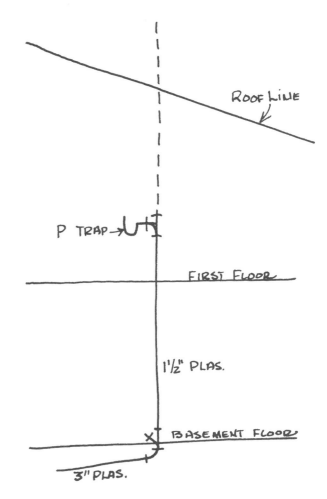

Fig. 7-18. Simple riser diagram of DWV piping.

sketches, it may be helpful to think of vertical pipes always being drawn vertically on the sketch and all horizontal pipes being drawn at 120 deg. angles to the vertical lines. Some

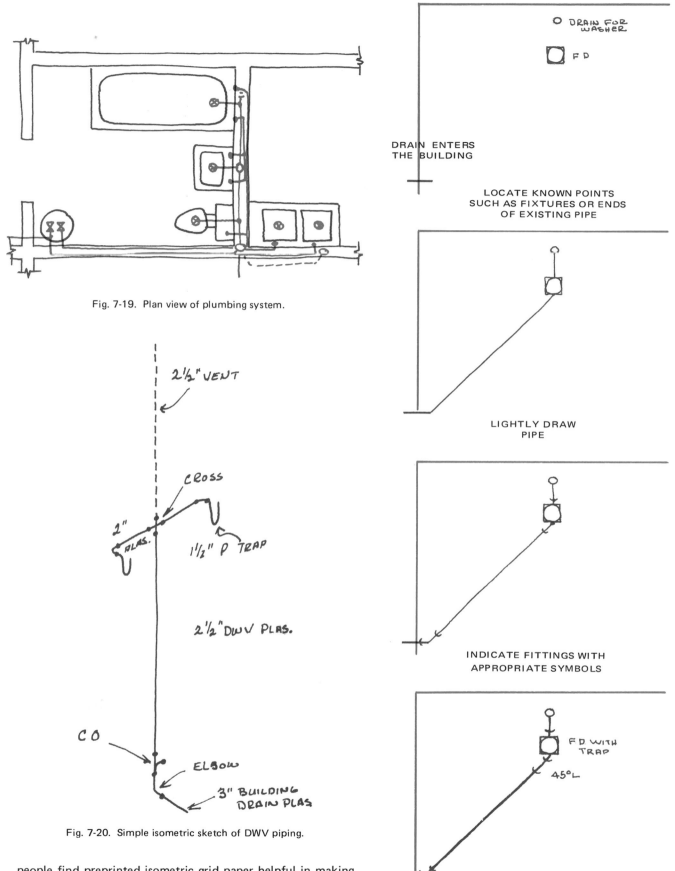

Fig. 7-19. Plan view of plumbing system.

2½" VENT

CROSS

2"
PLAS.

1½" P TRAP

2½" DWV PLAS.

CO

ELBOW

3" BUILDING
DRAIN PLAS

Fig. 7-20. Simple isometric sketch of DWV piping.

DRAIN FOR
WASHER

FD

DRAIN ENTERS
THE BUILDING

LOCATE KNOWN POINTS
SUCH AS FIXTURES OR ENDS
OF EXISTING PIPE

LIGHTLY DRAW
PIPE

INDICATE FITTINGS WITH
APPROPRIATE SYMBOLS

FD WITH
TRAP

45°L

45°L

DARKEN LINES AND ADD NOTES
AFTER ALL MODIFICATIONS
HAVE BEEN MADE

Fig. 7-21. Steps in making a plan view sketch. The colored lines
represent the outline of the building.

people find preprinted isometric grid paper helpful in making isometric sketches. The DWV piping drawing shown in Fig. 7-23, illustrates how the grid can be used. Fig. 7-24 shows a pictorial drawing of a residential DWV and water supply system. The steps in making a simple isometric sketch are shown in Fig. 7-25. Fig. 7-26 illustrates how the system pictured in Fig. 7-24 would look in an isometric view.

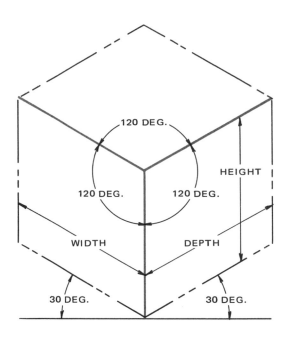

Fig. 7-22. In isometric axis, converging lines form angles of 120 deg. or 30 deg. from horizontal.

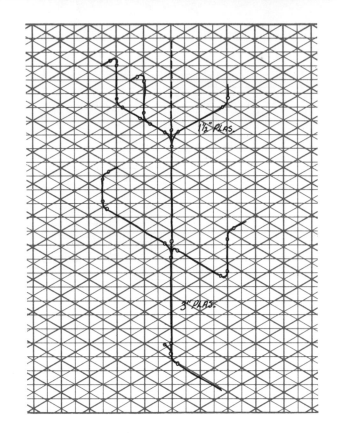

Fig. 7-23. Sketching on isometric grid paper.

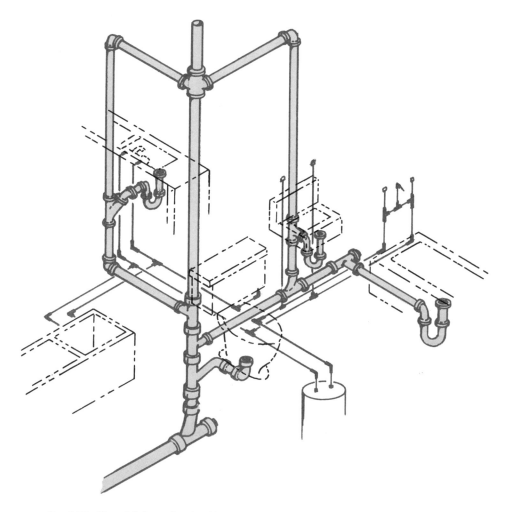

Fig. 7-24. Pictorial view of a plumbing system as an artist or drafter might produce it.

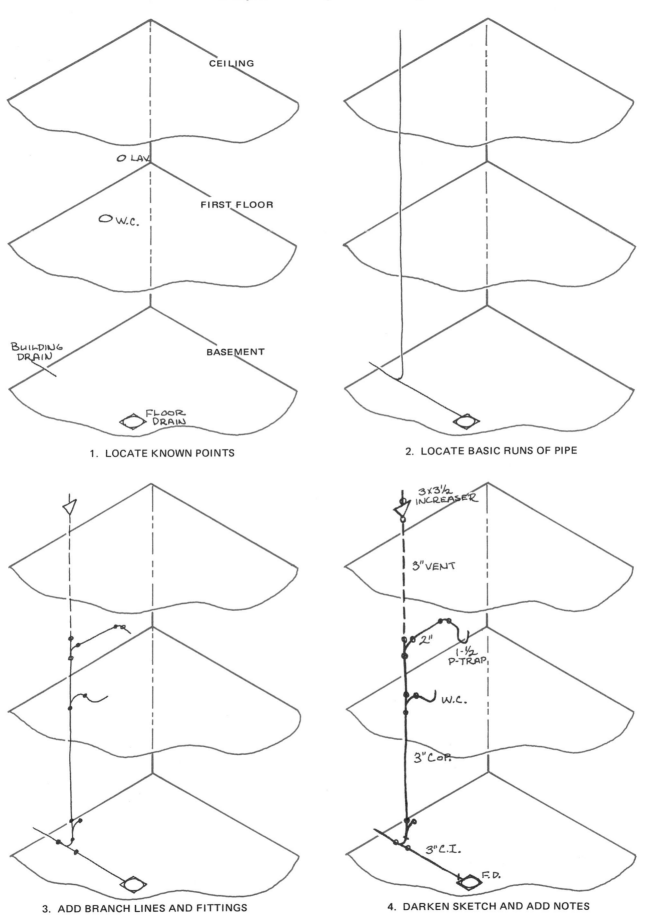

CEILING

O LAV

FIRST FLOOR

O W.C.

BUILDING
DRAIN

BASEMENT

FLOOR
DRAIN

1. LOCATE KNOWN POINTS

2. LOCATE BASIC RUNS OF PIPE

3. ADD BRANCH LINES AND FITTINGS

3x3½
INCREASER

3" VENT

2"

1-½
P-TRAP

W.C.

3" COP.

3" C.I.

F.D.

4. DARKEN SKETCH AND ADD NOTES

Fig. 7-25. Steps in making an isometric sketch. Colored lines indicate the basic outline of the building which may or may not be required to make the drawing understandable.

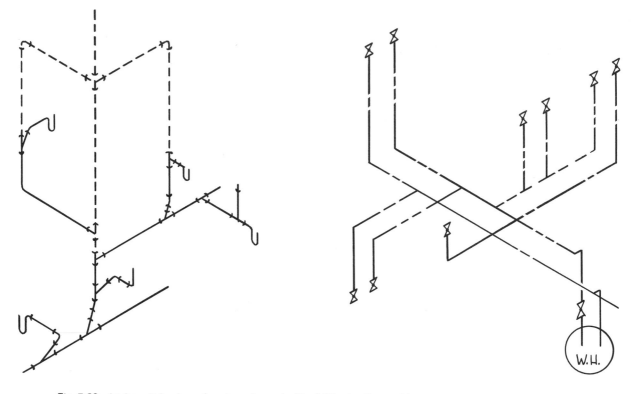

Fig. 7-26. An isometric view of system shown in Fig. 7-24 using line and fixture symbols. Drainage system has been separated from the water supply system.

## TEST YOUR KNOWLEDGE — UNIT 7

1. Referring to Figs. 7-2 through 7-6 answer the following questions:
   a. What fixtures are included in the smaller bathroom?
   b. How many hose bibs are required?
   c. What type sink is required in the kitchen?
   d. The hot water heater is located in the _____ room.
   e. Assuming that the water supply pipe enters the building approximately three feet to the right of the front stoop, how many feet of pipe will be required to supply the hose bib outside the utility storage room with water?
   f. The center-to-center dimension between the brick piers supporting the floor is _____ ' — _____ ".
   g. The vertical height of the crawl space under the floor is _____.
   h. The DWV piping for the structure is Schedule 80 _____.

2. Sketch the correct symbol for each of the following:
   a. Drain and waste piping above ground.
   b. Vent.
   c. Cold water.
   d. Hot water.
   e. 90 deg. elbow — cast iron.
   f. Reducing elbow-copper.
   g. Concentric reducer — galvanized iron.
   h. Connector-plastic.
   i. Gate valve-brass.

3. What are the abbreviations for each of the following terms?
   a. Cast iron _____.
   b. Cleanout _____.
   c. Galvanized iron _____.
   d. Hose bib _____.
   e. Plumbing _____.
   f. Water closet _____.
   g. Water softener _____.

## SUGGESTED ACTIVITIES

1. Using a set of residential house plans and specifications, determine the type and quantity of fixtures required and type of pipe and fittings necessary.

2. Using the plans in Activity 1, prepare an isometric sketch of the DWV piping system. This sketch should be made on isometric grid paper. Include appropriate symbols for all pipe and fittings.

3. On the same grid, sketch the hot and cold water piping.

# Unit 8
# DESIGNING PLUMBING SYSTEMS

## Objectives

In this unit you will find the principles for designing plumbing systems that will provide long and satisfactory service.

After studying this unit you will be able to:
- Apply various tables for determining adequate size of drainage and water supply piping.
- Plan efficient plumbing systems which conserve plumbing materials.
- Design plumbing systems that are easily serviced.
- Explain the relationship of height to water pressure.
- Recognize cross connections and explain how to avoid them.
- Describe systems for collecting and draining storm water.

Every successful plumbing system design begins with identification of needs. The owner, architect, or some other responsible person must indicate:
1. The number and type of bathrooms.
2. Fixtures, water supplies and drainage facilities needed in the kitchen and laundry rooms.
3. Type of hot water supply.
4. Whether or not a water softener and/or a filter will be included.

In commercial buildings, the number of people who will use restroom facilities must be considered. In addition, water supply and drainage will be needed for:
1. Food preparation areas.
2. Water fountains.
3. Cooling towers for air conditioning equipment.
4. Locations where manufacturing processes require liquids or gases.

In this unit, attention centers chiefly on designing residential plumbing systems. However, many of the same design considerations are appropriate for small commercial installations.

## BASIC DESIGN CONSIDERATIONS

All plumbing designs must consider these basic needs:
1. Ample facilities for those who live or work in the building.

2. Adequate piping for water supply and drainage to each fixture.
3. Easily cleaned fixtures, fittings, cabinets, walls and floors in the bathroom, kitchen and laundry.
4. Drainage, supply and vent piping that is free of leaks for the life of the structure. This precaution will prevent damage to the building and assure that unpleasant and potentially harmful gases will not enter the building.
5. Drainage piping systems with traps that are adequately vented to insure an airtight seal.
6. Economical plumbing installation. Fixtures should be clustered as much as is practical. This reduces cost.

With these basic considerations in mind, it is possible to make decisions about the plumbing system for different parts of the dwelling.

## RELATIONSHIP OF ROOMS

Since plumbing materials should be conserved, it is good economics to place rooms needing plumbing near each other. Such rooms can be grouped in different ways. In well-planned buildings, the kitchens, bathrooms and laundry areas are often placed back to back, above and below each other or arranged so that plumbing runs are considerably reduced.

This aspect of planning is the job of the architect or the person who has drawn up the plan for the structure. *The plumber's first step is to study the plans to identify the fixtures needed in kitchen, bath and utility rooms. Skill is required in locating piping to take advantage of grouping these fixtures.*

Fig. 8-1 is an example of how the room arrangement in a two-story home can make efficient plumbing installation possible. Because the soil stack is the most expensive part of the plumbing system, considerable savings can be realized if fewer fittings and less pipe is used.

Large apartment buildings with many stories are a good example of this principle. In nearly all cases they contain a kitchen-bathroom core which runs vertically up the building. Thus, one soil stack serves many apartments. See Fig. 8-2.

## ARRANGING THE ROOM FIXTURES

Another important step in designing plumbing systems is locating fixtures inside the room. Again, proper grouping of

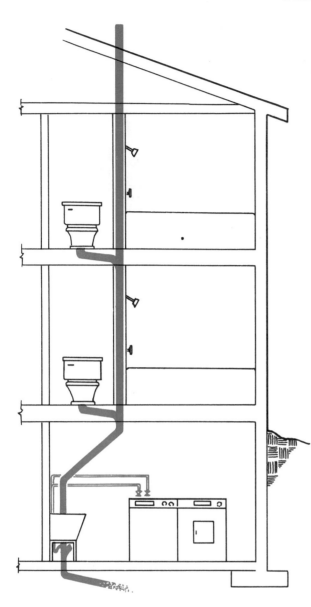

Fig. 8-1. This is an efficient layout of utility room and bathrooms. Since bathrooms and laundry area are aligned vertically they are served by one soil stack.

the fixtures will save materials and pipe fittings. *The fixtures, ideally, should be placed so that all drainage piping is in or near one wall.*

Several conventional bathroom layouts are shown in Figs. 8-3 through 8-9. They differ in size and type of fixtures included. Some layouts succeed in keeping piping in one wall. Other layouts are acceptable but will require more materials.

In laying out bathrooms, care should be used to provide ample room for using the facilities.

Maximum and minimum dimensions for fixtures and accessories are given in Fig. 8-10.

## KITCHENS

In kitchens, as in bathrooms, it is desirable to keep piping to a minimum. However, the planner must also be concerned about the work area created by the relative positions of the stove, refrigerator and sink. See Fig. 8-11. If this work

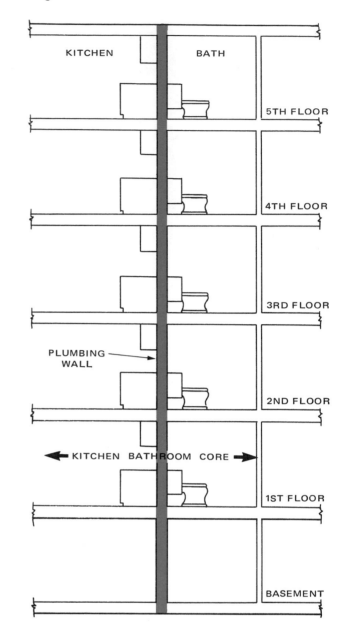

Fig. 8-2. By stacking the bathroom-kitchen core in multistory buildings, it is possible to place all plumbing lines in a single plumbing wall.

Fig. 8-3. This arrangement is suitable for a small bathroom. All plumbing lines are contained in the plumbing wall. Door does not provide for greatest privacy.

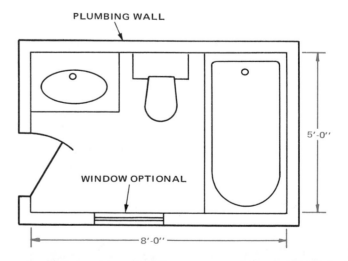

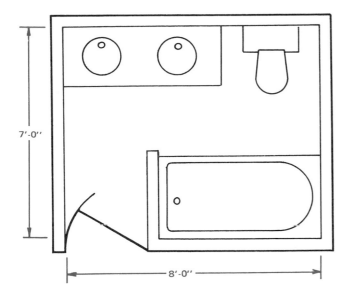

Fig. 8-4. This arrangement for a nearly square room will require more piping to reach tub.

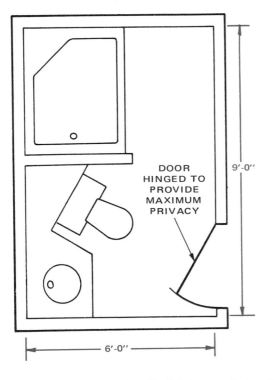

DOOR HINGED TO PROVIDE MAXIMUM PRIVACY

Fig. 8-5. Slightly larger bathroom than Fig. 8-4 also uses single wall for plumbing. Greater privacy is provided at doorway and through wall between tub and water closet.

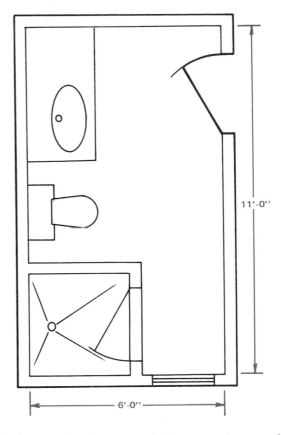

Fig. 8-6. Layout of a long narrow bathroom requires use of more piping but fixtures are grouped for greatest saving in waste piping.

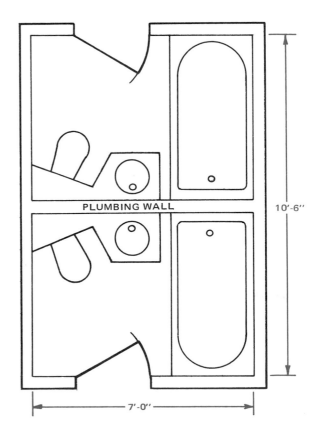

PLUMBING WALL

Fig. 8-7. Back-to-back bathrooms make maximum use of common plumbing wall. From standpoint of using least DWV piping can you determine where the soil stack should be in the plumbing wall?

"triangle" is kept small and free of cross traffic there will be fewer steps and interruptions while moving from any one of these points to the others.

*When the basic layout of the kitchen has been determined, the plumber must identify the fixtures and accessories that are to be installed.* Water supply and/or drainage pipe may be needed for:

1. Sink.
2. Garbage disposal.
3. Dishwasher.
4. Ice maker.

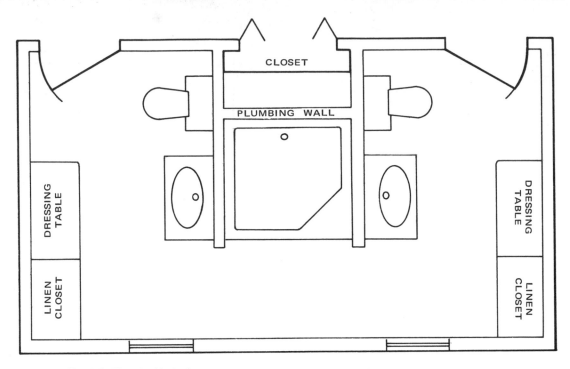

Fig. 8-8. This double bathroom arrangement uses common bathtub and provides excellent opportunity for economizing on piping.

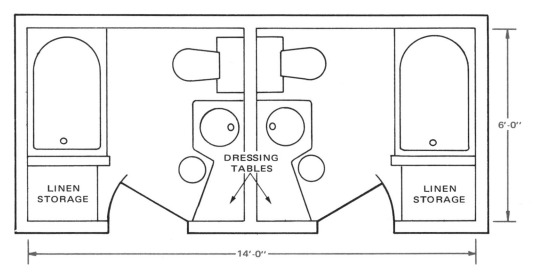

Fig. 8-9. This arrangement, while acceptable, will require more piping material than the double baths in Fig. 8-7 and Fig. 8-8.

In the utility room, Fig. 8-12, plumbing is usually needed for a washer, water heater, water softener and water filter. It is also common practice to place a floor drain in this area. In addition, it may be necessary to include connections and lines for water supply to a furnace humidifier and drainage for an air conditioner evaporator in the furnace bonnet.

As with the bathroom and kitchen, fixtures in the utility room should be located as close to one another as is practical in order to keep the cost down.

## SELECTING SPECIFIC FIXTURES

Before the complete details of the plumbing system can be worked out, fixtures must be selected so that the "rough-in"

dimension of each can be determined. *The rough-in dimensions indicate where the DWV and water supply lines must be located in the wall and/or floor.* Because the drainage and water supply pipes may vary for different fixtures, it will be necessary for the plumber to know the sizes of the pipe on each fixture.

## DESIGNING THE PIPING SYSTEMS

In a typical residence, plumbing consists of:

1. A drain-waste-vent piping system which empties into the sanitary sewer, a septic tank, a cesspool or a lagoon.
2. A water supply piping system which is fed by a municipally owned water main or a privately owned water supply.

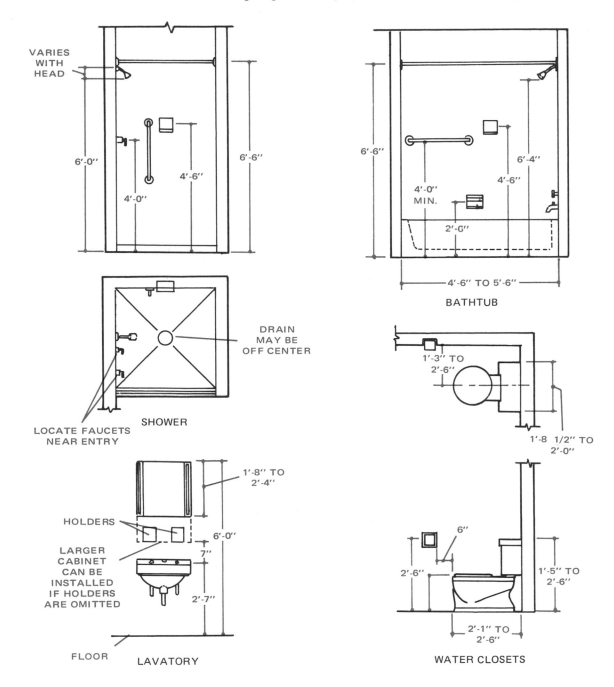

Fig. 8-10. Dimensions and location for bathroom fixtures and accessories.

3. A storm water piping system which empties into a storm sewer, a drywell or a nearby stream.

The DWV piping system and the storm water piping systems are sometimes combined into one sewer piping system. However, this type of installation is undesirable because heavy rainfall can overload the system and cause drains to back up into buildings. Furthermore, a combined piping system imposes an extremely heavy volume of water on sewer treatment facilities. This, in turn, increases cost of construction and operation.

The STORM DRAINAGE piping system is different from the DWV piping system. It collects only water which is unpolluted and carries it to the storm sewer. Downspouts, driveway drains and sump pumps which remove ground water

from below basement floors are examples of piping connected to a storm sewer.

Other than the fact that the storm water piping system and the DWV piping system must be kept separate, they are very similar. They both depend on gravity to function. Both are constructed from similar materials.

## DESIGNING THE DWV PIPING SYSTEM

DWV piping systems operate on gravity. Proper slope of horizontal runs is very important. A drop of 1/8 to 1/2 in. per foot is considered adequate. This will provide for good drainage but will not move liquids along so rapidly that solid waste is left behind to clog the drain.

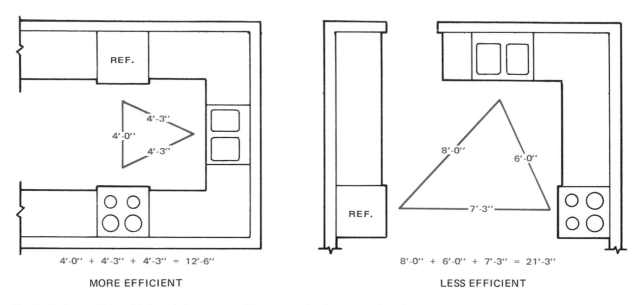

$4'\text{-}0'' + 4'\text{-}3'' + 4'\text{-}3'' = 12'\text{-}6''$

MORE EFFICIENT

$8'\text{-}0'' + 6'\text{-}0'' + 7'\text{-}3'' = 21'\text{-}3''$

LESS EFFICIENT

Fig. 8-11. In an efficient kitchen design, stove, refrigerator and sink are placed so that traffic between them forms a tight triangle. Some planners recommend that combined distance between the three should not exceed 22 ft.

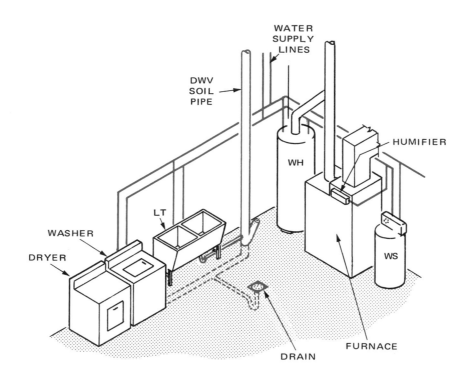

Fig. 8-12. Efficient layout in utility room groups appliances close to water supply and drainage.

*Drainage piping is designed before the water supply system because it is the most expensive and most important part of the plumbing system.* It should be kept simple not only to avoid drainage problems but to keep down its cost. In determining the location of bathroom fixtures, for example, the designer should try to place the water closet where it requires the least material and fewest turns. Tub, shower and lavatory location would be considered after the water closet, since the waste piping is less expensive than soil pipe. In Fig. 8-13, the water closet has a direct, short run into the soil stack.

The type of material selected for the DWV system (ABS,

PVC, copper or cast iron) will depend upon local plumbing code requirements, cost and preference of the owner. (Unit 4 has additional information about piping materials.)

## Size of pipe

*Selecting the correct DWV pipe size for each fixture drain and for the soil stack is extremely important.* Pipe which is too small will not permit waste to flow properly from the fixture. It tends to clog. If the pipe is too large, several other disadvantages become apparent:

1. Considerably more space is needed for installation. Studs must be wider in plumbing walls in order to conceal the

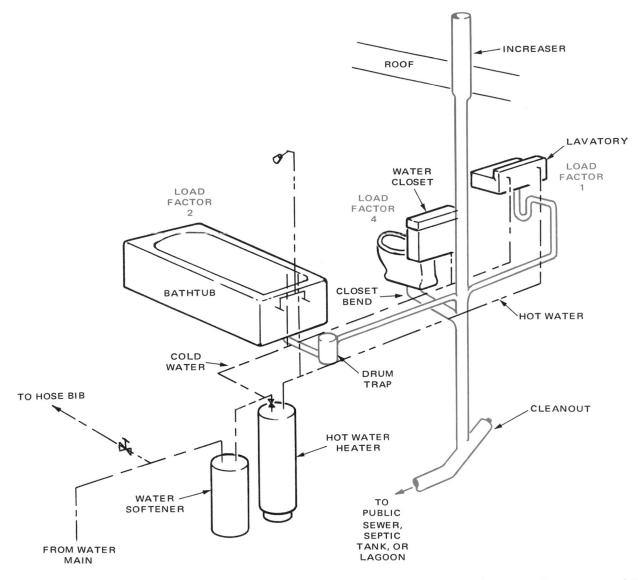

Fig. 8-13. A well-planned bathroom layout. Note that planner has given each plumbing fixture a factor according to amount of liquid the fixture discharges during a given interval. The larger the number, the greater the volume.

large-diameter pipe.
2. Large piping is extremely expensive.
3. Large pipe is more difficult to install.
4. Oversize pipe is not efficient in carrying away solid waste. Scouring action of the water is reduced while flow is too shallow to carry solids along. They tend to settle in the drain where they soon cause a stoppage.

## Drainage studies

Studies have been made to determine:
1. The smallest pipe size that can be satisfactorily used for various types of fixtures.
2. What size the pipe must be if a given combination of fixtures is installed on a branch line or on a soil stack.

The study group, made up of representatives of management, labor and governmental agencies, tested plumbing fixtures measuring the amount of liquid each could discharge over a measured time interval. A lavoratory, they found, could discharge about 7 1/2 gallons of water a minute. This was so close to a cubic foot of water that it became the basis of the

measurement system. Since a lavatory can discharge about 1 cu. ft. of water per minute it receives a number 1. This is called its LOAD FACTOR.

## Load factors

Thus, a load factor indicates how many cubic feet of water a fixture discharges into the drain. Fig. 8-13 is a sketch of a typical bathroom plumbing system. It shows load factors for each fixture. A factor of 4 indicates that the fixture discharges 4 cu. ft. of water in a minute. Standard charts have been developed so that these factors are now given for many fixtures. They can be used by plumbers to properly determine drain sizes. See Fig. 8-14. Factors not listed in load factor chart can be estimated by using the values given in Fig. 8-15.

## Computing sizes

Consider the DWV piping for the bathroom shown in Fig. 8-16. Branch A receives a load factor of 4 from the tank-operated water closet. Branch B also receives a load factor of 4 from a bathtub and large lavatory. The soil stack

| FIXTURE TYPE | LOAD FACTOR | MINIMUM TRAP SIZE |
|---|---|---|
| BATHROOM GROUP<br>  WATER CLOSET, LAVATORY<br>  BATHTUB OR SHOWER | 6 TANK TYPE<br>WATER CLOSET<br><br>8 FLUSH VALVE<br>WATER CLOSET | SEE INDIVIDUAL<br>FIXTURES<br><br>BELOW |
| BATHTUB W/NO SHOWER<br>BIDET<br>DRINKING FOUNTAIN<br>DISHWASHER | 2<br>3<br>1/2<br>2 | 1 1/2<br>1 1/2<br>1<br>1 1/2 |
| FLOOR DRAIN<br>KITCHEN SINK — DOMESTIC<br>KITCHEN SINK — DOMESTIC<br>  W/GARBAGE DISPOSAL | 1<br>2<br>3 | 2<br>1 1/2<br>1 1/2 |
| LAVATORY — SMALL<br>LAVATORY — LARGE<br>LAUNDRY TRAY<br>  1 OR 2 COMPARTMENTS | 1<br>2<br>2 | 1 1/2<br>1 1/2<br>1 1/2 |
| SHOWER — DOMESTIC<br>WATER CLOSET<br>  TANK OPERATED<br>  FLUSH VALVE OPERATED | 2<br><br>4<br>8 | 2<br><br>3<br>3 |

Adapted from Table 11.4.2 National Plumbing Code

Fig. 8-14. Load factors and trap sizes for many plumbing fixtures.

| FIXTURE<br>DRAIN SIZE (IN.) | LOAD<br>FACTOR |
|---|---|
| 1 1/4 & SMALLER<br>1 1/2<br>2 | 1<br>2<br>3 |
| 2 1/2<br>3<br>4 | 4<br>5<br>6 |

Table 11.4.3 National Building Code

Fig. 8-15. Table for estimating load factors for other types of fixtures not listed in Fig. 8-14. (National Building Code)

must be able to handle a load factor of 8 (water closet, tub and lavatory combined). The chart in Fig. 8-14 gives the branch lines sizes required for these fixtures: bathtub 1 1/2 in., water closet 3 in. and lavatory 1 1/2 in. However, the combined load factor for the branch to the lavatory and the bathtub is 4. Therefore, the pipe connecting the lavatory drain to the soil stack must be 2 in. in diameter, Fig. 8-17. (Since the maximum load for a 1 1/2 in. horizontal branch is 3, the next larger size must be selected for the run of pipe from the stack to the lavatory drain.)

As a general rule, stacks must never be smaller than the largest branch entering them. They must also be designed to accommodate the load factors indicated in Fig. 8-14. In the simple example just given, a 3 in. stack is required.

Building drains — these are the drainage pipe connecting the soil stack to sewer or septic tank — must be sized and sloped to carry the load of waste received from the dwelling. Fig. 8-18 gives load factors for various combinations of sizes and slopes.

The foregoing example and the tables should be used only as a guide in designing a system. The plumbing code for a given area may contain somewhat different requirements. *In every case, the local plumbing code is the final authority and must be followed.*

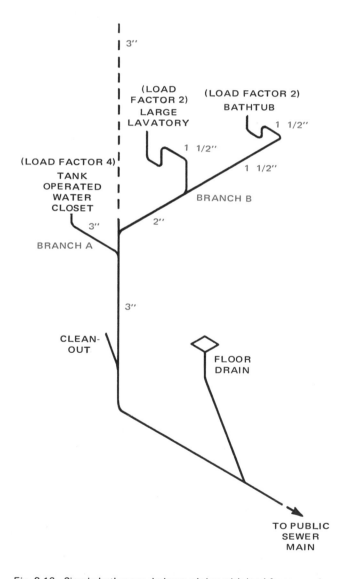

Fig. 8-16. Simple bathroom drainage piping with load factors and sizes determined.

| DIAMETER OF PIPE (IN.) | MAXIMUM LOAD FACTOR THAT MAY BE CONNECTED TO: | | | |
| --- | --- | --- | --- | --- |
| | ANY HORIZONTAL FIXTURE BRANCH | ONE STACK OF 3 STORIES OR LESS | ONE STACK OF MORE THAN 3 STORIES | |
| | | | TOTAL | MAXIMUM PER BRANCH INTERVAL |
| 1 1/4 | 1 | 2 | 2 | 1 |
| 1 1/2 | 3 | 4 | 8 | 2 |
| 2 | 6 | 10 | 24 | 6 |
| 2 1/2 | 12 | 20 | 42 | 9 |
| 3 | 20 | 30 | 60 | 16 |
| 4 | 160 | 240 | 500 | 90 |
| 5 | 360 | 540 | 1100 | 200 |
| 6 | 620 | 960 | 1900 | 350 |

Adapted from Table 11.5.3 National Plumbing Code

Fig. 8-17. Plumbing codes use charts like this to guide plumbers in designing adequate size into drainage branches and soil stacks.

| DIAMETER OF PIPE (IN.) | MAXIMUM LOAD FACTOR THAT MAY BE CONNECTED TO ANY PORTION OF A BUILDING DRAIN OR SEWER | | | |
| --- | --- | --- | --- | --- |
| | FALL PER FOOT | | | |
| | 1/16 | 1/8 | 1/4 | 1/2 |
| 2 | | | 21 | 26 |
| 2 1/2 | | | 24 | 31 |
| 3 | | 20* | 27* | 36 |
| 4 | | 180 | 216 | 250 |
| 5 | | 390 | 480 | 575 |
| 6 | | 700 | 840 | 1000 |
| 8 | 1400 | 1600 | 1920 | 2300 |

Adapted from Table 11.5.2 National Building Code
*Not more than two water closets

Fig. 8-18. National Building Code recommended sizes and slopes for building drains. Note that 1/16 in. fall is not recommended with drain pipe smaller than 8 in. diameter.

## Selecting DWV fittings

The drain-waste portion of the DWV system must be installed with fittings that encourage smooth flow of waste through the pipes. In Unit 4, DWV fittings were compared to pressure fittings. Three major characteristics of drain-waste fittings are:

1. The smooth large-radius curves produced at pipe joints.
2. The shoulders inside the fittings which prevent an offset from being created when the pipe is installed.
3. The built-in slope of 1/8 to 1/4 in. per foot in horizontal runs. (This is shown in Fig. 4-31, Unit 4.)

Fig. 8-19, is the same DWV piping system discussed in Fig. 8-16. The fittings necessary to install the system have been identified. Note that sanitary Ts are used at all branch intersections and that a cleanout is provided at the point where the soil stack changes directions.

## DESIGNING THE VENTING SYSTEM

Vents, properly installed as a part of the DWV system, permit air to circulate through the waste piping. Their function is to maintain air inside the waste piping at a nearly constant pressure. Further, vents exhaust sewer gas buildup above the roof. Venting is important. Without it, the waste piping system will not function as intended.

## MAINTAINING ATMOSPHERIC PRESSURE IN THE WASTE PIPING

Traps installed in the waste piping prevent sewer gas from entering the building by providing a water seal. A primary function of vents is to prevent loss of this seal. SIPHONAGE, BACK PRESSURE, EVAPORATION, CAPILLARY ACTION, and WIND can cause the loss of water normally in the traps. The following discussion will help you understand how each of these conditions occur.

### Siphonage

Siphonage happens for two different reasons:

1. If a partial vacuum is created in the waste piping, water may be siphoned (sucked) out of one or more traps, Fig. 8-20. The installation of a vent, as shown in Fig. 8-21, will prevent this from occurring because air, at atmospheric pressure, enters the waste piping near the trap.
2. INDIRECT or MOMENTUM SIPHONAGE. Takes place when the discharge of one fixture causes the water to be drawn from another fixture. Fig. 8-22, illustrates how this can occur. Negative pressure at the lower trap is caused when the flow of water drags air with it through the vertical pipe. This problem can be overcome by installing the proper size stack and by adding a vent near the trap on the

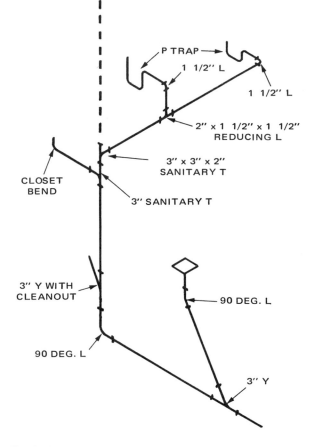

Fig. 8-19. Isometric diagram showing plumbing fittings needed.

lower level. If the stack is large enough, the water flowing through it will not completely fill the pipe. Thus, air may enter or escape while water is being discharged. A vent near the lower trap, however, would bring in air at atmospheric pressure to protect the trap seal.

## Back pressure

Back pressure can cause the loss of a trap seal when air pressure builds up in the system. The high-pressure air blows through one or more traps. Back pressure becomes a greater problem as buildings become taller. For example, if the water flowing through the waste piping completely fills the stack, air pressure will build up ahead of the slug of water, Fig. 8-23. Unless this pressure can escape directly to the atmosphere it will blow through a trap. This problem can be prevented by the installation of a vent near the fixture traps or where piping changes direction.

## Evaporation

Evaporation of water in the traps can also cause the seal to be lost if the plumbing system is not used for extended periods. If this is likely to happen, deeper traps can be installed.

## Capillary action

Capillary action is still another cause for loss of water in a trap. Again, this is unlikely to happen. However, foreign

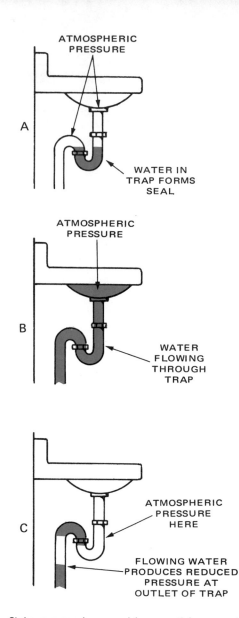

Fig. 8-20. Siphonage can be caused by a partial vacuum in the waste piping. A—With atmospheric pressure equalized, trap remains filled. B—Discharging waste sets up conditions for forming a vacuum in waste pipe. C—Without a vent to equalize pressure on both sides of trap, water is siphoned destroying trap seal.

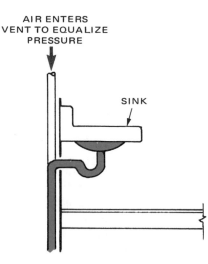

Fig. 8-21. Vent installed in line near trap equalizes air pressure to stop siphonage from the trap.

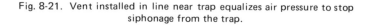

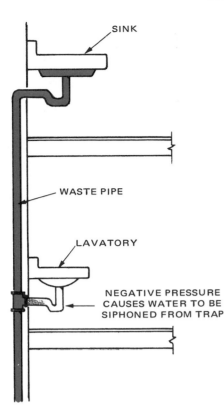

Fig. 8-22. Water flow through the stack can cause reduced pressure at the outlet of the trap. The vacuum thus created "sucks" the water out of the trap.

materials, such as string or a rag, can lodge in the trap as shown in Fig. 8-24. Capillary action then could cause the water to leak from the trap through the fibers of the material. The string or rag will absorb water and act much like a hose to siphon away the water. This process takes very little time. Generally, the next use of the fixture will flush the string or rag through the trap and the problem ceases.

## Wind

Wind blowing across the stack can cause a downdraft in the stack. The downdraft may act as back pressure. Placing stacks away from roof valleys and roof ridges tends to eliminate this problem.

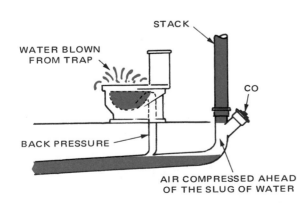

Fig. 8-23. Too much air pressure ahead of slug of water can cause water to be blown out of trap.

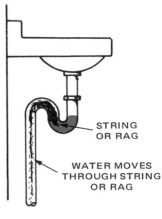

Fig. 8-24. Rag or string caught in trap can trickle away enough water to destroy trap seal.

## VENTING METHODS

Several different methods of venting will maintain atmospheric pressure at all points in the waste piping. Residential and light commercial plumbing may use one or more of the following type vents:

1. Individual.
2. Unit.
3. Circuit.
4. Wet.
5. Looped.

INDIVIDUAL VENTING is best because it vents every trap individually, Fig. 8-25. The vent must be installed as near the trap as possible. Few residential plumbing installations are individually vented because of the cost. Much experimentation has shown that other forms of venting are acceptable under certain conditions.

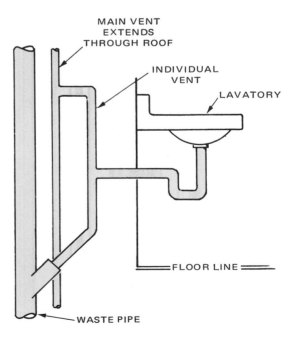

Fig. 8-25. Individual vents nearly eliminate the possibility of trap seals being lost.

UNIT VENTS can be installed where two similar fixtures discharge into the waste piping, Fig. 8-26. The installation is made with a double combination Y and 1/8 bend with deflectors. This fitting produces the smooth waste flow required for proper functioning of the waste piping.

CIRCUIT VENTS are installed on two or more fixtures which discharge into a horizontal waste branch, Fig. 8-27. The vent is placed between the last two fixtures on the branch. Discharge from the third fixture tends to wash away any material which could otherwise block the vent.

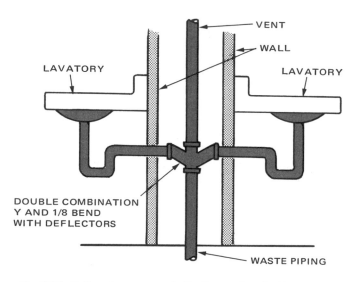

Fig. 8-26. Unit vents serve two similar fixtures installed back to back.

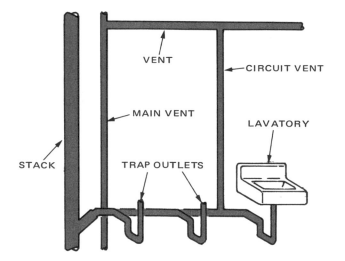

Fig. 8-27. Circuit vents are installed on horizontal branches serving two or more fixtures.

RELIEF VENTS were mentioned earlier in discussion of back pressure. At any point in the waste piping where the direction of the pipe changes it is desirable to install a relief vent. It helps eliminate back pressure buildup, Fig. 8-28. This is important when the waste piping is more than three stories high, Fig. 8-29.

WET VENTS are that portion of the vent piping through which liquid waste from another fixture also flows. Only fixtures discharging liquid waste can be safely connected to a

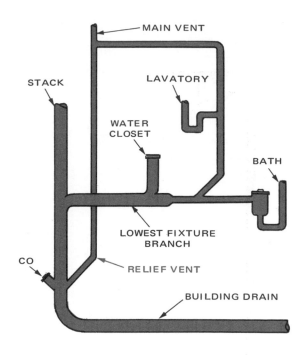

Fig. 8-28. Relief vents are placed where waste piping changes direction.

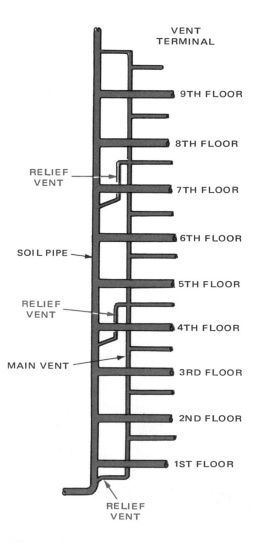

Fig. 8-29. Relief vents are even more important in plumbing for tall buildings.

wet vent. Connections from bathtubs, showers, lavatories and drinking fountains are generally acceptable. Kitchen sinks, water closets, washing machines and dishwashers should not be connected to wet vents. A typical example of a wet vent is shown in Fig. 8-30.

A LOOPED VENT is one which dips below the flood rim of the fixture before rising to connect with the vent stack. It is sometimes installed in remodeling work or in special installations where the extension of a vertical stack near the fixture is impractical. Fig. 8-31 is an example of a looped vent. This is not the most desirable venting method but it will work satisfactorily if the pipe is large enough.

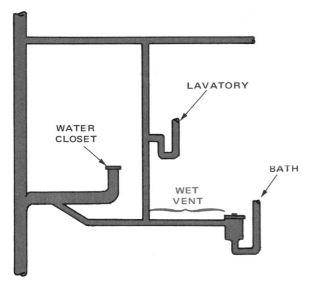

Fig. 8-30. Wet vents can be installed to serve fixtures which discharge only liquid waste.

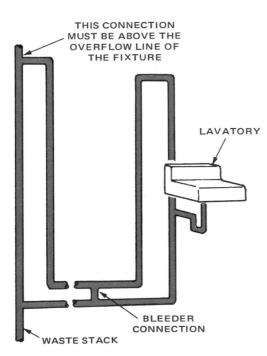

Fig. 8-31. Looped vents are sometimes installed when vertical stacks would be objectionable. Bleeder connection is needed to drain collected moisture.

## SIZING VENT PIPING

*As a general rule, vents may never be smaller than the largest branch entering them.* They must be designed to handle the load factors indicated in Fig. 8-15.

Maximum acceptable horizontal distance between the trap and the stack for different sized waste piping has been established by the National Building Code. By applying the data in Fig. 8-32, the need for a relief or circuit vent can be determined.

The piping system in Fig. 8-33, illustrates the use of a relief vent. The pipe fitting used to install vent piping are the same as the waste piping. All joints must be sealed and the pipe and fitting must be free of defects. Leaks in the vent piping could cause undesirable odor to escape into the building.

| SIZE OF FIXTURE DRAIN (INCHES) | MAXIMUM HORIZONTAL DISTANCE FROM TRAP TO VENT (FEET/INCHES) |
|---|---|
| 1 1/4 | 2-0 |
| 1 1/2 | 3-0 |
| 2 | 5-0 |
| 3 | 6-0 |
| 4 | 10-0 |

Fig. 8-32. Specifications for venting lavatories which are long distances from the soil stack.

## DESIGNING THE WATER SUPPLY PIPING SYSTEM

The water supply piping system is designed to function under a pressure of 40-60 psi within the building. This means that different needs determine its design:

1. It is extremely important that supply piping not contaminate the water running through it.
2. Each fixture must be served by an adequate supply of water.
3. Provision must be made to prevent water from freezing in the pipe.
4. Cross connections between the water supply and DWV piping systems must be prevented.
5. Special provision must be made to relieve excessive pressure which can build up in water heaters.
6. Valves need to be installed so that portions of the piping system can be isolated while repairs are made.
7. Provision needs to be made to reduce noise and avoid damage from water hammer and vibration.

## SUPPLY PIPING

Based on these considerations, materials, fittings and valves must be selected for trouble-free service during the expected life of the piping system. PVC, CPVC, copper, or galvanized iron pipe and fittings are suitable for installation of the water supply system. (See Unit 4 for additional information about these materials.) *In general, black iron and lead pipe should not be used in the water supply system.*

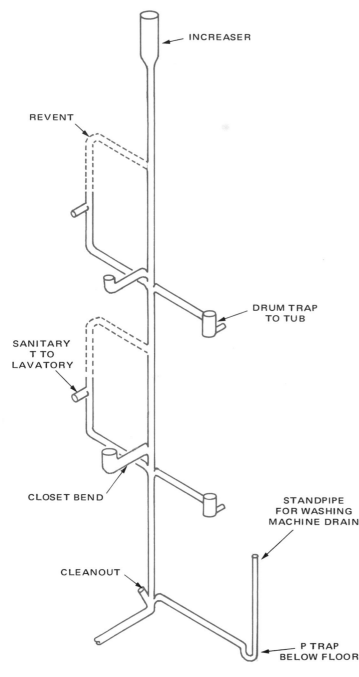

INCREASER

REVENT

DRUM TRAP
TO TUB

SANITARY
T TO
LAVATORY

CLOSET BEND

STANDPIPE
FOR WASHING
MACHINE DRAIN

CLEANOUT

P TRAP
BELOW FLOOR

Fig. 8-33. A more complex system is needed for two-story layout.

*deliver the proper flow.* Others have already predicted the probable need at each fixture and have recommended a suitable size of pipe. These are found on the chart in Fig. 8-34.

The calculations are based on the same discharge rate or load factors used to determine sizes of drainage pipe. This load factor was divided by 4 since no fixture nor group of fixtures is used to capacity at all times.

For example, a water closet equipped with a flush tank has a load factor of 4 because it discharges several gallons of water into the trap within a few seconds. Yet, the supply pipe required is only 3/8 in. because, in normal usage, this size pipe would refill the tank rapidly enough so that the fixture could be flushed again in a few minutes.

Following the recommendations in Fig. 8-34, the branches to the lavatory and the tub in Fig. 8-35 should be 1/2 in. diameter. The water closet needs only a 3/8 in. line. (In practice, though, the water closet branch would be roughed in with 1/2 in. pipe. Only the finished, exposed pipe would be 3/8 in.)

There is a rule of thumb for sizes of pipes which supply two or more branch lines:

1. Up to three 3/8 in. branches can be supplied by a 1/2 in. pipe.
2. Up to three 1/2 in. branches can be supplied by a 3/4 in. pipe.
3. Up to three 3/4 in. branches can be supplied by a 1 in. main.

For small installations like the one in Fig. 8-35, a 1/2 in. main is sufficient. If galvanized iron pipe is used for the installation, it would be desirable to use 3/4 in. because galvanized pipe has a tendency to collect mineral deposits, particularly in the hot water piping.

In residential and small commercial installations an adequate water supply can generally be obtained by following the procedures just outlined. However, there are three other considerations which need attention when selecting pipe sizes:

1. Height of the installation above the entry of the water supply into the building.
2. Length of pipe.
3. Number and type of fittings installed.

## BRANCH PIPE SIZES

| FIXTURE | PIPE SIZE (IN.) | FIXTURE | PIPE SIZE (IN.) |
|---|---|---|---|
| BATHTUB | 1/2 | SHOWER | 1/2 |
| DISHWASHER | 1/2 | URINAL — FLUSH VALVE | 3/4 |
| DRINKING FOUNTAIN | 3/8 | — FLUSH TANK | 1/2 |
| HOSE BIB | 1/2 | WASHING MACHINE | 1/2 |
| KITCHEN SINK | 1/2 | WATER CLOSET — FLUSH VALVE | 1 |
| LAUNDRY TRAY | 1/2 | — FLUSH TANK | 3/8 |
| LAVATORY | 1/2 | WATER HEATER | 1/2 |

Fig. 8-34. Testing and experience has shown which size of pipe is adequate for each kind of fixture.

## Selecting pipe sizes

*Selecting pipe to provide adequate water supply is a matter of determining or predicting demand at each fixture and matching the probable demand with the ability of piping to*

HEIGHT OF INSTALLATION. Pressure to force water through the water supply piping is produced in two different ways:

1. By pumping water into a water tower and permitting it to

110

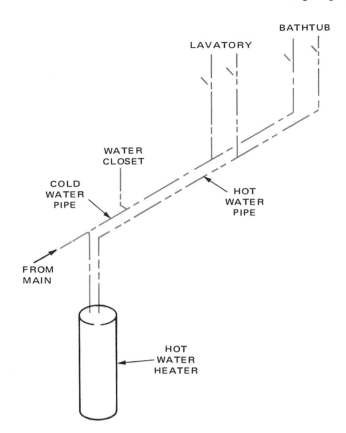

Fig. 8-35. Simple bathroom water supply system.

feed the water mains using the force of gravity. This system is typical of publicly owned water supply systems.

2. Using a pump to push the water through the water supply piping system. This method is frequently used in privately owned residential water systems. The water tower system has the advantage of providing almost constant pressure while storing a reserve of water for peak use periods.

To understand how the water pressure at a given outlet will be affected by height, study Fig. 8-36. The water pressure at the base of the water tower has a direct relationship to the height of the tower. For each foot of height the pressure increases 0.433 psi.

Assuming that the first floor of building B is 15 ft. above the base of the water tower and each of the floors in the building is 10 ft. apart, the theoretical water pressure on the third floor is determined by the height of the water in the tower above the third floor (15 ft.).

Further study of the illustration will show that the difference between the theoretical water pressure available to the third floor and that available in the basement is about 11 psi.

*It is very important that this factor be considered when designing a water supply system.* Upper floors of tall buildings may be nearly as high or higher than the water tower. It may be necessary to install a water storage tank on the roof and pump water into this tank in order to adequately supply the upper floors. On the other hand, in buildings that are

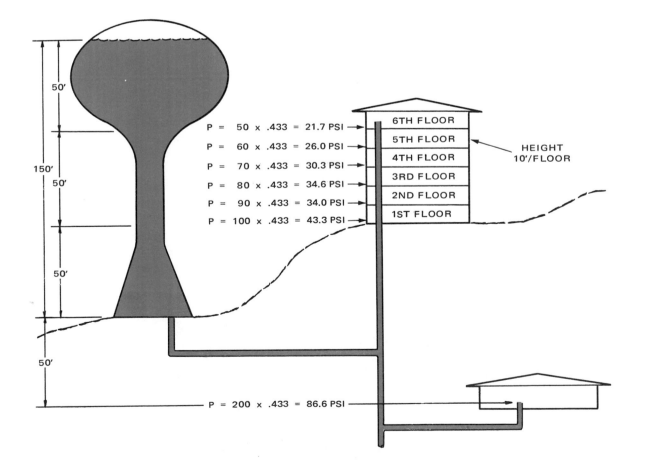

Fig. 8-36. Simplified diagram of gravity pressure system shows relationship of water pressure to height of water tower.

considerably below the water tower, it may be necessary to install control devices to reduce the water pressure.

LENGTH OF PIPE. Water passing through a pipe is slowed because of friction. The longer the pipe the more the pressure will drop. The amount of friction per foot of pipe is determined by the inside diameter of the pipe and the roughness of the inside wall. Calculation of the amount of pressure loss due to friction within a pipe is rather complicated and need not be considered except on relatively large piping jobs. For residential work, it will suffice to install the next larger diameter pipe on runs of 50 ft. or more. This will compensate for the friction loss. For additional information on water pressure losses through pipe friction, see Useful Information, pages 261 to 265.

NUMBER AND TYPE OF FITTINGS. This factor will have considerable effect on the loss of water pressure. See Fig. 8-37. Piping, therefore, should be installed with the least number of fittings possible. It will be necessary to use pipe one size larger, where there must be many fittings, in order to maintain water pressure.

A combination of errors or problems in height, pipe size and number of fittings can have extreme effects on water pressure. It is absolutely necessary that plumbers be aware of the problem. Equally important, they should be able to apply principles discussed in this unit in solving particular problems.

## PREVENTING FREEZING

In northern climates where temperatures fall below 32 F (0 C), exposed pipe with water standing in it will freeze. Since water expands as it turns to ice, the added pressure may burst water pipes and fittings. Replacing damaged piping is expensive. An added inconvenience is the loss of the use of all or part of the system until the pipe is thawed or replaced.

Freezing is likely to occur in the following situations:
1. When a water supply line is not installed below the frost line.
2. When supply lines must be installed in crawl spaces rather than in heated basements.
3. When piping is installed in outside walls in northern climates.
4. When outdoor faucets are provided for hose connections.

While planning installations, plumbers can take certain percautions to avoid the inconvenience and expense of repairing frozen plumbing:
1. Consult local plumbing codes for depth of frost line. Install

outside lines below that level. (Although this is generally done when the structure is first built, subsequent landscaping or remodeling may change the contour of the ground causing pipe to be nearer the surface.)
2. Exposed pipe in crawl spaces can be wrapped with insulation or with thermostatically controlled heating tape.
3. Avoid placing plumbing in outside walls in colder climates. Try, instead, to place pipes in partitions. If they must go in outside walls, insulate them well.
4. Hose bibs should be of the freeze-proof type. This design is discussed in Unit 5. See Fig. 5-23. An alternative is to install a drain or waste cock inside the building. This type of valve will shut off the water and drain water standing in the pipe between the drain cock and the hose bib.

Before installing either type of fixture, make certain that the space inside the building, where the hose bib or drain cock is installed, is going to be heated or that temperatures will not drop to the freezing range. Faucets are freeze-proof only so long as they stop the water at points where it cannot freeze.

*If the foregoing precautions cannot be taken to protect pipe from freezing, a supply line should not be installed.* There is one exception to this general rule. Plumbing can be installed in buildings which will only be used during the summer. In such cases, a cutoff valve below frost line must be provided. In addition, it is absolutely necessary that provisions be made for draining the water line. This will generally mean that a valve for draining the system will be installed at the lowest point in the piping system. Long horizontal runs of pipe must be sloped toward the valve so the system will drain completely.

## CROSS CONNECTIONS

CROSS CONNECTIONS occur when the piping system containing sanitary drinking water (potable water) can become contaminated by the content of another piping system (DWV or unsanitary water supply used for irrigation or other purposes.)

*In all cases, the actual flow of contaminated fluid into the sanitary water supply occurs because a difference in pressure exists between the two piping systems.*

Fig. 8-38, view A, illustrates how a water closet with a flush valve could contaminate the sanitary water system. Favorable conditions for a cross connection are set up by a sequence of events:

## EQUIVALENT LENGTH OF PIPE ALLOWANCES FOR FRICTION LOSS ON THREADED FITTINGS

| DIAMETER OF FITTING | EQUIVALENT LENGTH OF PIPE FOR VARIOUS FITTINGS (FT.) | | | | | |
|---|---|---|---|---|---|---|
| | 90 DEG. L | 45 DEG. L | 90 DEG. T | COUPLING | GATE VALVE | GLOBE VALVE |
| 3/8 | 1 | 0.6 | 1.5 | 0.3 | 0.2 | 8 |
| 1/2 | 2 | 1.2 | 3.0 | 0.6 | 0.4 | 15 |
| 3/4 | 2.5 | 1.5 | 4.0 | 0.8 | 0.5 | 20 |
| 1 | 3 | 1.8 | 5.0 | 0.9 | 0.6 | 25 |

Fig. 8-37. Pressure losses due to fittings in the supply line can be measured in terms of equivalent pipe lengths.

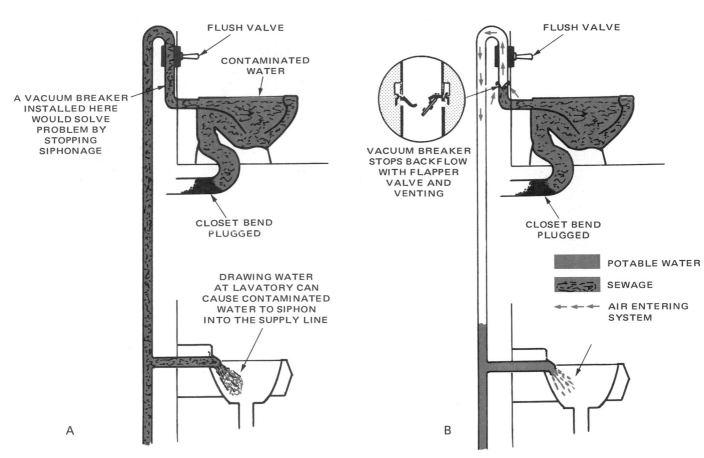

Fig. 8-38. A plugged water closet can introduce waste water into water supply unless furnished with a vacuum breaker. A—Installation is incorrect. Contamination results when water is drawn from the fountain on the floor below. B—Correct installation. Vacuum breaker flapper valve closes to stop backflow of contaminated water. Air enters vacuum breaker vents to prevent formation of vacuum in water supply line.

1. The water closet is clogged.
2. The water supply is turned off or there is a break in the water main which causes loss of water pressure in the supply pipe.
3. The sanitary water flows out of the riser and creates a vacuum in the branch line.
4. Because the water closet is flooded to the rim, its contents are drawn into the sanitary water supply, thus contaminating the water.

The extent of the contamination depends on how long the negative pressure lasts. The water coming from the lavatory on the floor below would be contaminated. This problem can be solved by installing a vacuum breaker on the water closet. The vacuum breaker, Fig. 8-38, view B, will permit air to flow into the line rather than allow a vacuum which would siphon sewage into the water supply.

In some cases, poorly designed fixtures create a cross connection. The lavatory in Fig. 8-39, has an internal faucet which is under water when the lavatory becomes flooded. A cross connection could develop if a vacuum were created in the sanitary water supply piping system.

Fig. 8-40 illustrates a water closet with a built-in flush tank. Note that the bowl inlet is below the flood rim of the water closet. A cross connection is possible. These problems can usually be prevented by selecting fixtures having the proper air gap, between the flood rim and the sanitary water supply inlet.

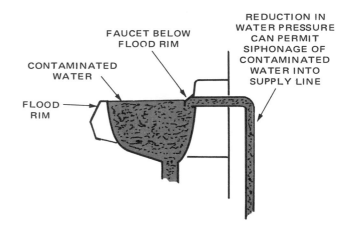

Fig. 8-39. Poorly designed lavatory with faucet below flood rim sets up ideal situation for cross connection.

Fig. 8-41 shows a properly designed lavatory.

Even in cases where the plumbing is correctly installed originally, it is possible for cross connections to occur if accessories are added to the sanitary water supply lines. As an example, consider the hose and spray nozzle which can be installed on some faucets for washing hair, Fig. 8-42. This type of cross connection could be particularly hazardous in a lavatory where strong chemicals or other poisonous materials are frequently present.

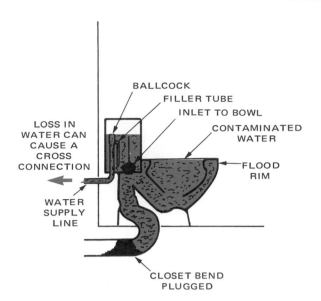

Fig. 8-40. In this water closet the supply tank is lower than the flood rim in the bowl. A plugged trap and a drop in water pressure would cause a cross connection.

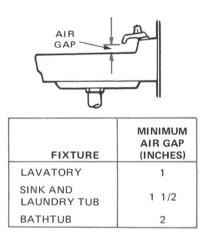

| FIXTURE | MINIMUM AIR GAP (INCHES) |
|---------|---------------------------|
| LAVATORY | 1 |
| SINK AND LAUNDRY TUB | 1 1/2 |
| BATHTUB | 2 |

Fig. 8-41. Properly designed fixtures have air gap which prevents cross connection during normal operation even if flooding occurs.

Designing a plumbing system which is free of cross connections requires two considerations:
1. Piping systems containing sanitary (potable) water must not interconnect with other piping systems.
2. Fixtures which provide appropriate air gaps and/or vacuum breakers must be selected.

Meeting these two conditions assures the design of a safe system.

## PRESSURE RELIEF VALVES

Pressure relief valves are devices which will prevent excessive pressure from being created in a hot water tank. It is general practice to have a dual safety system on hot water heaters.

One device shuts off the gas or electricity if the pressure exceeds a safe level. A second device permits hot water or steam to escape through a pressure relief valve, Fig. 8-43.

These devices may be a part of the hot water heater or they

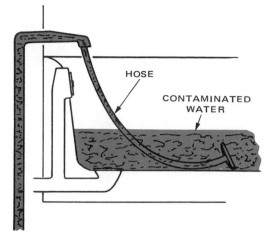

Fig. 8-42. Spray nozzle attached to open faucet and lying in tub can create cross connection if water pressure is lost in water supply system.

may need to be added to it. Their purpose is to prevent the hot water heater from exploding in case the thermostat ceases to function. In theory, either device would be satisfactory; however, these are mechanical attachments which may malfunction. It is essential that both be installed as a safety precaution.

## VALVES

*Shutoff valves should be placed in the water supply piping system at different points to isolate small parts of the system.* This makes it much more convenient to repair faucets or shut off the water supply when parts of the system develop leaks. For example, if the bathroom sink needs repair, the entire

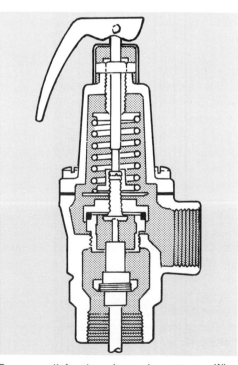

Fig. 8-43. Pressure relief valve shown in cutaway. When pressure becomes too high, spring at top is compressed allowing excess pressure to bleed out the overflow outlet at right.

water supply system need not be turned off to fix a faucet.

For greatest convenience, a cutoff valve should be installed on each fixture branch. Thus, hot and cold water supplies can be turned off independently at each fixture. However, this type of installation may add too much to the cost. The second alternative is to install cutoff valves in the supply lines leading to each room. This permits the isolation of that particular room while repairs are made.

The absolute minimum is to install a valve on the cold water main and another at the inlet to the hot water heater. Thus, the entire system or only the hot water may be cut off. Fig. 8-44, illustrates the water supply piping for the DWV system pictured in Fig. 8-33. Note that valves have been included to cut off the water supply to each room.

## REDUCING WATER HAMMER AND VIBRATION

The banging sound sometimes heard when faucets are turned off quickly is called WATER HAMMER. It is caused by the sudden stopping of the flow of the water through the piping system. To understand how this noise is created, it may be helpful to think of the water as a piston moving rapidly through a cylinder, Fig. 8-45. If the piston suddenly encounters a wall across the cylinder, a collision of considerable force takes place. In the case of water flow (piston) through a pipe (cylinder) this force has been known to break pipes. Water hammer can be avoided by installing air chambers near faucets. These devices, Fig. 8-46, provide a chamber of air which will absorb the energy causing the water hammer.

A second source of noise in the water supply piping system is vibration from the flow of water through pipe and fittings. This problem can generally be solved by securely anchoring the pipe to the frame of the building.

## DESIGNING THE STORM WATER PIPING SYSTEM

Disposing of heavy rain fall can be a major problem. As was mentioned earlier in this unit, it is common practice to provide

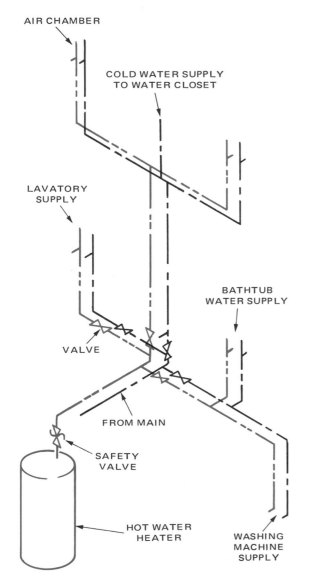

Fig. 8-44. This water supply piping system was designed for a two-story house.

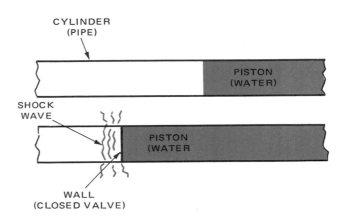

Fig. 8-45. Water hammer caused by sudden stopping of water flow.

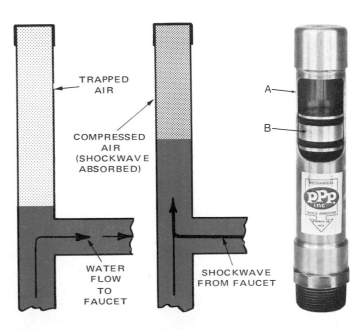

Fig. 8-46. Short lengths of capped pipe or special shock arrestors stop water hammer in water supply pipes. A—Air chamber. B—Movable piston. (Precision Plumbing Products, Inc.)

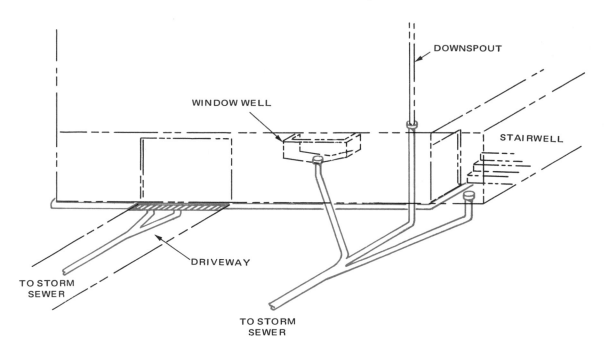

WINDOW WELL

DOWNSPOUT

STAIRWELL

DRIVEWAY

TO STORM
SEWER

TO STORM
SEWER

Fig. 8-47. Storm water drainage takes water away from foundation, roof, stairwells and window wells.

a separate storm water piping system.

A typical system is shown in Fig. 8-47. Note that the outlet of this piping system places the water on the street where it can flow into a storm sewer.

Two alternatives to this system are used where a storm sewer is not available:

1. Storm water is directed into a nearby stream or lake.
2. A drywell is constructed where the water can be stored until it seeps into the ground.

Referring again to Fig. 8-47, a FOUNDATION DRAIN collects ground water and drains it into a basin or sump. From the sump, the water is pumped to a level where it can flow away by gravity. The foundation drain is made from pipe which has openings in it, permitting water to enter, Fig. 8-48. This pipe may be plastic or clay. It may be installed inside or outside of the foundation wall depending upon the local code requirements.

The purpose of the foundation drain is to remove the water which tends to collect under the basement floor and outside the foundation wall. If this water is not removed, the basement walls and floor are likely to be damp. Even worse, the walls may be damaged by hydraulic pressure from the water on the outside.

In locations where the storm sewer entry is above the foundation drain it is necessary to install a sump pump to lift water from the sump into the storm sewer. The sump pump is turned on automatically when the water in the basin reaches a certain level. When designing the foundation drainage system it will be necessary to consult the electrical plans to make certain that electrical service has been provided to power the sump pump.

When the sewer drain is at a lower level than the foundation drain, no sump pump is needed. The ground water will flow into the sewer by gravity.

Other provisions for storm water removal include:

1. Installing drains to each gutter downspout.
2. Providing drains in stairwells and in some cases, window wells.
3. Installing drains outside of garage doors if the driveway slopes toward the building.

## SIZE OF STORM DRAINS

The appropriate size for vertical and horizontal pipes which connect roof drainage systems to the storm sewer can be determined by referring to Figs. 8-49 and 8-50. For example,

Fig. 8-48. Foundation drain pipe is perforated to allow it to collect ground water.    (Hancor, Inc.)

| DIAMETER OF DRAIN (IN.) | ROOF AREA (SQ. FT.) |
|---|---|
| 2 | 720 |
| 3 | 2200 |
| 4 | 4600 |
| 5 | 8650 |
| 6 | 13,500 |

Fig. 8-49. Pipe sizes for vertical storm drains.

| DIAMETER OF DRAIN (IN.) | ROOF AREA DRAINED BY PIPES HAVING VARYING SLOPES (SQ. FT.) | | |
|---|---|---|---|
| | 1/8 IN. | 1/4 IN. | 1/2 IN. |
| 3 | 822 | 1160 | 1644 |
| 4 | 1880 | 2650 | 3760 |
| 5 | 3340 | 4720 | 6680 |
| 6 | 5350 | 7550 | 10,700 |
| 8 | 11,500 | 16,300 | 23,000 |

Adapted from National Plumbing Code Table 13.6.2.

Fig. 8-50. Pipe sizes for horizontal storm drains.

if the drain serves a roof area of 2000 sq. ft., a 3 in. vertical and a 4 in. horizontal pipe which slopes 1/4 in. per ft. of run will be required. The same tables could be used to determine the size of the driveway drain pipe, provided an allowance is made for any water which drains from the surrounding ground sloping toward the driveway.

## TEST YOUR KNOWLEDGE — UNIT 8

1. For a plumbing installation to be economical, the kitchen, bathroom and utility room must be _____ _____.
2. The critical design factor for a kitchen is the _____ triangle.
3. Name the three basic piping systems found in most houses.
4. The required size of drainage piping is determined by adding the load factors of all the fixtures that empty into the pipe. True or False?
5. A horizontal fixture branch which has a combined load factor of 5 would require pipe of the following diameter:
   a. 3/4 in.
   b. 1 in.
   c. 2 in.
   d. 5 in.
6. A _____ in. diameter building drain with a 1/8 in. fall per ft. of run can carry a maximum load of 180 fixture units.

7. What type of fitting is used to connect horizontal branch drain lines to the DWV stack?
8. Venting systems are designed for one main purpose:
   a. Keep pipes from freezing.
   b. Maintain atmospheric pressure inside waste piping.
   c. Provide other routes for waste flow.
   d. Seal out sewer gases.
9. List the six methods of venting plumbing systems and describe each.
10. The size of water supply piping varies with the volume of water required by the fixture it serves. What diameter pipe is generally appropriate for each of the following fixtures?
    a. Flush tank water closet.
    b. Lavatory.
    c. Bathtub.
    d. Kitchen sink.
11. Explain why water pressure is typically higher in the basement of a multistory building than on the tenth floor.
12. The water pressure loss due to friction is greater through a coupling than through a 45 deg. L of the same diameter. True or False?
13. When a piping system containing sanitary water becomes contaminated by the content of another piping system a _____ _____ has occurred.
14. Plumbing codes generally require that hot water heaters be equipped with dual safety devices. One of the devices turns off the heat source and the other device known as a _____ _____ valve permits steam or hot water to escape if the pressure inside the tank becomes excessive.
15. Name the piping system designed to dispose of excessive amounts of rain water.
16. The banging sound sometimes heard when faucets are closed rapidly is known as _____ _____.
17. Explain how to eliminate water hammer.
18. A vertical pipe which carries the water from a 3000 sq. ft. area of roof should be _____ in. in diameter.

## SUGGESTED ACTIVITIES

1. Using a set of blueprints for a single family residence, prepare an isometric sketch of the DWV or water supply piping system. The sketch should:
   a. Make use of appropriate symbols.
   b. Provide adequate valves and safety devices.
2. Inspect the plumbing system in a home under construction to determine if it meets the criteria mentioned in this unit.
3. While referring to piping drawings for a particular installation, select pipe sizes and fittings which conform to local building code requirements.

# Unit 9
# PIPE AND FITTING INSTALLATION

## Objectives

In this unit the three stages of plumbing installation are detailed: first rough, second rough and attaching of fixtures.

After studying the unit you will be able to:
- Plan the steps involved in bringing water and sewer service into the building.
- Describe proper procedures for locating and installing DWV and water supply systems using various materials.
- Use the prescribed techniques for working with vitrified clay, cast iron, galvanized and black iron, copper and plastic plumbing materials.
- Compare and use the three methods of measuring pipe length between fittings.
- Describe or demonstrate methods for testing and inspecting completed plumbing systems.

When plumbing is installed in a building, the work is done in two or three stages. The first stage, called ROUGH-IN, may be divided into two parts known as first rough and second rough. In the first rough stage, the building sewer and water supply are installed. These are pipes which extend from the water and sewer mains to the inside of the building. The second rough stage is the installation of all plumbing which will be enclosed in the walls or which must be installed before the walls are finished. This includes DWV, hot and cold water supply piping, bathtubs and shower bases.

In the final stage, lavatories, water closets, sinks, faucets, hose bibs and shower heads are installed. These units are attached after walls are closed in.

The need for dividing the plumbing installation into two or three stages becomes readily apparent when the coordination of the plumber's work with other construction workers is considered. The first rough is frequently necessary even before the foundation is completed. This way there will be water to mix mortar for concrete block foundations. Having the sewer line extended into the basement before the concrete footing is poured also saves a considerable amount of hand digging.

## FIRST ROUGH

Responsibility for installation of sewer and water service from the street into the building is divided between municipal workers and plumbing contractors. While each municipality may differ slightly, the following is typical:
1. The contractor or plumber is responsible for trenching between street connection and building foundation. See Fig. 9-1.
2. City workers will tap the water main, install the corporation stop, the curb stop and the "B" box. (The "B" box or

Fig. 9-1. Private contractor will use heavy equipment to dig trenches and provide access to sewer and water mains.

"Buffalo" box, shown in Fig. 9-2, is a pipe open at one end that slips over the top of the curb stop. It extends vertically to ground level. The top is capped. When water is to be turned on or off, a long "key" inserted in this housing turns off the water.)

3. The plumber will attach pipe to water service at the curb stop and bring it to the house supply inlet pipe or the water meter.

4. The plumber opens up the sewer main, installs the sleeve connector and the lateral line to the building.

Fig. 9-2. The long tube is called a Buffalo box. It provides access to curb stop (also shown) when water service to building needs to be turned on or off.

## ATTACHING LATERALS TO WATER MAIN

Two different methods are used to attach the corporation valve to the water main:

1. If the main is large enough, the side wall is drilled and tapped. The corporation valve is threaded into the tapped hole.

2. A saddle is attached to the main and the corporation stop is attached to the saddle. This method must be used when the water main is 4 in. in diameter or less and the water supply line to the house is an inch or more in diameter.

### Installing the corporation stop

Installation of the corporation valve (stop), Fig. 9-3, requires some skill in the use of special drills. Careful work will

Fig. 9-3. Saddle is securely attached to main with U-bolts. This type of installation is permanent.

insure a tight seal between the stop and the main. The following procedures should be followed when a saddle is used:

1. Remove soil from around the water main to provide access for the U-bolts. Carefully remove dirt and loose rust from surface where the saddle must seal.

2. Position U-bolts and place saddle on top of the main. Adjust U-bolts to align with the holes in the saddle plate. Secure saddle so that rubber or plastic gasket seals the pipe and saddle surfaces.

3. Attach the corporation stop to the collar and turn the valve to the open position, Fig. 9-4. Check U-bolts again to insure tightness.

Fig. 9-4. Corporation stop is in place ready for the drill.

4. Insert the drill bit through the opened corporation stop. Attach housing of drill to the corporation stop by screwing it onto the threaded end.

5. Drill the hole in the water main. Twist the drill housing clockwise to maintain pressure on the bit. Drill can be turned with a ratchet wrench, Fig. 9-5, or with an electric drill.

Fig. 9-6. As drill is withdrawn, wrench is used to shut off valve on corporation stop.

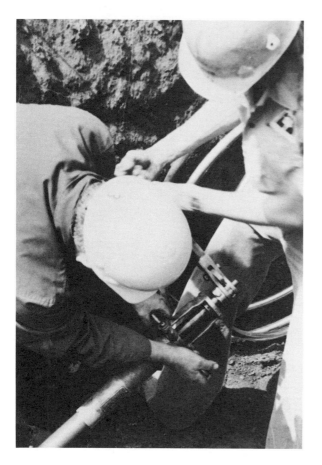

Fig. 9-5. Bit is turned with ratchet wrench to drill hole in water main.

6. When the hole is completed, withdraw the tool slowly allowing water pressure to flush out the drilling debris. Close the valve, Fig. 9-6.

7. Install a short section of water supply pipe or tubing from the corporation stop to the point where the Buffalo box is to be installed.

8. Attach the curb stop, Fig. 9-7.

9. Adjust the telescoping housing of the Buffalo box to ground level and set it squarely over the curb stop.

Normally, the plumbing contractor is responsible for installation from the curb stop into the house. A water line is laid between the curb stop and the point where the water meter will be installed. City workers will install the meter. Fig. 9-8, diagrams the water supply installation. It also shows tools and fittings used.

Openings in the foundation wall for water and sewer access may have been provided in the construction of the foundation. If not, the plumbing contractor will cut openings with a masonry drill or saw, Fig. 9-9.

Fig. 9-7. Curb stop is installed and checked for leakage.

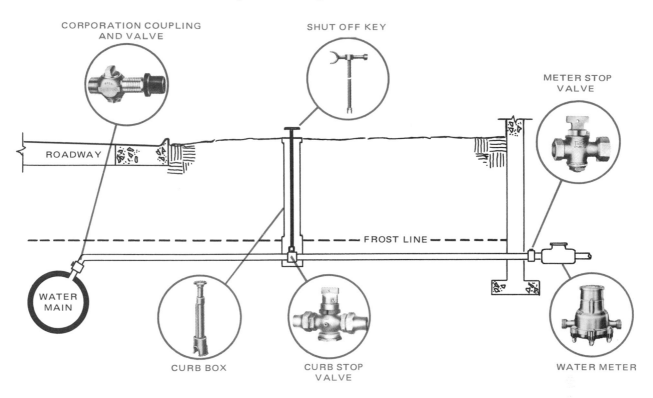

CORPORATION COUPLING AND VALVE

SHUT OFF KEY

METER STOP VALVE

ROADWAY

FROST LINE

WATER MAIN

CURB BOX

CURB STOP VALVE

WATER METER

Fig. 9-8. Complete municipal water supply hookup shows various connections and tools. Corporation valve allows tapping into water main without interrupting service. Curb stop permits shutting off water supply for service or emergencies. Meter stop valve allows cutoff of water service to entire building.

Fig. 9-9. Carbide blade on gasoline engine powered saw, cuts openings in foundation wall with minimal damage to the wall.

## Sewer line installation

The HOUSE SEWER is the drainage piping between the building's foundation and the sewer main or septic tank. Where swampy soil provides poor support for the pipe, cast iron must be used. Vitrified clay tiles are used where soil is firm enough to provide support. Cast iron is considered superior in all cases except where acids in the soil or in waste material are likely to attack the metal.

In some communities, the plumbing contractor will lay the entire house sewer from the foundation wall to the main sewer line. Before water lines or sewer lines are covered, they must pass inspection by the plumbing inspector.

## Procedure

After the sewer trench has been dug and the sewer main exposed, the plumber can install the sewer. Following procedure is typical:

1. Attach supportive banding. This protects sewer pipe from damage during drilling.
2. Attach rotary saw to sewer main and cut opening as shown in Fig. 9-10. Use water to lubricate the cutting operation.
3. Attach sleeve fitting over hole and cement in place with concrete or special plastic materials, Fig. 9-11. If additional reinforcement of the joint is desired, more concrete may be placed around it. When the connection has hardened, the rest of the sewer line may be laid. Care must be taken to provide proper slope of 1/4 in. per foot. Bell joints of vitrified clay pipe are sealed with special gaskets or with oakum and a special mixture of cement. A good seal is important so that tree roots do not enter and block the drain.

ADJUSTING
WHEEL

ROTARY
BLADE

Fig. 9-10. Cutting opening in sewers with rotary saw. Top. Drill is clamped in place and beginning to cut. Adjusting wheel raises and lowers rotary blade onto sewer pipe. Bottom. Sewer opening completed and ready for connector.

## SECOND ROUGH

Obviously, it would be difficult to install more of the piping system until the foundation is completed, the floor framed, subflooring installed, walls erected and the roof on. However, when the carpenters have completed framing the building, the plumber must return and complete the second rough stage of the plumbing installation.

In many ways, the second rough is the most critical and most difficult stage of plumbing. Pipes and fittings must be placed accurately so that fixtures can be hooked up easily during installation. Plumbers must make allowances for finished walls and flooring that will be installed later by other workers in the construction trades. If measurements are taken accurately and allowances are made for the thicknesses of materials still to be put in place, there will be few if any problems in connecting fixtures.

## LOCATING FIXTURES

*In plumbing,* LOCATING FIXTURES *means finding the exact spot in a room where a fixture is to be placed. It also includes marking the exact spot where pipes and fittings are supposed to be.* Locations are marked on the walls and floors before any actual installation begins.

After the frame of the building is completed the plumber will study the blueprints and the rough-in dimensions for each of the fixtures to be installed. The floor plan will show exactly where each fixture is to be installed, Fig. 9-12. The wall section shown in Fig. 9-12 will indicate thicknesses of wall and flooring materials. Still another measurement is needed before the plumber can determine where water supply and drainage piping will be located in the wall or floor. This is the rough-in dimension, Fig. 9-13. This measurement is always supplied by the manufacturer of the fixtures.

## LOCATING BATHROOM FIXTURES

Planning a rough-in should follow a step-by-step process. The layout of the fixtures for the bathroom plan in Fig. 9-12 will show how the process works.

From the floor plan, note that the distance from the wall stud (vertical framing members) on the right to the centerline of the water closet is 4'-6". (Before transferring any dimension to the bathroom, allow for the thickness of the finished wall.)

The next dimension is the distance from the rear wall to the center of the opening for the water closet drain. This is found to be 12 3/4 in. This distance is the rough-in dimension from Fig. 9-13 plus a 3/4 in. allowance for the thickness of the drywall and ceramic tile which will be installed over the studs later. Mark this point on the floor, Fig. 9-14.

The soil and vent stack in this installation are to be located directly behind the water closet. This is the most economical spot since it will require the least soil pipe from the water closet to the stack.

Depending upon the size of the stack and the type of pipe installed, the wall may need to be "furred-out" to make room

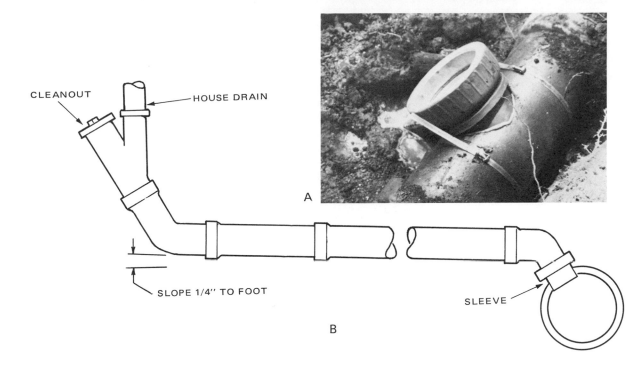

Fig. 9-11. Sewer connection completed. A—Sleeve connector is cemented in place with a quick-setting plastic. Later, cement will be placed around joint for added protection. B—Completed sewer line will slope from the house drain to the sewer main.

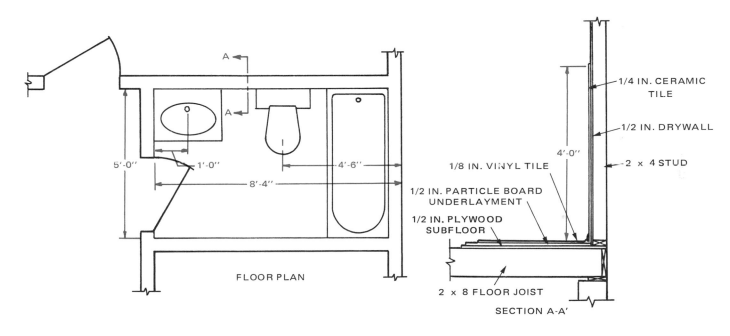

Fig. 9-12. Floor plans and wall section for bathroom give necessary information needed for placement of fixtures. The section is taken between lavatory and water closet.

for a large diameter pipe. Fig. 9-15 illustrates one method of furring. This is more likely to be needed if bell and spigot cast iron pipe is used for the DWV system.

Sometimes the carpenters have already framed the plumbing wall with 2 x 6 in. or 2 x 8 in. studs to provide space for the stack. In any case, this is the time to check if the wall will require furring. If so, allowances would need to be made for a thicker wall as the location of fixtures is marked on the floor.

If the water closet has a wall-hung tank it will need special blocking, Fig. 9-16. Its location will be marked on the studs so

that carpenters or plumbers will install the blocking.

Another style of water closet is a wall-hung unit supported completely by a bracket attached to the wall framing. See Fig. 9-17. Before this bracket is installed, cut openings in the framing for the DWV piping.

## Supply piping

The next step will locate cold water supply piping for the water closet. (Note, however, that supply piping is installed only after the DWV system is completed.) In this installation

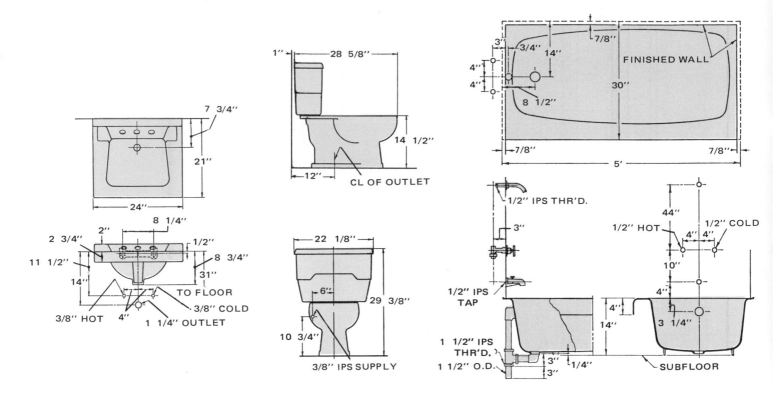

Fig. 9-13. Manufacturers supply rough-in dimensions for bathroom fixtures. (Kohler Co.)

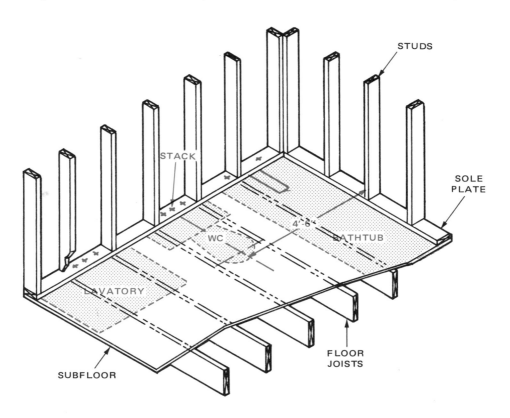

Fig. 9-14. Openings for DWV and water supply piping are marked on sole plate of wall framing.

the cold water supply pipe will come through the wall rather than up through the floor. Wall entry is generally better for two reasons:

1. Flooring is more easily installed.
2. The bathroom is more easily cleaned.

The correct location for cold water supply is found in the rough-in dimensions supplied with the water closet. (Review Fig. 9-13.) In this case, the pipe should come through the wall 10 3/4 in. above the finished floor and 6 in. to the left of the centerline.

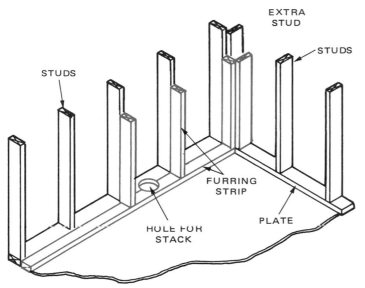

Fig. 9-15. Furring strips may be needed to make wall wide enough to conceal piping. In this example, 2 x 4 blocking doubles wall thickness.

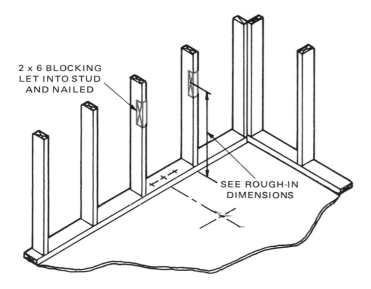

Fig. 9-16. Studs are marked for notches to let in 2 x 6 blocking.

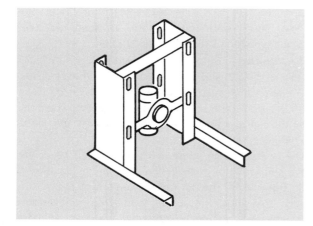

Fig. 9-17. Metal bracket is fastened to studs and supports wall-hung water closet.

The hole for the water supply line should always be drilled in the center of the sole plate. This reduces the likelihood of puncturing the pipe while nailing on drywall, Fig. 9-18.

As locations are marked for DWV and supply piping, note the sizes of holes. While marking holes for water supply piping, it is also helpful to indicate the distance from the finished floor to the point where the pipe must come through the wall. These notes will save time as the plumber will not need to return repeatedly to the rough-in dimensions while the pipes are being installed.

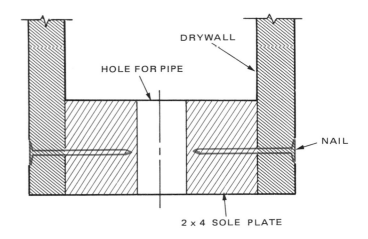

Fig. 9-18. When piping holes are centered, drywall nails are not likely to puncture the pipe.

## Cutting floor joists

In Fig. 9-15, the framing members (joists, studs and plates) of the building do not interfere with plumbing installation. This is not always the case. Sometimes, the location of the water closet shown by the blueprints will require cutting of floor joists. If this happens first consider moving the water closet one direction or the other. If this is impossible, change the floor framing as shown in Fig. 9-19. Doubling up joist and headers assures that the strength of the floor will remain

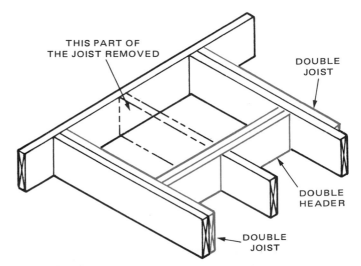

Fig. 9-19. When part of a floor joist must be removed, joists on either side are doubled and shortened joist is supported by a double header.

unaffected. Sometimes a joist can be notched as shown in Fig. 9-20, and then doubled to provide pipe clearance.

A similar but even more difficult problem arises when the floor joists run parallel to the wall against which the water closet is installed. Notching the joist (as shown in Fig. 9-21) to install the closet bend is not practical. It greatly weakens the

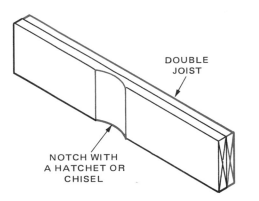

Fig. 9-20. Doubling up a notched floor joist.

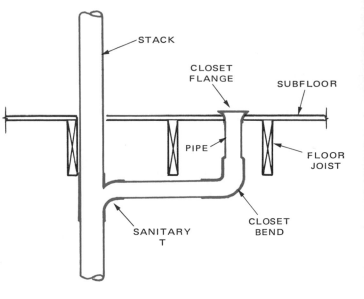

Fig. 9-22. Install closet bend below joist and connect it to closet flange with short length of soil pipe.

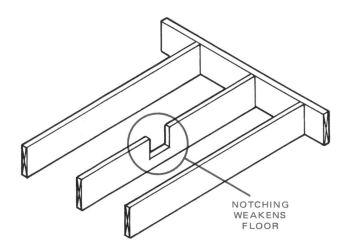

Fig. 9-21. Notching floor joists substantially weakens the floor. Result may eventually be a sagging floor.

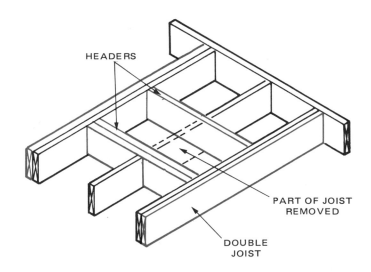

Fig. 9-23. Another example of joist doubled and headers installed. Header holding shortest joist stub need not be doubled as the joist needs very little support.

floor. The right solution is to install the closet bend below the joists. Use a short length of pipe to connect the closet bend with the closet flange, Fig. 9-22.

If the outlet of the water closet must be placed directly over a joist, install headers while doubling joists on either side, Fig. 9-23.

### Locating the lavatory

The floor plan, Fig. 9-12, indicates a distance of 7'-4" from the wall to the centerline of the lavatory. Making allowances for the thickness of the wall materials, this dimension becomes 7'-4 3/4".

Because the lavatory drain and water supply piping will come through the wall, holes for these pipes are laid out on the sole plate. The fixture rough-in dimensions, Fig. 9-13, indicate that the drain should come through the wall 17 in. above the floor (31 in. minus 14 in.). Water supply pipes

should be located 4 in. either side of the drain extending through the wall 19 1/2 in. above the floor (31 in. minus 11 1/2 in.). These requirements should be noted near the holes. They will be referred to during installation.

If a wall-hung lavatory is to be installed, locate blocking at the proper height and firmly secure it so that it will support the weight which may be placed on the lavatory during use, Fig. 9-24. Add more blocking near the points where the pipes come through the wall so that the pipes may also be securely fastened. See Fig. 9-25.

The lavatory drain in the floor plan, Fig. 9-12, can connect to the soil stack through a side inlet in the sanitary T, Fig. 9-26. But, when the lavatory drain must run horizontally through the studs, Fig. 9-27, to make a connection to the stack, the studs should be drilled or notched. If notching is required, reinforce the studs with metal or wood inserts as shown in Fig. 9-28.

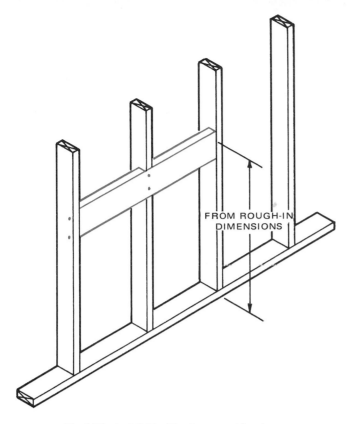

Fig. 9-24. Install blocking to support lavatory.

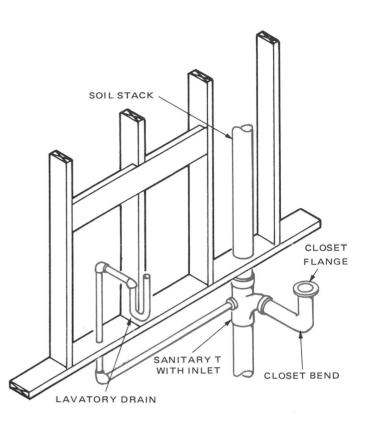

Fig. 9-26. Where floor joists do not interfere, lavatory drain can be connected directly to sanitary T below floor.

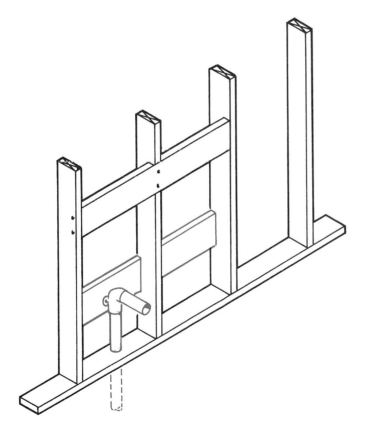

Fig. 9-25. Blocking installed to support water supply piping. Drop ear L can be attached to blocking with screws.

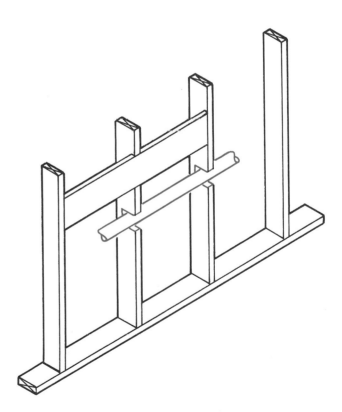

Fig. 9-27. Installing lavatory drain horizontally in wall requires notching or drilling of studs.

## Locating the bathtub

Locate the bathtub in the same manner as the water closet and the lavatory. Usually an opening is cut in the floor for access to the drainage fittings and stopper mechanism. Refer to the rough-in dimensions.

Next, locate the position of the water supply piping. As with the water closet, any joists cut must be reinforced by

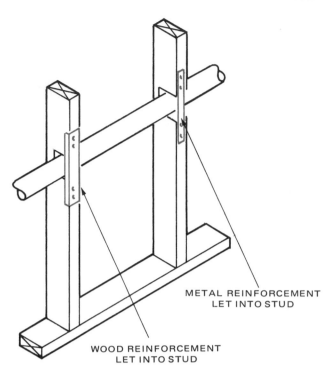

Fig. 9-28. Metal or wood reinforcing is installed flush with the outside of the studs.

doubling. This is important. A tub full of water weighs several hundred pounds. A recess tub generally requires blocking next to the wall to support the edge of the tub. See Fig. 9-29.

## Cutting openings

After all layout dimensions have been rechecked for accuracy, the drilling and cutting of openings can begin. A portable electric drill fitted with a spade bit, an auger or a hole saw will cut holes up to 2 or 3 in. in diameter. Larger holes will need to be cut with a saber saw. Unit 1 describes these tools.

Work can begin on the DWV piping system after the required blocking is installed. It is very desirable that blocking

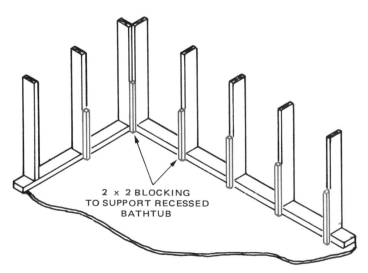

Fig. 9-29. Vertical blocking rests on floor and will support great loads from the tub.

be installed for soap dishes and towel bars before the wall is finished; however, it may be in the plumber's way if installed before pipes and fittings.

The steps for installing pipe and fittings are the same, whether the work is done in the first rough or second rough. The rest of this unit will concentrate on how to install different types of piping materials. Frequent reference will be made to the order or sequence in which work must be accomplished during the various stages.

## INSTALLING THE DWV PIPING SYSTEM

As mentioned in Unit 8, DWV piping is installed before the water supply piping because it must slope toward the sewer and because it is larger. It is much easier to fit the water supply piping around the DWV piping than to attempt the reverse.

The general procedure for installing DWV piping is the same regardless of the type of material used. Certain dimensions are more critical than others. Therefore, it is important to complete the work in an order which will assure their accuracy.

As noted earlier, the location of the closet flange is extemely important. Therefore, begin the installation with the closet flange. Support the closet flange on blocking equal in thickness to the finished floor material, Fig. 9-30. Check the distance from the wall to the center of the closet flange. Temporarily secure the closet flange with a weight.

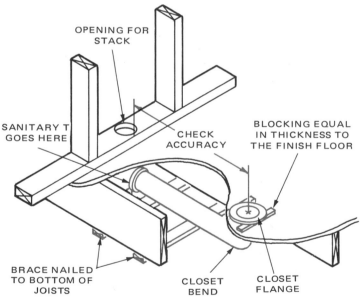

Fig. 9-30. Closet bend must be supported with wood braces, metal clamps or metal straps.

A closet bend or a 90 deg. L and a short length of pipe must be cut to connect the closet flange to the sanitary T. The T will be located in the stack which will be installed later. The length of the closet bend can be determined by temporarily supporting the sanitary T in its correct position while measuring from the lips on the inside of the spigot to the closet flange. See Fig. 9-31. Cut the closet bend, check it for accuracy by trial assembly and join it to the sanitary T. The closet flange will not be permanently secured until after the floor is installed.

Because the sanitary T accurately locates the stack, it is possible to suspend a plumb bob through the center of the sanitary T to locate other parts of the stack. Locate the fittings at the base of the stack next. In most installations with a basement, this will be a Y fitting and a 1/8 bend, Fig. 9-32. If necessary, excavate in order to position the Y at the correct elevation. The Y and the 1/8 bend can be permanently joined and temporarily positioned under the plumb bob. After carefully aligning and bracing, the Y and the 1/8 bend, must be set in concrete.

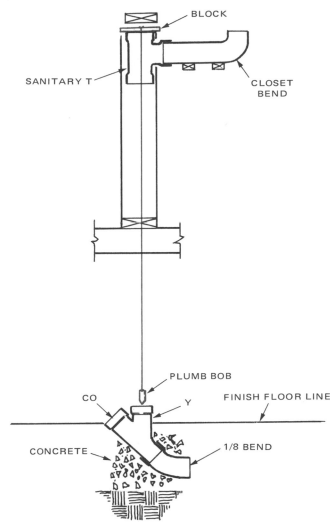

Fig. 9-32. To locate Y at base of stack, suspend plumb bob through center of sanitary T. Concrete can be poured later to provide solid support for weight of stack.

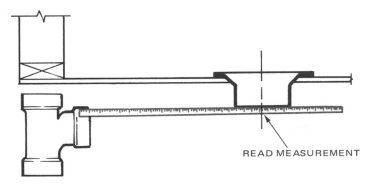

Fig. 9-31. Length of closet bend can be determined by measuring from the center of the closet flange to shoulder of sanitary T.

## Excavating for basement floor drains

In the basement, some digging is generally needed to provide the correct slope from the stack to the building drain and to floor drains. Before digging, locate the outlets of the floor drains and any secondary stacks. Then design a network of pipe to connect these points to the building drain, Fig. 9-33. Remove only the required amount of earth. This leaves a solid base upon which to lay the pipe. Loose dirt should not be used as fill under pipe. It tends to settle allowing the pipe to shift. Sand or brick are frequently used where fill is required.

## Installing the stack

When the excavation is complete, place fittings where branch lines intersect the main. Cut pipe sections to fit. Then run the branch lines.

Install the vertical soil pipe connecting the Y and the sanitary T. It is extremely important that the fittings be

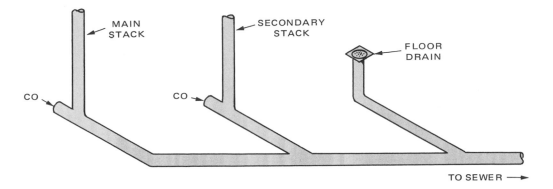

Fig. 9-33. Connecting main and secondary stack to drain line under basement floor.

correctly positioned so their outlet will correctly align with branch piping. Correct alignment can be assured if the pipe and fittings are trial-assembled and the position of the pipes marked at joints as shown in Fig. 9-34. Marking in this way permits the plumber to lay the pipe down or secure it in a vise while joining subassemblies.

The next critical element of the stack is the reducing sanitary T which connects the lavatory branch to the stack. The position of the T is important. Allow a fall of 1/8 to 1/4 in. per foot. Measure fall between point where the rough-in dimensions locate the lavatory drain to the inlet of the reducing sanitary T, Fig. 9-35. As a practical example, assume that the lavatory drain is 4 ft. from the stack and that the rough-in height of the lavatory drain is 24 in. The height of the inlet to the reducing sanitary T will be 23 in. above the floor if a slope of 1/4 in. per ft. is provided. Now cut the pipe which connects the sanitary T to the reducing sanitary T. Install it, making sure the alignment is correct.

Extend the stack through the ceiling and add an increaser and a larger pipe, Fig. 9-36. The increaser and larger pipe are required in colder climates to prevent blocking of the vent by frost. The stack is continued through the roof at least 6 in. and is flashed to prevent leaks.

Install the remainder of the pipes and fittings for the branch lines to the lavatory and the bathtub. Make sure that the drains come through the floor or walls at the spots indicated by the rough-in dimensions.

When all branch lines of the DWV system are installed, the DWV system is ready for inspection and testing.

A similar procedure is used to install water supply piping. In either case, start with the known (position of fixtures and location of building drain or water supply line) to locate other fittings. Then establish the length of pipes.

## MEASURING PIPE

Measurements to determine the length of pipe are taken in one of three ways:

1. Center-to-center.

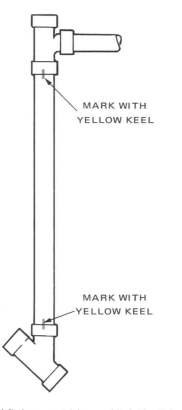

MARK WITH YELLOW KEEL

MARK WITH YELLOW KEEL

Fig. 9-34. Pipe and fittings are trial assembled. If satisfactory, position of fitting is marked as shown. Alignment is then certain to be correct when joists are made.

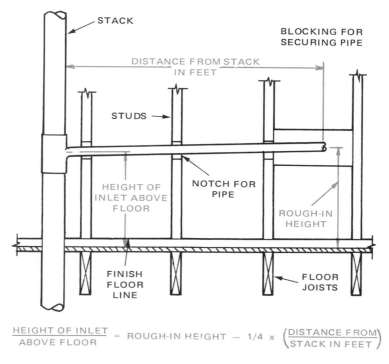

$$\frac{\text{HEIGHT OF INLET}}{\text{ABOVE FLOOR}} = \text{ROUGH-IN HEIGHT} - 1/4 \times \left(\frac{\text{DISTANCE FROM}}{\text{STACK IN FEET}}\right)$$

Fig. 9-35. Method of calculating location of sanitary T for lavatory branch drain.

## Pipe and Fitting Installation

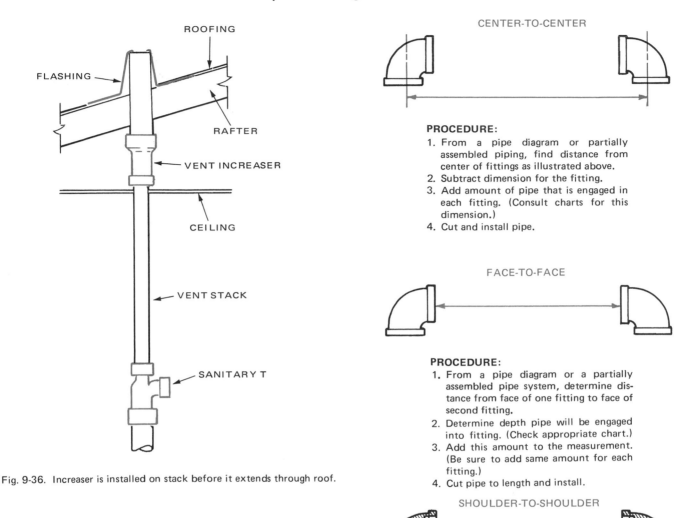

Fig. 9-36. Increaser is installed on stack before it extends through roof.

2. Face-to-face.
3. Shoulder-to-shoulder.

Fig. 9-37 illustrates the three methods. (Unit 4 has additional information on dimensions of fittings.)

Shoulder-to-shoulder dimensions provide a direct reading of the actual length of pipe required. Therefore, they are more accurate and save time because additional calculations are not required. However, this method is only useful with DWV fittings. Pressure fittings do not have shoulders.

### INSTALLING PLASTIC PIPE AND FITTINGS

DWV plastic piping is generally measured by the direct method. Plastic water supply lines are measured by the face-to-face technique with an allowance for the depth of the fitting socket, Fig. 9-38. For example, if the face-to-face measurement for a 3/4 in. pipe is 4'-6" and the pipe must engage a fitting at both ends, then the pipe must be cut 4'-7 1/4" long. See Fig. 9-39.

### CUTTING

Plastic pipe is cut with a fine-tooth saw and a miter box, Fig. 9-40, or a special plastic pipe cutter, Fig. 9-41. In either case the cut must be square so that a full joint will be made with the socket. All burrs on the cut end of the pipe should be removed with a reamer or abrasive paper.

CENTER-TO-CENTER

PROCEDURE:
1. From a pipe diagram or partially assembled piping, find distance from center of fittings as illustrated above.
2. Subtract dimension for the fitting.
3. Add amount of pipe that is engaged in each fitting. (Consult charts for this dimension.)
4. Cut and install pipe.

FACE-TO-FACE

PROCEDURE:
1. From a pipe diagram or a partially assembled pipe system, determine distance from face of one fitting to face of second fitting.
2. Determine depth pipe will be engaged into fitting. (Check appropriate chart.)
3. Add this amount to the measurement. (Be sure to add same amount for each fitting.)
4. Cut pipe to length and install.

SHOULDER-TO-SHOULDER

PROCEDURE:
1. Measure distance between fitting from shoulder-to-shoulder.
2. Cut pipe to this length.

Fig. 9-37. There are three ways to determine the length of pipe needed between two fittings.

### PLASTIC PIPE DEPTH OF ENGAGEMENT

| PIPE SIZE (IN INCHES) | ENGAGEMENT (IN INCHES) |
|---|---|
| 1/2 | 1/2 |
| 3/4 | 5/8 |
| 1 | 3/4 |
| 1 1/4 | 11/16 |
| 1 1/2 | 11/16 |
| 2 | 3/4 |
| 3 | 1 1/2 |
| 4 | 1 3/4 |
| 6 | 3 |

Fig. 9-38. Chart should be referred to in determining lengths of plastic pipe.

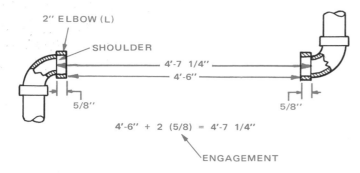

4'-6'' + 2 (5/8) = 4'-7 1/4''

ENGAGEMENT

Fig. 9-39. When computing pipe length, be sure to include allowance for depth of socket.

## JOINING

Before joining any type of plastic pipe, apply a solvent to both the pipe and fitting or use abrasive paper. This cleans the pipe and removes the gloss for better bonding.

Fig. 9-40. Miter box assures square cut on plastic pipe.

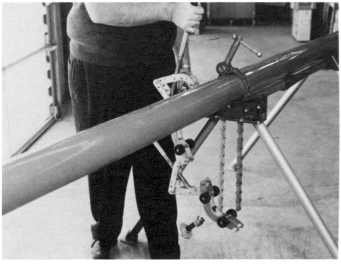

ATTACHING THE CUTTER

CUTTING THE PIPE

CLOSE-UP OF THE CUTTER

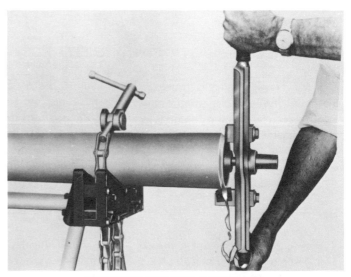

TRIMMING THE END OF THE PIPE

Fig. 9-41. Follow the procedure for cutting and trimming plastic pipe with a cutter. (Wheeler Mfg. Corp.)

Apply a light coat of the appropriate solvent cement, Fig. 9-42, to the fitting socket with a natural bristle brush. (Do not use the same brush to apply the cleaning solvent.)

Next, apply a heavy coat of cement to the pipe spigot, Fig.

| TYPE OF PLASTIC | CEMENT |
|---|---|
| ABS | ABS DISSOLVED IN METHYL ETHYL KETONE |
| CPVC | CPVC DISSOLVED IN TETRAHYDROFURAN |
| PVC | PVC DISSOLVED IN TETRAHYDROFURAN |
| PE | MECHANICAL JOINTS ONLY |
| SR | SR DISSOLVED IN TOLUENE |

Fig. 9-42. Several solvent cements are available for making joints in plastic piping.

9-43. Immediately insert the pipe all the way into the fitting socket while giving it a quarter turn, Fig. 9-44. The rotation assures that the solvent cement is evenly distributed in the joint. A bead of solvent cement completely around the fitting indicates that the proper amount of cement was applied, Fig. 9-45. No bead or a partial bead may indicate incomplete bonding which will result in a leaking joint.

The solvent cement will set in two to five minutes and can be handled with care at that time. Allow 24 hours before testing the pipe.

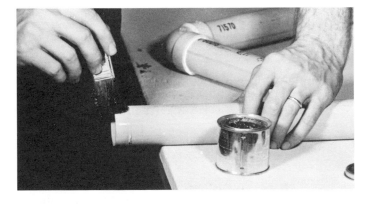

Fig. 9-43. Apply solvent cement to plastic pipe with brush or dabber.

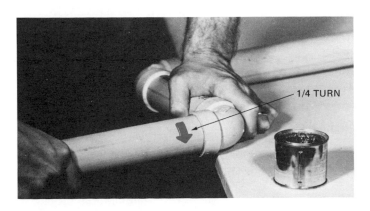

1/4 TURN

Fig. 9-44. Turn fitting 1/4 turn as pipe goes full depth into fitting socket.

Fig. 9-45. Full bead around socket means that the proper amount of solvent cement was used.

## SUPPORTING

Horizontal runs of plastic pipe should be supported every 3 to 4 ft. with metal hangers 3/4 in. or more in width. Stacks should be set in concrete at their base and secured to the building frame in order to maintain alignment.

## INSTALLING COPPER PIPE AND FITTINGS

Copper DWV pipe measurements can frequently be taken shoulder-to-shoulder. However, face-to-face measurements are frequently more convenient when pressure fittings are being installed. Appropriate allowances for fitting socket depth are shown in Fig. 9-46.

## CUTTING PIPE AND TUBING

Copper pipe and tubing should be cut with a tubing cutter, Fig. 9-47, or a fine-tooth hacksaw and a miter box. Slowly rotate the tubing cutter around the pipe. Tighten the cutter with each revolution until the pipe is cut. Ream the cut end of the pipe to remove the burr produced in cutting off the pipe, Fig. 9-48.

## JOINING

Solder type fittings are installed by cleaning, fluxing and heating the joint as described in Unit 10. Flare type joints are

SOCKET ALLOWANCES FOR COPPER FITTINGS

| PIPE SIZE (IN INCHES) | ENGAGEMENT (IN INCHES) | PIPE SIZE (IN INCHES) | ENGAGEMENT (IN INCHES) |
|---|---|---|---|
| 1/4 | 5/16 | 2 | 1 3/8 |
| 3/8 | 3/8 | 2 1/2 | 1 1/2 |
| 1/2 | 1/2 | 3 | 1 11/16 |
| 3/4 | 3/4 | 3 1/2 | 1 15/16 |
| 1 | 15/16 | 4 | 2 3/16 |
| 1 1/4 | 1 | 5 | 2 11/16 |
| 1 1/2 | 1 1/8 | 6 | 3 1/8 |

Fig. 9-46. Make allowance for these socket depths when measuring copper piping.

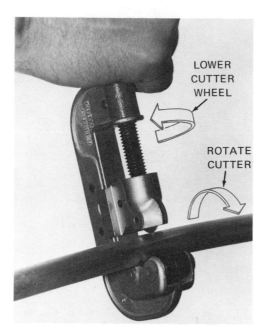

Fig. 9-47. Copper tubing is cut with light pressure on the cutting wheel.

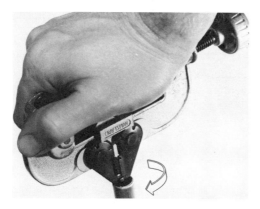

Fig. 9-48. Ream tubing with hand tool to remove ridge left by cutting.

Fig. 9-49. Top. Flaring tool and parts. (Imperial Eastman Corp.) Bottom. Insert tubing, end flush with top of die block, and tighten clamp.

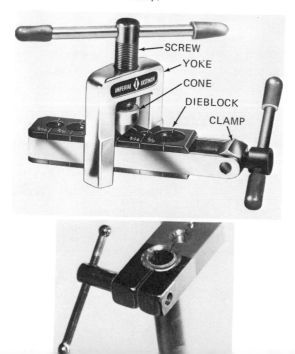

SCREW
YOKE
CONE
DIEBLOCK
CLAMP

made with a flaring tool. Be sure to slip the nut onto the tubing before flaring the end. Place tubing end into the die block, Fig. 9-49. Center yoke over tubing end and turn the cone into the tube, Fig. 9-50.

To assemble compression type fittings, first place the nut on the pipe followed by the compression ring, Fig. 9-51. No forming operation is needed. Insert the pipe into the fitting and tighten the nut.

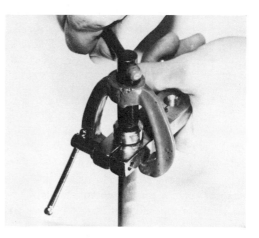

Fig. 9-50. Flare is formed on the tubing with the press.

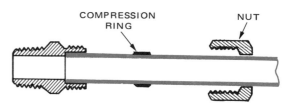

COMPRESSION
RING              NUT

Fig. 9-51. Proper assembly of compression type fitting. Compression ring is squeezed between male and female parts of fitting causing a tight seal.

## SUPPORTING

Horizontal runs of copper pipe should be supported every 4 to 6 ft. with copper pipe hangers. Vertical runs of pipe should be secured to maintain alignment.

### INSTALLING GALVANIZED AND BLACK IRON PIPE

Fig. 9-52 gives allowances for the engagement of threads into galvanized or black iron pipe fittings. These allowances must be added to all face-to-face measurements.

### CUTTING AND THREADING

A pipe cutter, described in Unit 1, is the best tool for cutting galvanized and black iron pipe. It works much like a tubing cutter. To operate, revolve the tool around the pipe and tighten the cutter wheel with each revolution until the pipe is sheared, Fig. 9-53. Once the cut is complete, the pipe must be

### DISTANCE IRON PIPE IS TURNED INTO STANDARD FITTINGS

| PIPE SIZE (INCHES) | ENGAGEMENT (INCHES) | PIPE SIZE (INCHES) | ENGAGEMENT (INCHES) |
|---|---|---|---|
| 1/8 | 1/4 | 1 1/4 | 11/16 |
| 1/4 | 3/8 | 1 1/2 | 11/16 |
| 3/8 | 3/8 | 2 | 3/4 |
| 1/2 | 1/2 | 2 1/2 | 15/16 |
| 3/4 | 9/16 | 3 | 1 |
| 1 | 11/16 | 3 1/2 | 1 1/16 |

Fig. 9-52. Dimensions in right-hand column are allowances to be added to face-to-face measurements of galvanized or black iron pipe. The allowance allows for the distance pipe is threaded into fitting.

reamed to remove the burr, Fig. 9-54.

Threads are cut with the correct size pipe die and a die stock, Fig. 9-55. A smooth sleeve guides the die onto the pipe to insure that it remains square with the end of the pipe, Fig.

Fig. 9-54. Reamer, chucked in hand brace, will remove burrs left from cutting of galvanized or black iron pipe.

Fig. 9-53. Views of cutter close up during process of cutting galvanized pipe.

Fig. 9-55. A pipe is threaded by hand with a die and die stock.

9-56. Cutting oil should be used to lubricate the die as threads are cut. Cutting, reaming and threading can also be done on a powered pipe machine, Fig. 9-57. A mobile shop unit, Fig. 9-58, can be trailered to the construction site.

## ASSEMBLY AND SUPPORT

Before putting the pipe and fittings together, apply pipe joint sealer to the threads. This seals the joint and prevents leaks. The threaded pipe or fitting is turned clockwise with a pipe wrench to tighten the joint, Fig. 9-59.

Galvanized and black iron pipe should be supported at 6 to 8 ft. intervals. Use metal hangers.

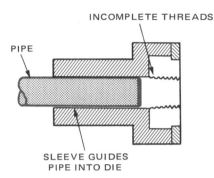

Fig. 9-56. Cutaway shows pipe being guided squarely into die.

## INSTALLING CAST-IRON PIPE

The correct allowances for the depth of the hub on common sizes of hub and spigot cast-iron pipe are given in Fig. 9-60. This allowance must be included when measuring pipe length.

Fig. 9-58. Gasoline engines power portable shop. Pipe can be cut and threaded. Unit is trailered to the job site. (Obear Industries)

## CUTTING

Cast-iron pipe is generally cut with a hydraulic pipe cutter, Fig. 9-61. This tool squeezes the chain tightly around the pipe. Small cutters in the chain bite into the pipe until it fractures. Hand methods give equally good results if these steps are followed:

1. Mark all the way around the pipe with yellow keel (grease pencil).

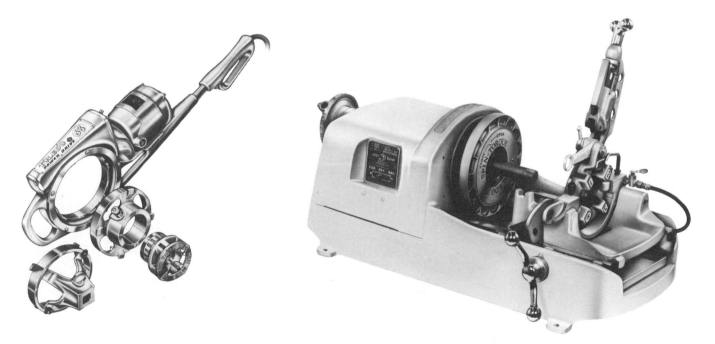

Fig. 9-57. Two types of power driven threaders. One is portable; the other is bench mounted.

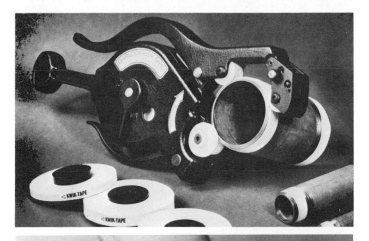

ALLOWANCES FOR HUB DEPTH

| PIPE SIZE (INCHES) | ENGAGEMENT (INCHES) |
|---|---|
| 2 | 2 1/2 |
| 3 | 2 3/4 |
| 4 | 3 |
| 5 | 3 |
| 6 | 3 |
| 8 | 3 1/2 |

Fig. 9-60. Add allowances above to the face-to-face measurements of cast-iron pipe.

2. Saw a 1/16 to 1/8 in. deep groove around the pipe, Fig. 9-62.
3. Roll the pipe and tap on it with a hammer near the groove until it breaks, Fig. 9-63. On extra heavy pipe, chisel around the groove until the pipe fractures.

## JOINING

Bell and spigot cast-iron pipe is joined with lead and oakum. Vertical joints are the easiest to make. Wipe away dirt and moisture from the inside of the socket and the outside of the spigot. This is important! Dirty surfaces will cause poor sealing and molten lead striking wet surfaces creates steam which expands with explosive force.

Carefully center the spigot in the socket and pack in oakum. (Oakum is a hemp treated with pitch to make it moisture proof.) Pack the oakum tightly using a yarning iron, Fig. 9-64. When the bell is half full the joint is ready to receive a pour of molten lead.

Melt about 1 lb. of lead for each inch of pipe diameter using a portable furnace, Fig. 9-65. Heat until cherry red. Pour the molten metal into the joint with a ladle. When the lead cools, caulk it to make the joint air and watertight. Use a standard caulking iron, Fig. 9-66. Move the iron slowly around the joint tapping it gently with a ball peen hammer. Use care, a hard blow could break the pipe.

When the lead takes on a dull gray appearance, the joint is properly caulked. It is ready for finishing. Using a wide iron,

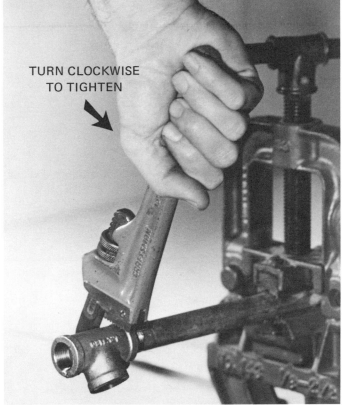

TURN CLOCKWISE TO TIGHTEN

Fig. 9-59. Threaded joints must be sealed and tightened to prevent leaks. Top. Teflon tape is applied as joint sealer. (Plastomer Products Div., Garlock Inc.) Bottom. Fitting is tightened with Stillson wrench.

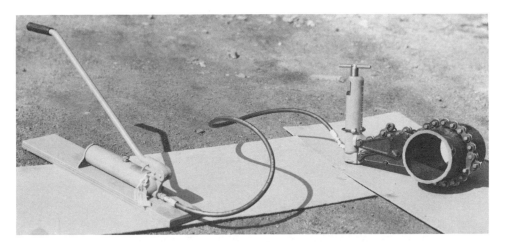

Fig. 9-61. Hand operated hydraulic cast-iron pipe cutter. (Wheeler Mfg. Corp.)

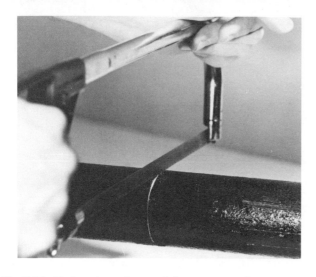

Fig. 9-62. Hacksaw is used to carefully cut groove all around pipe.

Fig. 9-63. Tapping along saw groove will cause pipe to break cleanly along saw line.

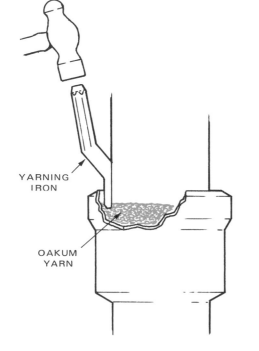

YARNING IRON

OAKUM YARN

Fig. 9-64. Yarning iron packs oakum into bell to form a seal so molten lead is contained in the bell.

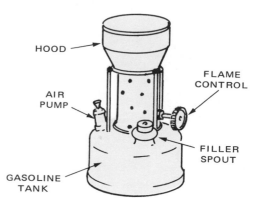

HOOD

FLAME CONTROL

AIR PUMP

FILLER SPOUT

GASOLINE TANK

Fig. 9-65. Furnace generates intense heat to melt lead.

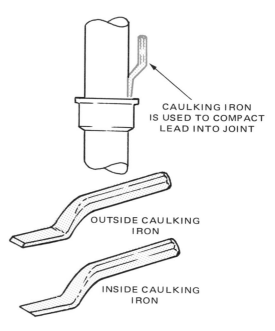

CAULKING IRON IS USED TO COMPACT LEAD INTO JOINT

OUTSIDE CAULKING IRON

INSIDE CAULKING IRON

Fig. 9-66. Tools with offset handles are used to caulk lead.

lightly tap the lead to make the surface smooth.

Horizontal joints are made in the same way except that a joint runner must be used to prevent the lead from flowing out of the hub before it can harden. This is a rope-like tool made of fireproof material, Fig. 9-67.

Hub and spigot cast-iron pipe may also be joined with a neoprene compression gasket as shown in Fig. 9-68. Such joints can be installed more easily if a rubber lubricant is applied before the pipes are forced together.

No-hub cast-iron pipe is joined with a neoprene gasket and a stainless steel clamp, Fig. 9-69. Slip the neoprene gasket onto the pipe as in Fig. 9-70. Position the clamp over the gasket, Fig. 9-71. Tighten the clamp screws, Fig. 9-72, to 60 inch-pounds. A torque wrench works best.

## SUPPORTING CAST-IRON PIPE

Cast-iron pipe, because of its weight, must be well supported while joints are being made. Permanent supports must be installed before the first rough is completed.

138

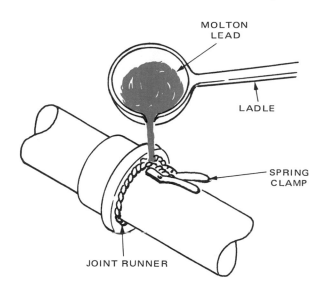

Fig. 9-67. Joint runner creates dam around top of bell when pouring lead into horizontal runs of soil pipe.

Fig. 9-68. Special gaskets can be used in place of oakum and lead. (E. I. du Pont de Nemours & Co.)

Fig. 9-69. First step in assembling gasket and clamp for no-hub cast-iron pipe.

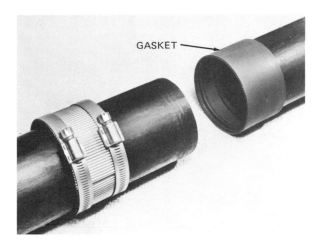

Fig. 9-70. Gasket and clamp are slipped onto the pipe.

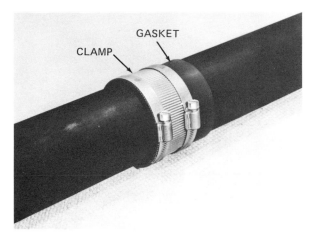

Fig. 9-71. Clamp is positioned on top of gasket.

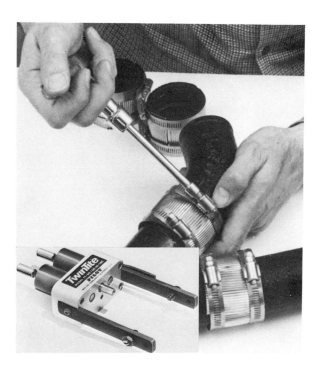

Fig. 9-72. Tightening clamps. Top. Special hand driver. Inset. Tool for tightening screws equally. (E.I. du Pont de Nemours & Co. and Pilot Mfg.)

It is generally recommended that these supports be placed at every joint on horizontal runs unless the distance between joints is less than 4 ft. In such cases, a support at every other joint is enough. Use strap iron or special hangers for this purpose, Fig. 9-73. Vertical runs of cast-iron pipe can be

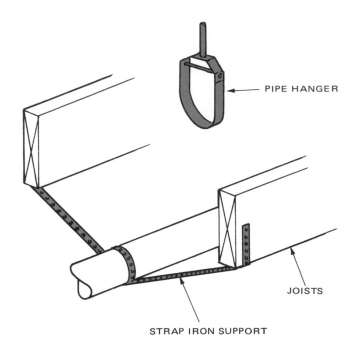

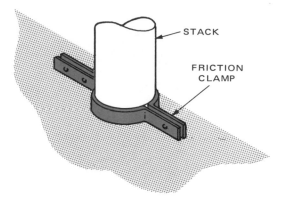

Fig. 9-75. Friction clamp should be used at every floor level to insure that each floor supports part of the weight of multistory stacks.

Fig. 9-73. Heavy straps or hangers are adequate support for soil pipe.

attached to the building structure with wire staples, vertical pipe brackets or pipe straps, Fig. 9-74. Friction clamps, Fig. 9-75, should support the weight of cast-iron pipe at each floor level.

## Masonry anchors

Many occasions will arise when the plumber must use fastening devices to attach pipe hangers or fixtures to concrete or masonry. In recent years, many different types of masonry fasteners have been developed and marketed.

The lag shield, Fig. 9-76, is made from lead. To install it,

1. Drill hole 1/4 in. deeper than anchor length and insert shield flush to surface of masonry.

2. Insert lag bolt or screw and tighten to expand anchor.

Fig. 9-76. Lag shields are commonly used to attach pipe hanger and fixtures to concrete and masonry supports.

drill a hole 1/2 in. in diameter or larger (depending upon the size of the anchor) into the concrete or masonry. Insert the lag shield. Insert and tighten the lag bolt. The shield expands and grips the concrete firmly. One advantage of the lag shield over caulking anchors is that lag shields do not have to be positioned as accurately.

Caulking anchors, Fig. 9-77, provide a fastener which is permanently attached to the concrete or masonry. They are internally threaded to accept machine screws and bolts. The caulking tool is specially made to permit the lead to be driven

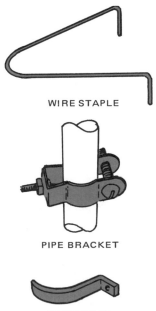

WIRE STAPLE

PIPE BRACKET

PIPE STRAP

Fig. 9-74. These supports are all suitable for securing vertical runs of cast-iron pipe.

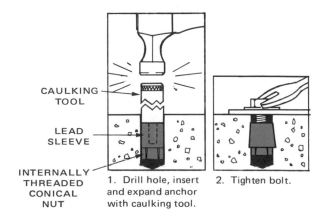

CAULKING TOOL

LEAD SLEEVE

INTERNALLY THREADED CONICAL NUT

1. Drill hole, insert and expand anchor with caulking tool.

2. Tighten bolt.

Fig. 9-77. Caulking anchors are securely fastened in concrete or masonry by driving the lead shield solidly around the conical nut. (Phillips Drill Co.)

firmly into the concrete without damage to the internally threaded cone-shaped nut.

Toggle bolts, Fig. 9-78, may be useful when attaching pipe hangers to hollow masonry units. The wings are spring operated. They open on the inside of the cavity. This makes it possible to tighten the screw. Note that the bolt must be long enough so the wings can swing free after the bolt is pushed

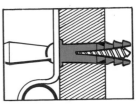
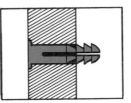

1. Drill hole 1/4 in. deeper than anchor length and insert anchor until flange is flush.

2. Fasten fixture by inserting screw through fixture and tightening.

Fig. 9-79. Plastic shields are generally easier to install because smaller diameter holes are required for their installation.

through the hole. Should it be necessary to remove the toggle bolt, the wings will be lost inside the wall.

Plastic anchors, Fig. 9-79, have become increasingly popular. They generally can be installed in a smaller hole than lag shields, caulking anchors or toggle bolts. They do not have as much holding power as other masonry anchors. But they are adequate for many installations where small diameter pipes are being suspended from hangers.

## INSPECTING AND TESTING PIPING SYSTEMS

After all drainage and water supply piping has been installed, it is the responsibility of the person to whom the building permit was issued to contact the plumbing inspector and arrange for the rough-in inspection. (See Unit 13 for additional information.) Generally, this inspection involves a water or air pressure test of each of the piping systems to determine if all joints are sealed.

The plumber is generally required to furnish all the equipment necessary to conduct the tests. This will include test plugs, Fig. 9-80, and a source of compressed air — if air

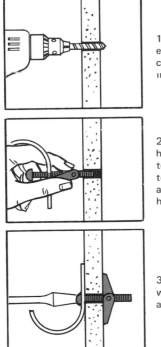

1. Drill hole large enough to permit closed wings to be inserted.

2. Insert bolt through hanger before threading toggle on bolt. Squeeze toggle wings together and insert in predrilled hole.

3. Tighten bolt. Spring wings will snap open after insertion.

Fig. 9-78. Toggle bolts can be installed in hollow masonry units and other walls which are hollow.

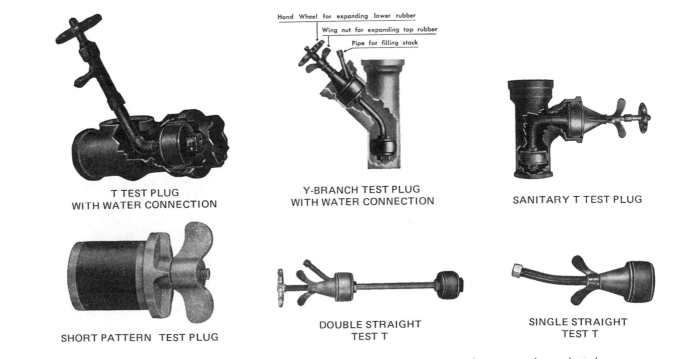

T TEST PLUG
WITH WATER CONNECTION

Hand Wheel for expanding lower rubber
Wing nut for expanding top rubber
Pipe for filling stack

Y-BRANCH TEST PLUG
WITH WATER CONNECTION

SANITARY T TEST PLUG

SHORT PATTERN TEST PLUG

DOUBLE STRAIGHT TEST T

SINGLE STRAIGHT TEST T

Fig. 9-80. A variety of test plugs may be used to seal openings in the piping so that tests can be conducted.

testing is required. Water testing can be done with test plugs and a hose. To test, close all outlets to a piping system with test plugs, caps or plugs. Fill the piping with water or air under pressure and look for leaks.

## WATER TESTING

A minimum of 10 ft. of water head must be used to water test DWV piping. Generally, this can be accomplished by filling the vent stack completely. Once the DWV piping is filled, visually inspect the complete piping systems to determine if any leaks exist. This inspection must be completed before any piping is covered. The inspector is also required to check the piping systems for such hazards as cross connections, defective or inferior materials and poor work.

Water testing of the water supply piping is conducted by closing all outlets and filling the system with water from the main. Again, visual inspection will locate leaks and other potential problems.

## AIR TESTING

Air testing is effective in detecting leaks. These tests are similar to the water tests, except that the piping system is filled with compressed air. A pressure gauge at the test plug will allow the inspector/plumber to determine if pressure is being lost at any point in the piping. For DWV piping, a pressure of 5 psi is generally adequate. Water supply piping is tested at a pressure at least equal to local water pressure or as much as 50 percent greater than the pressure in the water main. Soap suds are helpful in detecting leaks. The suds are brushed onto joints and any escaping air will form bubbles.

## ONE-LINE WATER SUPPLY SYSTEM

An alternative to the typical hot and cold water supply system has been introduced in recent years. This system has a centralized valve unit, Fig. 9-81, to control the water temperature and rate of flow of water to each individual spout. The valve unit is installed near the water heater. A

single pipe runs from the outlet of the valve unit to spouts at the lavatory, sink, tub and shower. Installation of this system also rquires connection of an electrical system which controls solenoids that turn the valves. Apart from these differences, installation is similar to that used for conventional hot and cold water systems. (See Unit 5 for additional information.)

## TEST YOUR KNOWLEDGE — UNIT 9

1. That part of the plumbing installation which extends the water and sewer lines into the building is known as the
   _____ _____.

2. Installation of all piping within the walls, floors and ceilings of the finished building is called:
   a. First rough.
   b. Fixture installation.
   c. Rough-in.
   d. Second rough.
   e. Trenching.

3. Describe the two methods used for attaching the corporation valve to the water main.

4. The _____ _____ is the drainage piping between the

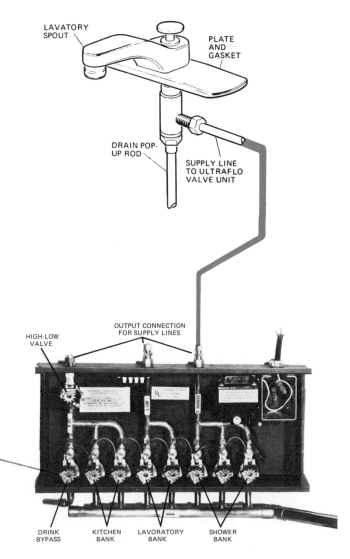

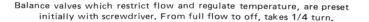

Balance valves which restrict flow and regulate temperature, are preset initially with screwdriver. From full flow to off, takes 1/4 turn.

Fig. 9-81. Newly developed one-line plumbing system requires only one pipe from the central control unit to the spout.

building's foundation and the sewer main or septic tank.

5. When locating fixtures, allowance must be made for the thickness of the finished _____ and _____ materials.

6. Holes for water supply pipes are made near the edge of wall plates and studs to reduce the likelihood that the pipe will be punctured by nails which secure the finish wall material. True or False?

7. Explain why blocking is installed in the wall.

8. Holes for pipe up to 3 in. in diameter can be drilled with a portable _____ _____ which is fitted with a _____ bit, an _____ bit or cut with a _____ saw.

9. The DWV piping system is installed before the water supply piping. True or False?

10. A _____ _____ can be used to transfer the center point of pipe and fittings in the stack from one floor level to another.

11. When measuring to determine the length of a pipe required to connect two fittings, the only method which does not require making allowances for fittings is:
   a. Center-to-center.
   b. Face-to-face.
   c. Shoulder-to-shoulder.

12. The first step in joining plastic pipe to a plastic fitting is to _____ both the pipe and the fitting.

13. When making connections in plastic piping, what operation insures even distribution of the solvent cement?

14. The plumber can be assured that a plastic pipe joint is correctly made if a _____ of cement forms completely around the fitting.

15. If the face-to-face distance between a 1/2 in. copper elbow and a 1/2 in. copper T is 6 1/2 in., the length of the connecting copper pipe will be:
   a. 6 1/2 in.
   b. 7 in.
   c. 7 1/4 in.
   d. 7 1/2 in.

16. If the face-to-face distance between a 3/4 in. galvanized iron T and a 3/4 in. elbow is 18 in., what length of 3/4 in. galvanized iron pipe will be required to properly connect the two fittings?
   a. 19 in.
   b. 19 1/8 in.
   c. 19 1/4 in.

17. Pipe _____ _____ is applied to threaded pipe before assembly to prevent leaks.

18. Name the tool used to pack oakum into the bell of a cast-iron pipe joint.

19. On horizontal joints of hub and spigot cast-iron pipe, a _____ _____ is used to prevent the molten lead from running out of the joint as it is poured.

20. Testing of the completed DWV and water supply piping must be done at the completion of the _____ stage.

## SUGGESTED ACTIVITIES

1. Working as a group, make a set of sample pipe assemblies with each of the different types of pipe commonly used in your area. Each of these completed exercises should be capped and fitted with adapters or test plugs so they can be tested using standard procedures.

2. In a model framed structure, while working in small groups, install the DWV and water supply piping for a kitchen, utility room or bathroom. The completed piping system should be tested using the testing procedure(s) found in your community.

# Unit 10
# SOLDERING, BRAZING, WELDING AND LEAD WIPING

## Objectives

This unit describes the making of watertight pipe joints using heat and various filler materials.

After studying it you will be able to:
- Identify the solders, plastics, leads and fluxes needed for succesfully joining all kinds of pipe and fittings.
- Describe or demonstrate the processes by which pipe, fittings and filler materials are joined.

Water supply pipe and fittings which are not threaded are joined by soldering, brazing, welding, lead wiping or cementing. Cementing is used with plastic pipe and does not require the application of heat.

## SWEAT SOLDERING

Soldering is a method of using heat to form joints between two metallic surfaces using a nonferrous filler material. (A nonferrous metal is one which contains no iron and is, therefore, nonmagnetic.) Soldering is generally used by plumbers to join rigid copper pipe and fittings. The filler material is distributed evenly between the close-fitting surfaces of the joint by CAPILLARY ATTRACTION. This is the tendency of a liquid to be drawn to the surface of solids in a kind of "soaking" or "spreading" action.

## SOLDERS

Soft solder which is suited for joining copper pipe is generally made up of equal parts of tin and lead. Its melting temperature 427 F (219 C) is well below the 800 F (427 C) maximum suitable for soldering. It is sold in spools. It will probably carry a designation such as "50A" indicating the percentage of tin in its composition. It should be used only where pipe temperatures will not exceed 250 F (121 C). Thus, it is suited to low-pressure steam applications too.

Hard solders are made up of various percentages of copper and zinc alloys. They are used in brazing of cast iron, iron and steel, brass and sometimes copper.

## FLUXES

A soldering flux performs several functions:
1. It protects the surface from oxidation during heating.

(Oxidation is the process of picking up oxygen which produces tarnish and rust in metals.)
2. It helps the filler metal flow easily into the joint.
3. It floats out remaining oxides ahead of the molten solder.
4. It increases the wetting action of the solder by lowering surface tension of the molten metal.

Highly corrosive fluxes contain inorganic acids and salts such as zinc chloride, ammonium chloride, sodium chloride, potassium chloride, hydrochloric acid and hydrofluoric acid.

Less corrosive fluxes contain milder acids such as citric acid, lactic acid and benzoic acid. Though briefly very active at soldering temperatures, their corrosive elements are driven off by the heat. Residue does not remain active and is easily removed after the joint is cool.

Noncorrosive fluxes — the only type suited for plumbing work — are composed of water and white resin dissolved in an organic or benzoic acid. Its residue does not cause corrosion. These fluxes are effective on copper, brass, bronze, nickel and silver. This type is recommended for joining copper pipe.

The best fluxes for joining copper pipe and fittings are compounded of mild concentrations of zinc and ammonium chloride. These are mixed with a petrolatum base.

## SWEAT SOLDERING PROCEDURE

Soldering is not difficult. However, it is important that each operation be carefully completed for satisfactory results. Carefully study the following procedures before attempting to make a solder joint:
1. Cut the copper pipe with a tubing cutter, Fig. 10-1.
2. Ream the ends of each pipe to remove metal burrs, Fig. 10-2.
3. Cleaning is a very important part of making good solder joints. Use a copper cleaning tool, Fig. 10-3, abrasive paper (fine grit), emery cloth, or No. 00 steel wool to clean the copper pipe ends and the socket or cup of the fitting. Do a thorough job. After the scale and dirt are removed, brush away any loose abrasive particles. Avoid touching the clean metal with your fingers.
4. Immediately apply the proper flux to all pipe and joint areas with a clean brush as shown in Fig. 10-4.
5. Assemble the fluxed pipes into the fitting, push and turn until the pipes are bottomed against the inside shoulders of the fitting.
6. Select the proper solid core solder.

Fig. 10-1. A tubing cutter is recommended for cutting copper pipe. It produces a square cut that needs little dressing.

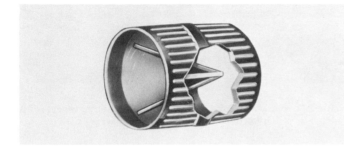

Fig. 10-2. A reamer is used to remove the wire edge or burr formed on the inside of copper tubing during cutting. (Imperial Eastman Corp.)

Fig. 10-3. This special cleaning tool is used to prepare copper tubing and fittings for soldering. It will polish the inside diameter (ID) or end of fittings and the outside diameter (OD).

Fig. 10-4. Soldering flux is applied to pipe and joint areas to reduce oxidation during the heating cycle.

7. Light a small portable propane gas torch, Fig. 10-5, for heating the pipe and fitting.

When lighting the torch, always use a spark lighter and hold the torch so it points away from you and any flammable material. See Fig. 10-6.

Fig. 10-5. Gas torches provide sufficient heating capacity for soldering copper tubing and fittings.

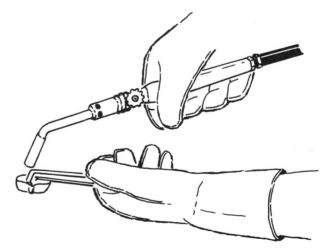

Fig. 10-6. When lighting the torch, direct the tip away from you and any flammable material. Use a spark lighter to ignite the fuel.

8. Direct the heat on the copper pipe before heating the fitting. This procedure heats the pipe (which generally dissipates more heat than the fitting) to the correct temperature without overheating the fitting. The fitting can be heated very quickly once the pipe has reached the right temperature. Hold the torch so the inner cone of the

flame touches the metal, Fig. 10-7.

9. Slowly touch the end of the solder wire to the joint area to check for proper temperature. Feed the solder into the joint as you move the torch flame to the center of the fitting. Do not melt the solder in the flame. In Fig. 10-8, a pipe joint is being soldered with an electric soldering gun designed especially for plumbing work.

10. The joint can be wiped with a clean cloth while hot, Fig. 10-9, to remove excess solder and any remains of the flux.

11. Secure the propane torch and other equipment.

Be certain the valve of a propane torch is closed after use. Store the torch in a cool place away from any unusual source of heat.

Fig. 10-9. To produce a joint that is smooth and neat, wipe with a burlap or denim cloth while still hot.

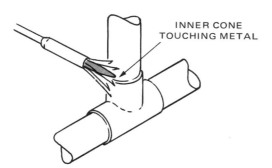

Fig. 10-7. The inner cone of the torch flame should touch the metal. This is the hottest part of the flame.

INNER CONE TOUCHING METAL

Fig. 10-8. Feed the solder into the joint when the pipe and fitting are hot enough to make solder flow. (Nibco Inc.)

## BRAZING

Like soldering, brazing uses a nonferrous filler material to join base metals. However, brazing is done with temperatures above 800 F (427 C). In brazing, the melting point of the filler metal is below that of the base metals being joined.

In plumbing, brazing is used for joining pipe and fittings in saltwater pipelines, oil pipelines, refrigeration systems, vacuum lines, chemical handling systems, air lines and low-pressure steam lines. Cast bronze fittings are commonly silver brazed to pipe and tubes of copper, brass, copper nickel and steel.

Braze welding of cast-iron parts by a metal worker should not be confused with the brazing done on piping. They are different processes.

Brazing of pipe is an adhesion process. The metals being joined are heated, but not melted. The joint formed by such a process is superior to soldering. It is used where mechanical strength and pressure-proof joints are needed.

The strength comes from the ability of the brazing alloys or silver braze to flow into the porous grain structure of the pipe and fitting. See Fig. 10-10. However, this excellent bond is only possible if:

1. The surface is clean.
2. Proper flux and filler rod is used.
3. The clearance gap between the outside of the pipe and the bore is only .003 to .004 in.

## BRAZING MATERIALS

Filler metal for brazing is available in different shapes: wires, rods, sheets and washers. The classifications, each with special uses, include:

1. Aluminum-silicon, used for brazing aluminum.
2. Copper-phosphorus, for joining copper and copper alloys or other nonferrous metals.
3. Silver, for joining virtually all ferrous and nonferrous metals except aluminum and several metals with low melting

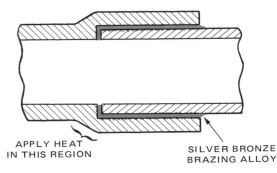

APPLY HEAT
IN THIS REGION

SILVER BRONZE
BRAZING ALLOY

Fig. 10-10. A cross-sectional enlargement of the clearance gap between a pipe and a bronze pipe fitting.

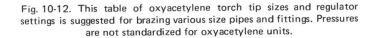

| TIP SIZE (NO.) | GAS PRESSURE REGULATOR SETTINGS | | ROD SIZE (INCHES) | PIPE AND FITTING DIA. (INCHES) |
| --- | --- | --- | --- | --- |
| | OXYGEN | ACETYLENE | | |
| 4 | 4 | 4 | 3/32 | 1/4 — 3/8 |
| 5 | 5 | 5 | 1/8 | 1/2 — 3/4 |
| 6 | 6 | 6 | 3/16 | 1 — 1 1/4 |
| 7 | 7 | 7 | 1/4 | 1 1/2 — 2 |
| 8 | 8 | 8 | 5/16 | 2 — 2 1/2 |
| 9 | 9 | 9 | 3/8 | 3 — 3 1/2 |
| 10 | 10 | 10 | 7/16 | 4 — 6 |

Fig. 10-12. This table of oxyacetylene torch tip sizes and regulator settings is suggested for brazing various size pipes and fittings. Pressures are not standardized for oxyacetylene units.

points.

4. Copper and copper-zinc, suited for joining both ferrous and nonferrous metals. This compound is used in a 50/50 mixture for brazing copper. A 64 percent copper — 36 percent zinc compound is used for iron and steel.

5. Nickel, used when extreme heat and corrosion resistance is needed. Applications include food and chemical processing equipment, automobiles, cryogenic and vacuum equipment.

Fluxes, considered so important in the soldering process, are even more necessary in brazing. In addition to protecting the surface from oxidation and aiding the flow of filler material, brazing flux serves to indicate the temperature of the metal. Without flux it would be almost impossible to know when the base metal reaches the correct temperature. Fluxes are produced in powder, paste and liquid form. Six different types of flux are commercially available.

## SUPPLYING HEAT

Most brazed joints are made at a temperature of 1400 F (760 C) or higher. Because of the higher temperature, an oxyacetylene torch, Fig. 10-11, is commonly used.

Fig. 10-11. An oxyacetylene torch unit can efficiently provide the higher temperature necessary for brazing.
(Linde Air Products Co., Div. of Union Carbide)

Correct torch tip size and the appropriate oxygen and acetylene regulator settings are shown in the table in Fig. 10-12. For example, to braze 1/2 or 3/4 in. pipe, a No. 5 torch tip is recommended. This tip requires an oxygen pressure of 5 psi and an acetylene pressure of 5 psi. See Fig. 10-13.

## PROCEDURE FOR BRAZING

When using a welding torch always wear welding goggles and protective clothing. Shut off tank valves when finished. Think "Safety."

Fig. 10-13. Regulator for controlling oxygen and acetylene pressure.
(Linde Air Products Co., Div. of Union Carbide)

1. Assemble the correct tip on the torch.
2. Make sure the regulator valves are closed. Open the tank valve. At this point the tank pressure gauge on the oxygen tank may read as much as 2000 psi and the acetylene tank pressure gauge may read up to 250 psi.
3. With the valves on the torch closed, adjust the regulator valve to the correct setting as indicated in the previous step.
4. Open the acetylene torch valve 1/4 turn. Holding the torch away from you or any flammable material, light the gas with a spark lighter.
5. Adjust the oxygen and acetylene torch valves until a neutral or carburizing (excess of acetylene) flame is produced, Fig. 10-14.
6. Apply flux to the pipe with a brush, and assemble the pipe and fitting.
7. Heat the pipe first. Watch the flux. It will first turn to a white powder. Then when the correct brazing temperature is reached, it will become liquid. At this time, shift the flame to the bronze fitting, Fig. 10-15.
8. At this point the brazing rod can be preheated by introducing it into the flame as the fitting is being heated. In a few seconds the rod will be hot enough that it can be inserted into the flux container. A coating of flux will melt onto the rod, Fig. 10-16.

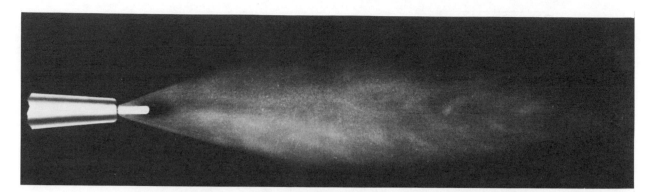

NEUTRAL FLAME — EQUAL AMOUNTS OF OXYGEN AND ACETYLENE

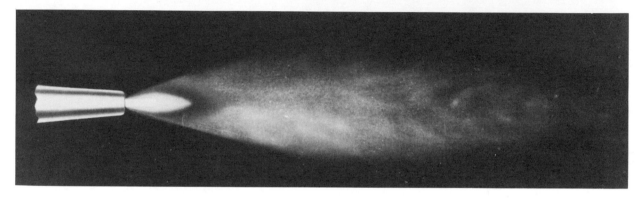

CARBURIZING FLAME — EXCESS ACETYLENE

Fig. 10-14. A neutral or carburizing flame is necessary for proper brazing.

Fig. 10-15. When the flux becomes liquid, shift the cone of the flame to the fitting.

9. Feed the brazing rod into the joint as the flame is moved back and forth between the pipe and the fitting, Fig. 10-17.
10. Allow the pipe and fitting to cool before moving or testing the joint.

## WELDING

Welding, which involves the melting of the parent material in order to form a bond, can be done on steel or plastic pipe and fittings. In the plumbing industry, welding is generally limited to repair work on thermoplastic pipe systems. The

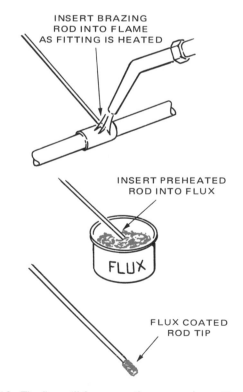

INSERT BRAZING ROD INTO FLAME AS FITTING IS HEATED

INSERT PREHEATED ROD INTO FLUX

FLUX

FLUX COATED ROD TIP

Fig. 10-16. The flux will form a coating on a preheated brazing rod.

surface or pipe to be welded is heated by an electrically operated welding unit which forces 500 F to 700 F (260 C to 371 C) air from a blowpipe nozzle.

Fig. 10-17. Feed the brazing rod into the joint as the flame is moved back and forth between the pipe and fitting. (Rigid Tool Co.)

## PROCEDURE

To understand how this tool is used to repair plastic pipe or fittings, study the following procedures for repairing a small fracture in a piece of polyethylene thermoplastic pipe:

1. Clean the welding surface to remove dirt, oil and loose particles. Use fine abrasive paper, detergent cleaner and a cloth.
2. Place the pipe on firebrick or other heat resistant material for welding, as in Fig. 10-18.
3. The welding unit must be capable of heating the surface to a weld temperature of 550 F (288 C). Position the welding filler rod at an angle of about 75 deg. to the weld surface. Weld one or two beads over the hole in the pipe.
4. Allow weld to cool completely before testing with water pressure.

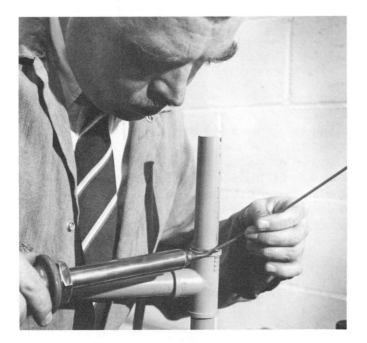

Fig. 10-18. This welder blows heated air to join thermoplastics. (Laramy Products Co., Inc.)

Electric arc welding is done on metal natural gas pipelines, pressure vessels, and storage tanks. The American Welding Society (A.W.S.) has rigorous welding performance tests and information for pipeline welders. Even though pipe, valves and joints are used, this type of special work is not usually done by a residential or commercial plumber.

For additional information on welding of metals refer to:

1. Walker, John R., ARC WELDING. Goodheart-Willcox Co. Inc., South Holland, IL.
2. Baird, Ronald J., OXYACETYLENE WELDING, Goodheart-Willcox Co., Inc., South Holland, IL.
3. Althouse, Andrew D., and others, MODERN WELDING. Goodheart-Willcox Co., Inc., South Holland, IL.

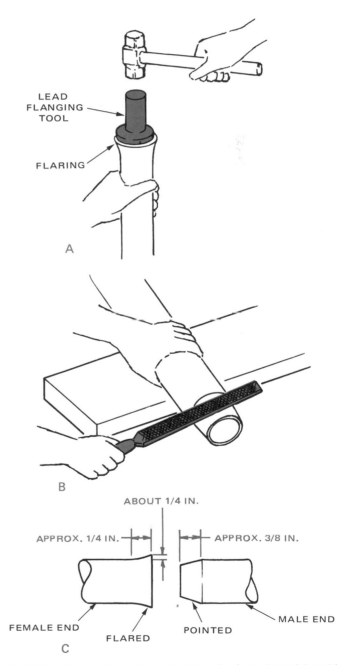

Fig. 10-19. Forming the male and female ends of a lead pipe joint with a lead flanging tool and a rasp. A—Flanging female end of pipe. B—Shaping the male end of the pipe. C—Final shape and dimensions of female and male ends.

## LEAD WIPING

Lead wiping is sometimes required on special jobs or where older plumbing installations need repair. At one time lead was used for DWV piping. It was considered an excellent plumbing material because of its corrosion resistance and its ability to be formed into almost any shape. However, copper and plastic have virtually replaced lead because of the ease with which they can be joined.

Solder for lead wiping is a 2 to 1 mix of lead and tin to which a small amount of black rosin has been added. This mixture will melt at 400 F (227 C).

Tin, being lighter than lead will float to the top in the molten state. Therefore, frequent stirring is necessary. The solder is hot enough to pour when it will char a wooden stick. If the stick ignites, it is too hot.

With experience the plumber is able to judge the correct temperature of the solder by its color or "bloom."

The following steps are basic for successful lead wiped joints:

1. Form the female and male ends of the lead pipe using a flanging tool and a rasp, Fig. 10-19.
2. Clean the pipe joint with chalk. Apply a coat of "soil" — this is a mixture of lamp black, glue and water — at locations where solder should not stick. See Fig. 10-20.
3. Support the lead pipes with bricks or blocks so your hand can go completely under the joint.
4. Prepare the lead in a furnace pot.
5. Using a ladle, pour a small amount of lead on the joint. Hold a moleskin wiping pad under the joint and roughly shape the lead as it is poured onto the joint, Fig. 10-21. A

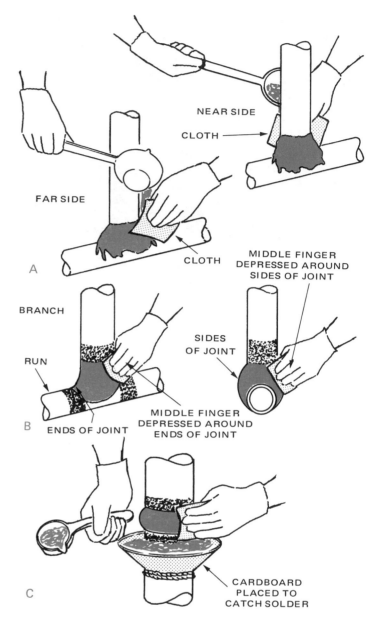

Fig. 10-21. Pour molten lead from a ladle and shape the lead around the joint with a moleskin pad. A—Using pad, shape lead roughly while pouring. B—Final shaping is done by wiping with the pad in smooth, even strokes. C—Cardboard apron catches lead falling from joint.

bulb-like joint is then formed by wiping while the solder is still in a plastic condition. The finished joint should be about 2 in. long, Fig. 10-22.

Considerable practice will be required to develop the skill necessary to consistently make good lead joints. The beginner should remember that the joint must be clean and the lead the correct temperature for good results.

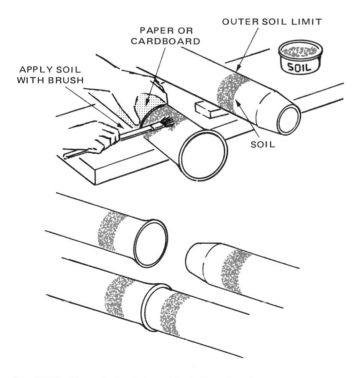

Fig. 10-20. Clean the lead pipe with chalk and apply "soil" at locations where solder is not wanted.

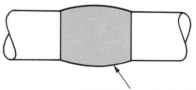

Fig. 10-22. The finished joint will be about 2 in. long and bulb shaped.

## TEST YOUR KNOWLEDGE — UNIT 10

1. Thermoplastic piping systems can be repaired by _____.
2. The operation which removes metal burrs from the inside of pipes is known as:
   a. Brazing.
   b. Chamfering.
   c. Deburring.
   d. Reaming.
   e. Routing.
3. Flux is a chemical which is used to prevent metal from oxidizing when it is being soldered or brazed. True or False?
4. The three conditions necessary for good solder joints are: clean _____, proper _____ and correct amount of _____.
5. List the four functions of flux.
6. The solder used for copper pipe and fittings is composed of:
   a. 50 percent tin and 50 percent lead.
   b. 100 percent tin.
   c. 100 percent lead.
   d. 30 percent tin and 70 percent lead.
7. _____ generally requires a lower temperature than brazing.
8. When heating joints that are to be soldered or brazed it is best:
   a. To heat the fitting first since it is heavier and requires more heat.
   b. To heat the pipe and fitting at the same time.
   c. To heat the pipe first since it dissipates heat faster than the fitting.
9. _____ requires a temperature high enough to melt the parent material.
10. For successful lead joints the pipe must be _____ and the molten lead at the correct _____.

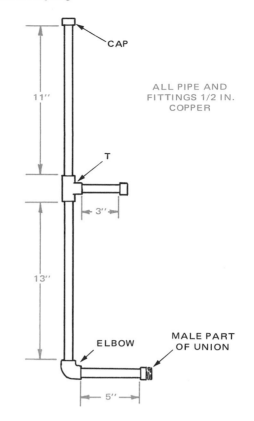

Fig. 10-23. Pipe soldering/brazing activity.

## SUGGESTED ACTIVITIES

1. Assemble the copper pipe and fittings shown in Fig. 10-23. When the assembly is completed, attach it to a water supply. Have your instructor test and inspect your work.
2. Make the same assembly described above from brass or bronze pipe and fittings. Have your instructor test and inspect your work.
3. Weld a break in a plastic pipe.

# Unit 11
# LEVELING INSTRUMENTS

## Objectives

This unit introduces two precision instruments used by plumbers to transfer heights accurately over long distances.

After studying it, you will be able to:
- Explain the operation of the builders' level and the cold beam laser.
- Explain and demonstrate the basic techniques for using these instruments to find levels and properly slope drainage pipe.

When the leveling job becomes too big for the level, straightedge, chalk line and square, the plumber must use a different kind of instrument for maintaining accuracy. Optical instruments are designed for long distance leveling. They work on the principle that a line of sight is always straight. It does not dip, sag or curve. If the line of sight is level, any point along the line will be the same height as any other point.

## BUILDERS' LEVEL OR TRANSIT

The SURVEYORS' LEVEL or BUILDERS' LEVEL as shown in Fig. 11-1, mounts on a tripod. It can be swung to the left or to the right 360 deg. but does not move from the horizontal position. It is used to check level and measure angles in the horizontal plane.

It is a very useful instrument for the plumber when installing sewer and septic tank lines. Since it can locate distant points that are level, it can also be used to measure the difference in elevations between two distant points.

A long graduated stick called a STADIA ROD is used along with the builders' level to find elevations. Its graduations are in feet, inches and fractions of inches. The stadia rod is rested in a vertical position on top of the spot where the elevation is to be measured. It has a marker which slides up or down to record the height sighted by the builders' level. See Fig. 11-2.

To see how this tool is used, consider the problem of installing a branch sewer line which connects a new house to the sewer main, Fig. 11-3. The branch line from the house, as

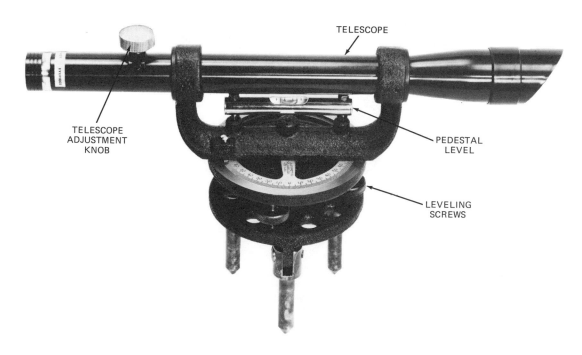

Fig. 11-1. Builders' level is a precision measuring instrument. It is used to establish heights of piping at high and low points.
(The L. S. Starrett Co.)

TELESCOPE

TELESCOPE ADJUSTMENT KNOB

PEDESTAL LEVEL

LEVELING SCREWS

# Leveling Instruments

Fig. 11-2. A stadia rod or a folding rule can be used to make measurements. Stadia rod marker provides a "target" for the transit.

rotated. Align the telescope over two opposing adjusting screws and move the screws in opposite directions until the bubble in the telescope level is centered. Rotate the telescope until it aligns with the second set of adjusting screws. Again, level the telescope. Repeat this adjusting procedure once more. Once the leveling operation is completed, extreme care must be taken not to disturb the instrument. Periodically check the pedestal level to insure that the instrument is reading correctly.

3. While one person holds the stadia rod or some other suitable measuring device, the person operating the builders' level adjusts the telescope adjusting knob until a

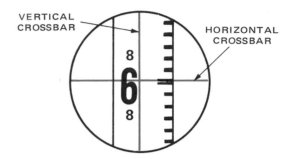

Fig. 11-4. Stadia rod as it looks through the telescope of the transit.

well as the branch tap into the sewer main, have been installed.

The procedure for laying out the trench for the branch sewer line follows:

1. Set up the tripod where it will be about the same distance from the tap and the branch. Stay 20 ft. or more away from the excavation for the branch line. Set the feet of the tripod firmly into the ground. Be careful not to touch or jar the legs.
2. Attach the builders' level to the tripod. Adjust the four leveling screws, Fig. 11-1, until the pedestal level indicates the pedestal is level in all positions to which the level is

reading, Fig. 11-4, can be made on the stadia rod. Note the cross hairs that appear in the telescopic site. *The vertical cross hair will help align the stadia rod. If it is not vertical, the measurement will be incorrect.* The horizontal cross hair indicates the point at which the reading should be made. In this case the reading is 8'-6". This is the elevation taken from the branch outlet at the house.

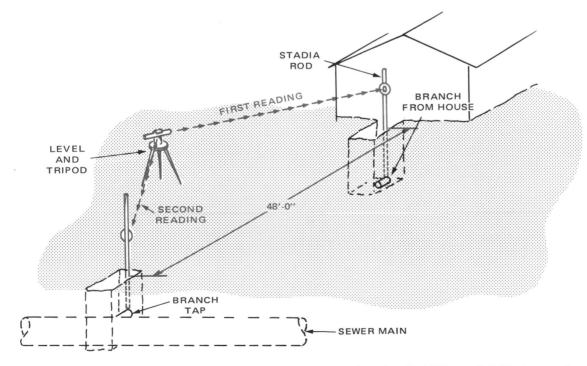

Fig. 11-3. First step in installing a branch sewer line is to sight through level to find difference in height between branch from building and sewer tap.

FALL = 1/2 IN. PER FT.

2 FT.

| | |
|---|---|
| SECOND READING | = 10'-6'' |
| FIRST READING | = 8'-6'' |
| DIFFERENCE IN ELEVATION | = 2'-0'' |

$$\frac{\text{DIFFERENCE IN ELEVATION}}{\text{LENGTH OF RUN}} = \text{RATE OF FALL}$$

$$\frac{2'\text{-}0''}{48'\text{-}0''} = \frac{1}{24} = 1/2 \text{ IN. FALL/FOOT OF RUN}$$

Fig. 11-5. Computing rate of fall on branch line using mathematics.

4. Repeat Step 3 with the stadia rod positioned at the base of the branch tap. This step produces a reading of 10'-6''.
5. Compute the rate of fall required in the branch sewer line as shown in Fig. 11-5.
6. As the trench is dug, check its depth by taking a reading on the stadia rod. For example, the reading from the bottom of the trench should be 8'-9'' at a distance of 6 ft. from the house (in 6 ft. of run, the trench should go down 3 in. or 8'-9'').
7. When the trench is completed, lay the pipe and check the fall with the builders' level and stadia rod. Use the same procedure as in Step 3. This will insure that the rate of fall in the branch line is uniform, reducing the likelihood of the line clogging from sediment.

The builders' level is useful anytime it is desirable to measure heights (elevations) or to transfer the measurements from one point to another. Care in setting up the instrument and in taking readings helps to produce accurate work.

## COLD BEAM LASER

A laser, Fig. 11-6, is an instrument that amplifies or strengthens light, projecting it as a thin beam. Plumbing contractors installing gravity flow pipelines can use this principle of alignment by laying pipe along this pencil thin beam of light.

Two principle types of lasers have been developed. The "hot beam" laser is usually operated for scientific research. The second type is termed a "cold laser" because it emits a

Fig. 11-6. "Cold beam" laser is used to establish slope in sewer pipe.

harmless beam of light with an output of .001 watt. This "working beam" can assist the contractor and crew when installing sanitary and storm drain pipelines.

To learn how this instrument is used, refer to Fig. 11-7. Use the following procedure for aligning pipe:

1. Mount a cold beam laser inside a manhole.
2. Connect the laser to a 12-volt storage battery.
3. Level to a grade stake and complete plumbing.
4. Use the laser to project a thin ray of red light through the inlet of the pipe or tile. See Fig. 11-7.

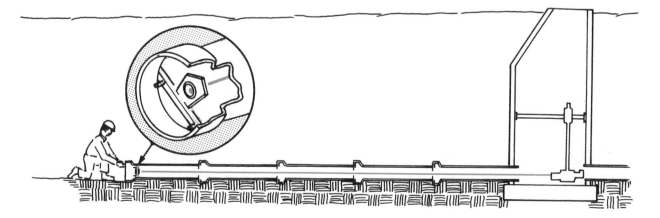

Fig. 11-7. Laser beam is projected through the sewer tile until it strikes target placed in the end of the pipe. This procedure is repeated as each section of pipe is laid.

5. Place a beam target in the pipe nearest the worker.

6. The target is self-aligned inside the pipe and is removed after each pipe is placed in position and aligned.

7. Pipes are thoroughly aligned when the laser beam produces a spot of light on the center of the target. The common working distance for a cold laser beam is 400 ft.

8. When the diameter of the pipe is small, mount the laser on a tripod, Fig. 11-8.

Cold beam lasers are rugged, compact instruments. Because of the low power output, they are not dangerous. However, never look directly into the working laser beam.

Fig. 11-8. Laser is mounted on a tripod when working with small diameter pipe.

## TEST YOUR KNOWLEDGE — UNIT 11

1. What is the name of the part of the builders' level which performs each of the following functions:
   a. Supports the level.
   b. Adjusts the pedestal so it is horizontal.
   c. Focuses the telescope.
   d. Helps the surveyor check the stadia and alignment.
   e. Indicates where the reading should be made.

2. If, in the problem discussed in the unit, the readings taken at the first position and the second position had been 9'-0" and 12'-4" respectively, what would the rate of fall have been if the sewer main was 160 ft. from the house?

3. Why is it important that a sewer line not change slope? (Select best answer.)
   a. Sediment tends to collect at points where slope changes.
   b. Less pipe is required to run a straight line.
   c. Pipe joints fit better.

4. Name the part or amount which completes the following statements about "cold beam" laser operation:
   a. Portable power supply for the laser.
   b. Placed inside each pipe or tile and is self-aligned.
   c. Common working distance of the laser beam.
   d. Supports laser when small pipe is being aligned.

## SUGGESTED ACTIVITIES

The following activities should be undertaken in the order indicated.

1. Working in groups of two or three, set up a level and make readings at a series of points. Record the readings on a chart. Compute the difference in elevation of each of these points from the predetermined elevation.

2. Transfer a known elevation from one point to another. Using the known elevation (bench mark) as the reference point, transfer this elevation to a stake driven at least 30'-0" from the bench mark.

3. Given two points of known elevation, one near a structure and one near a sewer main:
   a. Determine the rate of fall of a branch sewer line between the two points.
   b. Insert a row of stakes 2 ft. from the proposed excavation and spaced at 4 ft. intervals. The tops of all stakes should be 5 ft. (or any distance your instructor may specify) above the bottom of the trench to be dug.

# Unit 12
# RIGGING AND HOISTING

## Objectives

This unit reviews mechanical means and tools that may be employed to lift or hold heavy piping during installation.

After studying the unit you will be able to:
- Describe the kinds of rope suitable for use in mechanical lifting devices.
- List other tools for hoisting and describe how they are used.
- Demonstrate the methods of securing rope to piping.
- Describe different types of ladders and how to use them safely.

Plumbers engaged in residential and commercial plumbing work may need to raise or lower heavy pieces of pipe and secure them temporarily while pipe hangers are installed and fittings connected. Vertical stands of pipe may need temporary support while joints are being completed. Such lifting and supporting can be safely done with the use of the right equipment.

## ROPES

Ropes made of either natural or synthetic fiber may be used for hoisting. Natural fiber ropes are used extensively in construction work. They will vary in strength depending upon the quality of the fibers used in their manufacture.

MANILA ROPE, made from the fiber of a wild banana plant called abaca, makes the strongest natural fiber rope. Manila rope is sold in several grades which differ in strength and appearance, Fig. 12-1.

AMERICAN HEMP ROPE, is close to manila rope in appearance and is about 80 percent as strong as No. 1 manila. SISAL ROPE, is about 60 percent as strong, while cotton and jute rope are only 50 percent as strong.

Rope is made from manufactured fibers such as nylon, rayon, Dacron and glass. These fibers are stronger than hemp. They also last longer because they resist rot and deterioration.

## HANDLING ROPES

Ropes are made from individual fibers which have been spun together much like string or yarn. Fibers can be damaged through improper use. Natural fiber ropes should be kept dry because moisture hastens their decay. A wet rope should be hung loosely in an area where it can dry before it is used.

Avoid running rope over sharp edges. It causes fibers to break and will eventually destroy the strength of the rope. Dragging a rope over concrete, gravel and other rough surfaces will wear away the fibers and reduce the strength.

A frozen rope should not be used until it is thawed. Frozen fibers tend to break as the rope is flexed.

| GRADE | DESCRIPTION |
|-------|-------------|
| YACHT | HIGHEST QUALITY STRONGEST VERY SMOOTH APPEARANCE |
| BOLT | 10 – 15 PERCENT STRONGER THAN NO. 1 |
| NO. 1 | STANDARD HIGH GRADE VERY LIGHT IN COLOR |
| NO. 2 | SAME INITIAL STRENGTH AS NO. 1 LOSES STRENGTH MORE RAPIDLY |
| NO. 3 | ABOUT THE SAME INITIAL STRENGTH AS NO. 1 LOSES STRENGTH VERY RAPIDLY |

Fig. 12-1. Grades of manila rope.

## INSPECTING ROPES

Inspect ropes frequently for damage. Surface inspection will reveal broken or worn strands. For interior inspection, twist the rope in a direction opposite to the way it was spun. This will open up and separate the strands so that interior fibers can be examined.

Evidence of powder between the strands generally indicates excessive wear. Open one or more of the internal strands to determine the extent to which the fibers have been broken. If damage is extensive, it may be necessary to destroy the rope. Certainly damaged rope should not be used for hoisting loads.

The danger of a rope breaking under a load is much greater when the load is jerked or when obstructions temporarily stop the load from rising. This places excessive strain on the rope.

## KNOTS

The proper knot correctly tied can make the difference between a load being secure and a load that may fall causing

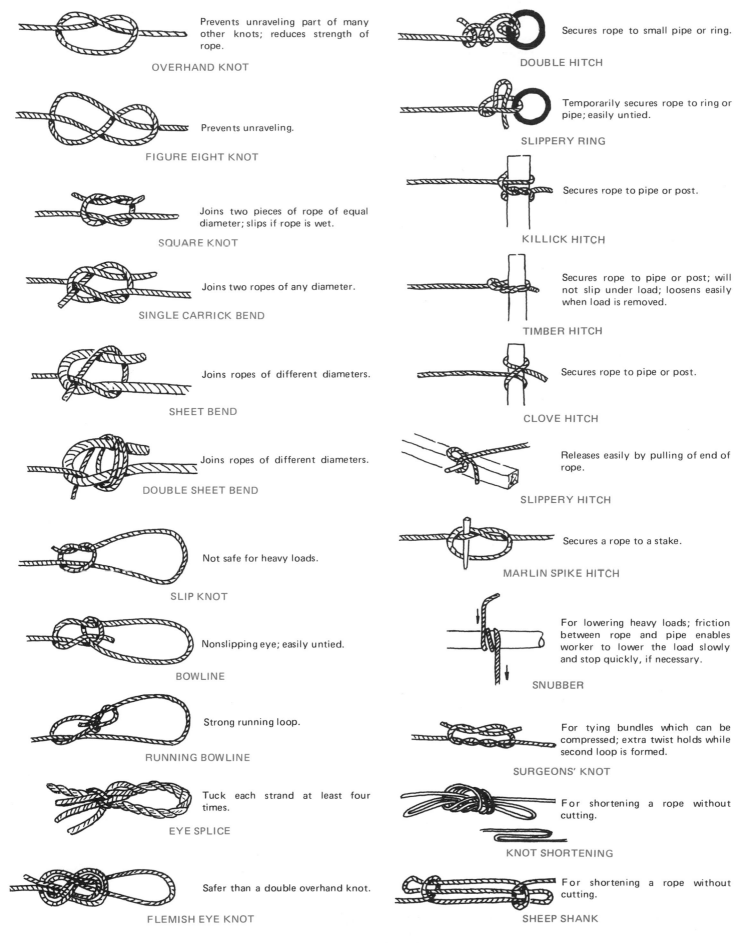

**OVERHAND KNOT** — Prevents unraveling part of many other knots; reduces strength of rope.

**FIGURE EIGHT KNOT** — Prevents unraveling.

**SQUARE KNOT** — Joins two pieces of rope of equal diameter; slips if rope is wet.

**SINGLE CARRICK BEND** — Joins two ropes of any diameter.

**SHEET BEND** — Joins ropes of different diameters.

**DOUBLE SHEET BEND** — Joins ropes of different diameters.

**SLIP KNOT** — Not safe for heavy loads.

**BOWLINE** — Nonslipping eye; easily untied.

**RUNNING BOWLINE** — Strong running loop.

**EYE SPLICE** — Tuck each strand at least four times.

**FLEMISH EYE KNOT** — Safer than a double overhand knot.

**DOUBLE HITCH** — Secures rope to small pipe or ring.

**SLIPPERY RING** — Temporarily secures rope to ring or pipe; easily untied.

**KILLICK HITCH** — Secures rope to pipe or post.

**TIMBER HITCH** — Secures rope to pipe or post; will not slip under load; loosens easily when load is removed.

**CLOVE HITCH** — Secures rope to pipe or post.

**SLIPPERY HITCH** — Releases easily by pulling of end of rope.

**MARLIN SPIKE HITCH** — Secures a rope to a stake.

**SNUBBER** — For lowering heavy loads; friction between rope and pipe enables worker to lower the load slowly and stop quickly, if necessary.

**SURGEONS' KNOT** — For tying bundles which can be compressed; extra twist holds while second loop is formed.

**KNOT SHORTENING** — For shortening a rope without cutting.

**SHEEP SHANK** — For shortening a rope without cutting.

Fig. 12-2. Knots commonly used to secure ropes.

injury and damage. The more common knots are shown in Fig. 12-2. Read descriptions and warnings carefully. Select only knots which suit the purpose.

## HOISTING DEVICES

A GIN BLOCK, Fig. 12-3, is a type of pulley. It provides leverage and allows loads to be lifted by an operator standing

Fig. 12-3. A gin block can be used to raise loads. A heavy rope is fed through the pulley.

on the ground or other solid footing. It can be hung off the side of scaffolding or from the building's frame. Simple, rapid operation is another of its advantages.

A LEVER HOIST, Fig. 12-4, will lift or pull much heavier loads. Because of its lever and ratchet mechanism, it moves the load more slowly than the gin block. It is suitable where heavy

Fig. 12-4. A lever hoist will lift or pull heavier loads for short distances.

loads need to be moved short distances. Most lever hoists will lift loads from 6 to 12 ft. The dogs on the ratchet must be kept in good condition to prevent slippage.

The portable WINCH CRANE, Fig. 12-5, is convenient in large buildings where large pipes must be lifted to considerable heights.

Fig. 12-5. The winch crane is handy when heavy pipe must be lifted long distances in large buildings.

## SECURING PIPES FOR HOISTING

Use great care in tying ropes to piping in preparation for hoisting. A rope wrapped twice around the pipe before tying will hold the pipe more securely. A single wrap is dangerous and may allow the pipe to slide out of the loop as it is being hoisted. If possible, tie the rope to the pipe in several places. See Fig. 12-6. Knot the rope with two half hitches and snug the knot close to the pipe.

## LADDERS

Plumbers will not often use ladders but they are needed for gaining access to roofs to complete the installation of the venting system. Less frequently, a ladder may be needed inside the building to reach points where a gin block will be attached or to get closer to certain tasks.

## LADDER CONSTRUCTION

Ladders are made of either wood or metal. Side rails of wood ladders are generally made of Douglas fir, spruce, fir or Norway pine. Rungs are second growth white ash, hickory or white oak. Whatever lumber is used, it should be sound, straight-grained, well seasoned and free of decay or knots.

Never paint a wooden ladder. Paint conceals defects and may lead to serious injury when wood members give way.

To preserve the wood, coat the ladder with spar varnish or a good clear all-weather sealer. This leaves the grain structure visible for inspection. Linseed oil is also a good perservative but adds to the weight of the ladder. Replace wood ladders which develop cracks or rot.

Other materials used for ladders include: aluminum, magnesium alloy, iron and steel. Aluminum and magnesium are preferred for plumbers because they are lighter weight than wood, iron or steel. Thus, they are more easily moved by one person.

## LADDER TYPES

Ladder types likely to be used by the plumber include:

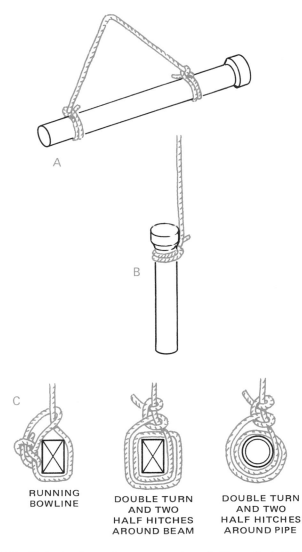

RUNNING
BOWLINE

DOUBLE TURN
AND TWO
HALF HITCHES
AROUND BEAM

DOUBLE TURN
AND TWO
HALF HITCHES
AROUND PIPE

Fig. 12-6. Methods of securing rope to pipe. A—If practical, tie the rope to both ends of the pipe. B—Rope wrapped twice around the pipe will hold the pipe more securely. C—Two methods of securing rope to pipe or beam. Knots on either hitches can easily be loosened when the rope is to be removed.

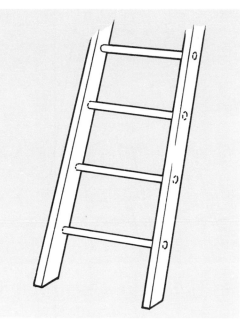

Fig. 12-7. Single ladder has only one section.

1. Single ladder.
2. Extension ladder.
3. Stepladder.

A single ladder, Fig. 12-7, has one straight section. Its length is determined by the length of the side rails. Single ladders over 30 ft. long are not considered safe.

An extension ladder, Fig. 12-8, has two or more sections which can be extended to adjust the length. The ladder size is determined by adding together the length of all its sections.

However, an extension ladder cannot be extended its total length. It is important to have sufficient overlap for safety. Two-section ladders measuring up to 38 ft. should have at least 3 ft. of overlap. Four feet of overlap is needed for 40 to 44 ft. ladders and a 5 ft. overlap is required for ladders 44 to 46 ft. Three-section ladders must overlap 4 ft. between each section. Two and three-section ladders must not exceed 46 ft. fully extended.

A stepladder is a self-supporting nonadjustable ladder, Fig. 12-9. It has flat steps and a hinged back which folds against the steps during transport. A rack near the top step will hold

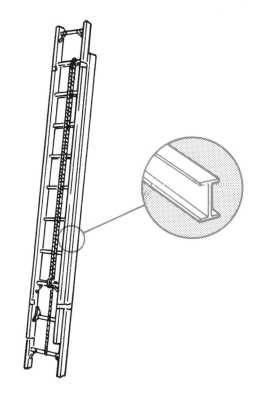

Fig. 12-8. Extension ladder has two or more sections. A rope to assist in extending the sections is desirable.

tools. Quality stepladders have steel spreaders which will not injure the hand while opening and closing the ladder. The legs should be sturdy and well braced. Sizes range from 4 ft. to 12 ft.

## TEST YOUR KNOWLEDGE — UNIT 12

1. The strongest type of natural fiber rope is _____.
2. No. 3 manila rope will retain its strength longer than No.

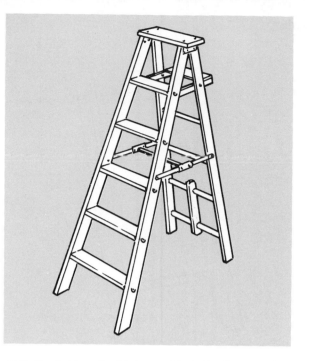

Fig. 12-9. Stepladders should be sturdily built. This one has metal struts beneath each step to keep rails from spreading.

1 manila rope. True or False?

3. Why is cotton fiber rope seldom used for hoisting work?

4. What precautions should be practiced when taking care of rope?

5. If the outside surface of a rope is sound, the rope is strong. True or False?

6. Name two knots which can be used to join two ropes of the same size.

7. Name three ways of making a loop on the end of a rope.

8. Name three knots that can be used to attach a rope to a pipe or post.

9. A _____ will lift or pull a heavier load but operates more slowly than a _____ block.

10. A rope wrapped _____ around the pipe to be lifted will hold the pipe more securely.

11. Name at least two materials used in constructing ladders.

## SUGGESTED ACTIVITIES

1. Practice tying 10 or more of the knots shown in Fig. 12-2 as directed by your instructor.

2. Attach a lever hoist to an overhead structural member and lift a length of cast-iron pipe. Make certain that the pipe is held at the proper slope to be installed as a part of a DWV system.

3. Attach a gin block to the side of a scaffold and practice using it to raise a variety of small parts.

# Unit 13
# BUILDING AND PLUMBING CODES

## Objectives

This unit outlines the functions and content of a building and plumbing code.

After studying it you will be able to:
- Explain the purpose of zoning laws and building codes.
- Cite typical examples of how codes are administered and enforced.
- List the specific points that a plumbing code should cover.
- Understand and apply code requirements to a plumbing installation.

Governmental units such as cities, counties or states make and enforce laws which specify the minimum standard for buildings erected within their jurisdiction. The purpose of these laws is to provide for the health and safety of the people who occupy these structures.

Laws regulating the type of structure which can be built in a given area are known as ZONING LAWS. In general, these laws serve to separate residential, office, light industrial and heavy industrial activities.

Building codes control such things as:
1. Quality of materials.
2. Loads the structural system must be able to support.
3. Number and type of exits required.
4. Quality of work required.

Plumbing and electrical codes are frequently separated from the building code. This is logical because the people who install and inspect these systems must be specially trained. In many cases, they are also licensed by the governmental unit to work in their jurisdiction.

## ADMINISTRATION OF CODES

Because codes are adopted and applied locally, it is generally impossible to describe precisely how codes are universally administered. However, some generalizations can be made concerning code enforcement. When plumbing, the individual must become thoroughly familiar with the codes applying to his or her specific area.

Typically, the building officials (titles will vary, but could include city engineer, building inspector or building commis-

sioner) have the responsibility to administer the codes established by the local government. In some cases, they are certified to administer state building codes if they exist. The chief building official will generally have a staff to study plans, inspect buildings and maintain records for all buildings under construction.

Frequently, the inspectors are required to obtain a license. This verifies that they are qualified to perform their work.

A brief review of the process used by the building officials will demonstrate the need for good planning and careful work:
1. A contractor will seek a permit to erect a certain kind of building on a certain plot of ground. To obtain a building permit, the contractor must submit two copies of plans and specifications to the building inspector along with an application for a building permit, a plumbing permit, Fig. 13-1, electrical permit, heating and ventilating permit and other permits as may be required.
2. The plans and specifications will be reviewed to determine if they meet the minimum standards specified by the code. If changes must be made, they will be noted on the plans by the building officials.
3. Assuming that changes are minimal, the plans will be approved and a permit issued. Along with the permits will be directions on the appropriate time to request inspection.
4. During the progress of the work, periodic inspections will be made. In the case of residential plumbing, this is done after roughing in is completed and before any pipes are covered. A final inspection is made at the completion of the job. Also, the connection of the house drain to the sewer will require inspection before it is covered.
5. The inspector will attach an approval or rejection notice, Fig. 13-2, to the building permit as may be warranted by the work done. If the work is rejected, the changes will need to be made before the inspector is asked to reinspect the work.
6. When the building has passed final inspection, it may be occupied.

## CODE ENFORCEMENT

In cases where the building officials have difficulty obtaining the quality of work they believe necessary, they may turn the case over to the prosecuting attorney. Court action can be initiated to stop work on the job until necessary

## Application for Plumbing Permit

PERMIT NO. _____

DATE _____

TO THE BUILDING INSPECTOR:          RECEIPT _____

The undersigned hereby makes application for a Plumbing Permit, according to the following specifications:

Name of Plumber _____

Name of Owner _____ Address _____

Location, House No. _____

Between _____ and _____

Kind of Building _____ Kind of Floor _____

Number Waterclosets _____

    "   Baths _____

    "   Washstands _____

    "   Kitchen Sinks _____

    "   Slop Sinks _____

    "   Laundry Trays _____

    "   Shower Baths _____

    "   Urinals _____

    "   Soda Fountain Wastes _____

    "   Fountain Cuspidor Wastes _____

    "   Refrigeration Wastes _____

    "   Drinking Fountains _____

    "   Cellar Drains _____

    "   Hot Water Installation _____

_____

_____

_____

Estimate Cost of Plumbing $ _____ Fee $ _____

Minimum Fee _____

In consideration of permission given _____ do hereby covenant and agree to construct said work in all respects in compliance with the Laws of the State of _____ and with the ordinances of the City of _____ _____ Code relating to Plumbing.

      Plumber _____

      Address _____

      Company _____

Fig. 13-1. Typical plumbing permit application. Fee paid to city depends on number of fixtures.

CITY OF _____

BUILDING DEPARTMENT

The Following Has Been Inspected and

# REJECTED

_____
_____
_____
_____
_____
_____
_____
_____
_____

Date_____ Inspector_____

CITY OF _____

BUILDING DEPARTMENT

The Following Has Been Inspected and

# APPROVED

_____
_____
_____
_____
_____
_____
_____
_____
_____

Date_____ Inspector_____

Fig. 13-2. Inspection stickers. Different colors are often used for rejection and approval.

changes are made. Frequently, it is impossible to obtain a connection to the water main until the plumbing has passed inspection.

## MODEL CODES

Today, each local government is permitted to adopt and to administer its own plumbing code. The advantage is that a community can have local control of its code. The disadvantage is confusion. Since each community in a large metropolitan area can have its own code, plumbing which is satisfactory in one community may not meet the requirements of an adjoining community. Another problem is that developing an adequate plumbing code is both difficult and expensive. Also, many rural areas have not adopted minimum standards for plumbing in their area. Plumbers, plumbing contractors, architects and suppliers are frequently confused by such inconsistencies.

In an effort to standardize plumbing codes, several model

| CODE | DATE |
|------|------|
| HOOVER CODE | 1928 |
| STANDARD PLUMBING CODE | 1933 |
| PLUMBING MANUAL | 1936 |
| UNIFORM PLUMBING CODE | 1938 |
| EMERGENCY PLUMBING STANDARD | 1941 |
| NATIONAL PLUMBING CODE | 1955 |

Fig. 13-3. Model code development dates back to 1928.

codes, Fig. 13-3, have been developed. These codes provide governmental units with carefully prepared, scientifically verified and workable standards which can be adopted or adapted to meet local needs. Because many cities have seen the advantage of adopting one of these codes, the inconsistencies in plumbing codes have been reduced in many areas.

These codes are not mandatory. They are models which a local government may use or ignore.

## CONTENT OF PLUMBING CODES

A well written plumbing code will generally contain the following types of information:

1. Assumptions or general principles upon which the specifics of the codes are based.
2. Definition of terms used in the code.
3. General regulations (type of structure to which the code applies, slope of drainage piping, quality of work, depth of building drains, etc.).
4. Quality and size limits on materials.
5. Requirements for joints.
6. Location of traps and cleanouts.
7. Plumbing fixture requirements.
8. Design requirements for water supply lines.
9. Design requirements for drainage system.
10. Design requirements for vents.
11. Requirements for storm drains.
12. Procedure for specific inspection and tests.

From this list, it is evident that design, material selection and installation practices are all controlled. It is the responsibility of the plumber to do the work in compliance with the code. Therefore, a thorough study of the existing code in the local area is absolutely essential before layout of the plumbing system is begun.

## TEST YOUR KNOWLEDGE — UNIT 13

1. Laws which regulate the type of structure which can be built in a given area are known as _____ laws.
2. Laws which control quality of materials, structural system design, etc., are known as _____ _____
3. The first inspection is generally made of a residential plumbing job:
   a. At the completion of the rough-in.
   b. After the drywalling is done.
   c. When the inspector has time.
   d. When the building permit is issued.

4. In cases where the building officials cannot agree with the plumber about how a job should be done, the officials have the authority to turn the case over to the _____ _____ who will take court action to stop work on the job.
5. The National Plumbing Code applies to all buildings built in the United States. True or False?
6. It is necessary for a plumber to be thoroughly familiar with the plumbing code in his or her area. True or False?

## SUGGESTED ACTIVITIES

1. Obtain a copy of the plumbing code for your area and identify the basic areas of work that it covers.
2. Ask a plumbing inspector to describe how the local plumbing code is administered. Arranging for the inspector to be a guest speaker in class may be most helpful.

Replica of early toilet design. Note gravity flush tank. (Heads Up, Inc.)

Environment, the newest bathroom fixture, provides a climate controlled atmosphere for the user. Heat, sun, rain and wind can all be preprogrammed to provide the desired cycle for relaxation and revitalization of the user.

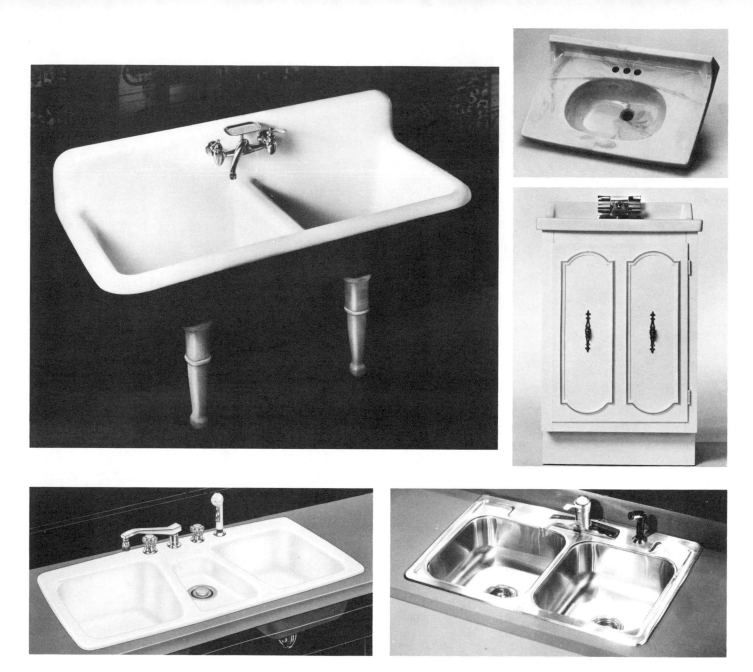

Fig. 14-1. There are a variety of shapes and sizes to choose from in selecting lavatories and sinks.

Fig. 14-2. Ledge type lavatory is hung on wall brackets.

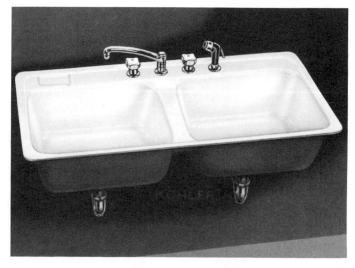

Fig. 14-3. Self-rimming sink rests on top of counter. (Kohler Co.)

# Unit 14

# PLUMBING FIXTURES

## Objectives

This unit describes many different fixtures designed for residential and small commercial buildings. The unit details procedures for proper installation.

After studying the unit you will be able to:
- Recognize the various types of fixtures and give them their correct name.
- Describe and demonstrate proper installation procedures for each fixture.
- Describe and demonstrate correct procedure for making water supply and drainage connections for each fixture.

Plumbing fixtures described in this unit include lavatories, sinks, bathtubs, shower stalls, water closets, bidets, urinals and water fountains. First, we shall take a look at the materials commonly used in the manufacture of the various fixtures. Then we will discuss each type of fixture and how it is installed. However, it should be understood that each fixture may have some unique characteristic. The plumber should study the manufacturer's instructions before attempting to install any fixture.

## MATERIALS USED IN FIXTURES

Plumbing fixtures may be made of a single material or combinations of materials. Common materials or combinations include:

1. Vitreous china (vitrified porcelain).
2. Steel coated with porcelain enamel.
3. Cast iron coated with porcelain enamel.
4. Stainless steel.
5. Acrylic plastic.
6. Fiber glass reinforced with plastic.

PORCELAIN, whether used alone or as a coating for steel or cast iron, produces a sanitary, easily cleaned surface. It is manufactured from a combination of materials that may vary in composition. Essentially, it is a mixture of a fine clay (called kaolin), quartz, feldspar and silica.

The most expensive fixtures are molded vitrified porcelain, also called china or vitreous china. The complete vitrification requires a temperature of 2600 F (1426 C) to fuse the mixture of materials.

Porcelain enamel is bonded to steel or cast-iron fixtures by fusion at a temperature above 800 F (427 C). Porcelain enamel is sometimes referred to as "glass lining" or VITREOUS ENAMEL.

Alloy sheet steels for plumbing fixtures such as lavatories, sinks and bathtubs are formed economically by the stamping process. Such fixtures are generally less expensive and less durable than those made of cast iron.

Gray iron is ideally suited for plumbing fixtures. It is low in cost and can be cast in a wide variety of shapes. Plumbing fixtures such as bathtubs, sinks and lavatories are frequently cast from it. Porcelain enamels are fused to the gray iron to provide a glossy, decorative, protective and sanitary coating.

Stainless steel is durable and has a good surface finish, two characteristics necessary for plumbing fixtures. The nickel-steel alloy has a silver, satinlike finish. No additional coating is required to produce a sanitary, easily cleaned surface. Unlike enameled surfaces, stainless steel does not chip. However, because of the difficulty of forming stainless steel, it is used only to make simpler fixtures such as kitchen sinks.

Plastics are the most recent material to be used for plumbing fixtures. Versatile and relatively low cost, they can be shaped to form nearly any fixture. One-piece construction, including fixtures and adjoining walls for bathtubs and showers, is possible using fiber glass reinforced plastic. Acrylic plastic sheets with marbleized colors are being used in lavatories. These fixtures are attractive and superior to marble fixtures because the plastic does not absorb water.

## LAVATORIES AND SINKS

LAVATORIES are designed for installation in bathrooms and other locations for washing hands and face. SINKS are designed to be used for food preparation and dishwashing. Because of the differences in their uses, the sizes and shapes of lavatories differ from sinks. See Fig. 14-1. However, they require similar installation procedures.

Lavatories or sinks fall into three basic types:

1. Ledge type, Fig. 14-2.
2. Self-rimming type, Fig. 14-3.
3. Built-in with metal rim, Fig. 14-4.

Each is available in a variety of shapes and styles. For example, a corner lavatory, Fig. 14-5, may be desirable where space is limited. Some lavatories are designed with a single supporting leg, Fig. 14-6.

Fig. 14-4. Built-in lavatory has metal rim. The rim supports the bowl in the countertop.

Fig. 14-5. Corner lavatory is a space saver in small bathrooms. (Eljer Plumbingware Div., Wallace-Murray Corp.)

Fig. 14-6. Lavatory is supported by vitreous china leg.

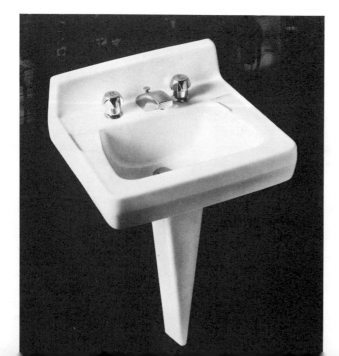

## INSTALLATION OF LAVATORIES AND SINKS

Ledge type lavatories and sinks are hung from the wall. A bracket is anchored to the studs as shown in Fig. 14-7. Position and install the bracket so that the top of the lavatory or sink is 30 or 31 in. above the floor line.

If possible, install faucets before installing the sink. This way it is much easier to reach the nuts on the underside. After installation, there will be little room between the bowl and rear of cabinet to manipulate tools. Follow procedure described later in this unit.

Self-rimming and metal rimmed lavatories and sinks are placed in countertops and vanities. These steps should be followed to install them:

1. Locate, mark, and cut the correct opening in the countertop, Fig. 14-8.
2. Place the lavatory or sink into the opening and check for proper fit.
3. Remove the lavatory or sink and correct any error in the opening size.
4. Apply mastic to the underside of the rim in self-rimming units. If unit is the rim type, put putty under the rim.
5. Replace the sink or lavatory, assemble and tighten the

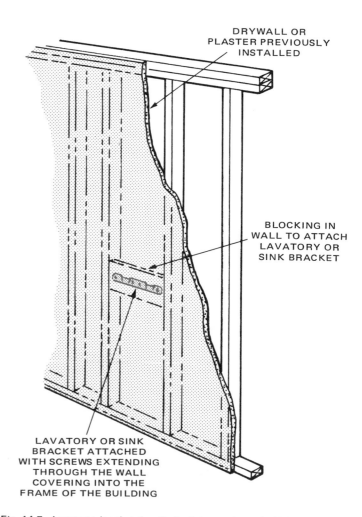

DRYWALL OR PLASTER PREVIOUSLY INSTALLED

BLOCKING IN WALL TO ATTACH LAVATORY OR SINK BRACKET

LAVATORY OR SINK BRACKET ATTACHED WITH SCREWS EXTENDING THROUGH THE WALL COVERING INTO THE FRAME OF THE BUILDING

Fig. 14-7. Lavatory bracket is attached to studs to hold ledge type lavatory or sink.

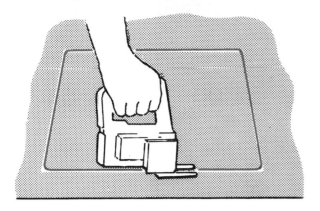

Fig. 14-8. Cut out sink opening using hole saw at corners and a saber saw or portable circular saw for straight cutting.

hold-down bolts and/or metal lugs. See Fig. 14-9 for installation of metal rimmed lavatory.

6. Remove excess mastic or putty.

## Installing drainage fittings

After placing the bowl, install the sink or lavatory drain. Fig. 14-10 illustrates the parts and their location.

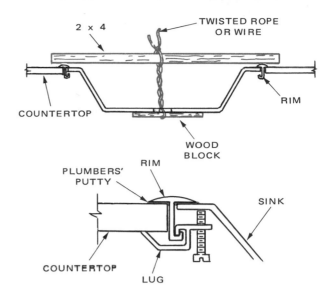

Fig. 14-9. Rim type sinks and lavatories are held in place by a rim and lugs. The rim fits around all four sides of the sink. Flange at top of rim conceals sawed edges of countertop and carries weight of sink. Top. Cutaway shows rim and sink in place being held temporarily with blocks and wire. Bottom. Lug hooks onto rim and bolt presses sink tightly against the rim.

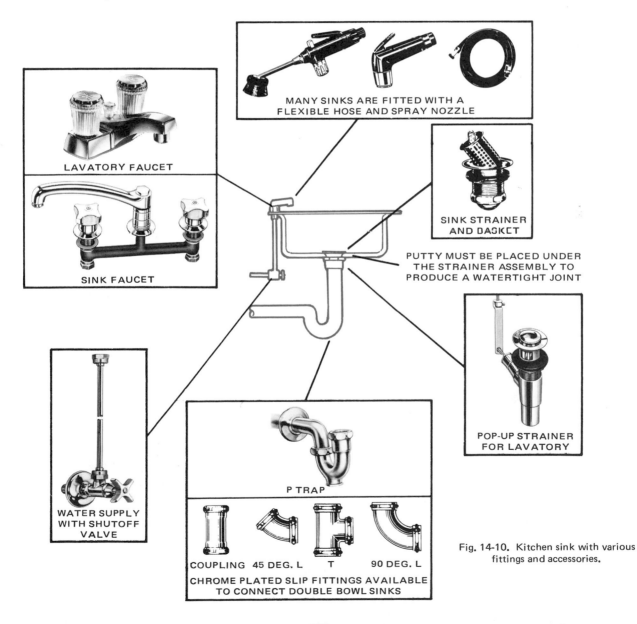

Fig. 14-10. Kitchen sink with various fittings and accessories.

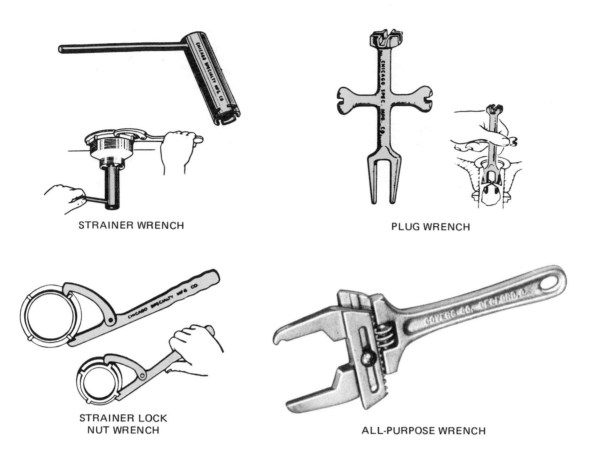

STRAINER WRENCH

PLUG WRENCH

STRAINER LOCK
NUT WRENCH

ALL-PURPOSE WRENCH

Fig. 14-11. These wrenches are commonly used to install strainers in lavatories and sinks.

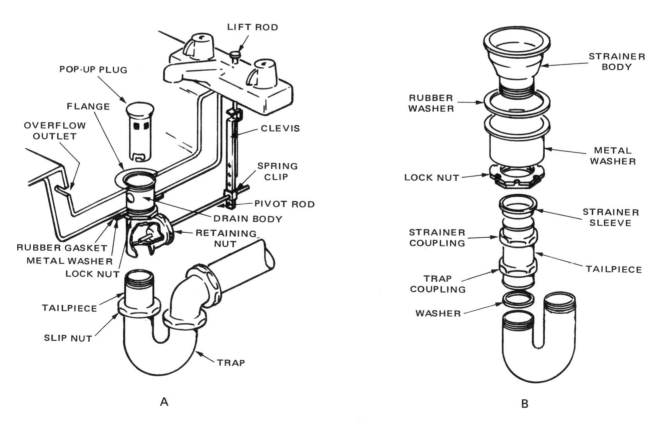

LIFT ROD

POP-UP PLUG

FLANGE

OVERFLOW
OUTLET

CLEVIS

SPRING
CLIP

PIVOT ROD

DRAIN BODY

RETAINING
NUT

RUBBER GASKET
METAL WASHER
LOCK NUT

TAILPIECE

SLIP NUT

TRAP

A

STRAINER
BODY

RUBBER
WASHER

METAL
WASHER

LOCK NUT

STRAINER
SLEEVE

STRAINER
COUPLING

TAILPIECE

TRAP
COUPLING

WASHER

B

Fig. 14-12. Details of complete assembly of drainage fittings.

Place a bead of putty at least 1/8 in. thick around the underside of the rim of the basket and strainer to form a watertight seal. Insert the strainer in the sink. Install strainer washer and any other washers furnished. Attach nut and tighten the assembly using one or more of the special tools pictured in Fig. 14-11. See Fig. 14-12 for proper order of assembling drainage fittings.

The strainer outlet is connected to the DWV piping with a tailpiece and a 1 1/4 in. or 1 1/2 in. diameter chrome plated P trap. Slip a rubber washer, a strainer slip nut, a trap slip nut and another rubber washer onto the tailpiece in that order. Slide the plain end of the tailpiece inside the P trap and push it down until the top end of the tailpiece will clear the end of the strainer. Lift it snugly against the strainer sleeve and attach the slip nut. Now attach and tighten the lower slip nut onto the P trap. Tighten all slip nuts by hand until snug. Then tighten them with a wrench that will not mar the chromed surfaces.

## Installing faucets

If they are not already in place, install faucets using a basin wrench, Fig. 14-13, to tighten the nuts.

Special 3/8 in. chrome plated, flexible copper supply tubes are used to connect the hot and cold water supply to the faucet. See Fig. 14-14 for installation. It is advisable to install shutoff valves in the water supply lines. When maintenance and repair work is needed on the sink or lavatory, it will not be necessary to shut off the entire water supply.

Kitchen sinks are frequently fitted with garbage disposals, as shown in Fig. 14-15, to assist in disposing of waste from food preparation. When a garbage disposal is desired, it is connected directly to the sink replacing the standard basket and strainer. The outlet of the garbage disposal is connected directly to the DWV piping with slip-joint pipe and fittings. The only other connection required is the electrical hookup for the electric motor.

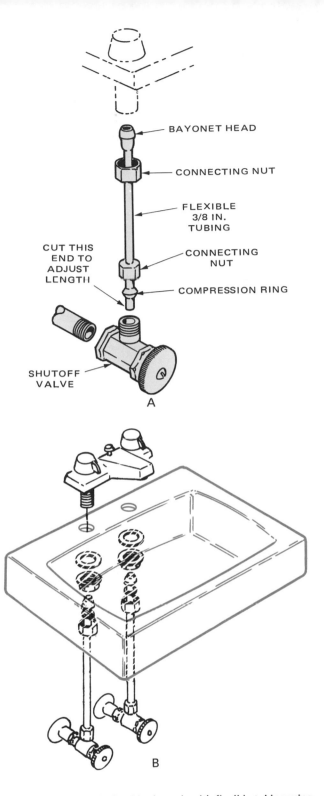

Fig. 14-14. Water supply hookup is made with flexible tubing using compression fittings. A—Cut tubing to length at bottom. Install connecting nut and compression ring. Ring will seal connection when nut is tightened. B—Exploded view shows typical connection for water supply.

## BATHTUBS

Common materials for bathtubs include:
1. Enameled cast iron.
2. Enameled steel.
3. Fiber glass reinforced plastic.

The exterior surface of a bathtub must be durable to

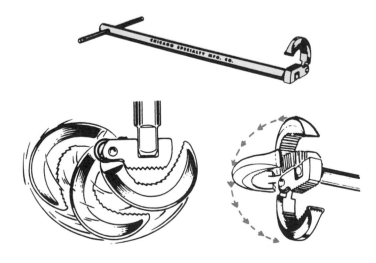

Fig. 14-13. Basin wrench makes tightening nuts on faucets easy. It is designed to work in cramped quarters. One jaw is fixed while the other swings freely. Fixed jaw is pinned to handle so it can swing 180 deg.

Fig. 14-15. Garbage disposals are designed to fit on drains of kitchen sinks. A—Mouth attaches to sink drain. B—Cutter. C—Motor for driving cutter. (A.O. Smith Corp.)

withstand frequent cleaning. A slope of 1/8 in. per foot to the drain outlet will provide proper sanitation and reduce maintenance.

Bathtubs are available in various sizes. Standard lengths are 4 1/2 ft., 5 ft. or 5 1/2 ft. Tubs in larger sizes are also available. A special safety feature is the nonslip or slip resistant bottom shown by the textured circles in the square tub in Fig. 14-16.

Fig. 14-16. Slip resistant materials in the bathtub reduce danger of falls. (Eljer Plumbingware Div., Wallace-Murray Corp.)

The bathing module, Fig. 14-17, features one piece of fiber glass reinforced plastic bath and shower unit. In this particular unit, walls 75 in. high eliminate the need for conventional wall materials such as ceramic tile. This has the added advantage of reducing the cleaning problems after installation since all joints are eliminated. Because of its size, installation is usually limited to new homes.

Another bathtub, shower and wall treatment combination is the four section cove, Fig. 14-18. The sections can be installed in an existing bathroom as part of a remodeling project.

Fig. 14-17. One-piece bath and shower unit is made of fiber glass reinforced plastic. Its use is restricted to new construction because of its size. (Kohler Co.)

Fig. 14-18. This bath and shower unit is made in four small sections so it can be used in a remodeling project. (Owens-Corning Fiberglas Corp.)

## INSTALLATION OF BATHTUBS

Bathtubs are installed in at least four different ways:
1. Recessed.
2. Sunken.
3. Corner.
4. Peninsular.

The recessed installation is common, Fig. 14-19. Rough-in dimensions for the length and width of the tub are known before interior walls are constructed. After the walls are framed and blocking installed, Fig. 14-20, set the tub. The wall covering — drywall, plaster or tile — is attached to the studs by the carpenters after the bathtub is in place. *Both tub and shower bases are installed during the early stages of construction and must be protected from damage while the interior of the building is being finished.*

Fig. 14-19. In a recess installation, walls enclose the tub on three sides.

During the final rough, the bathtub drain is connected to the DWV piping and the faucets are connected to the hot and cold water supply. Working through the bathtub access opening, the plumber installs the drain, overflow and stopper assembly, Fig. 14-21. Note that an overflow for the tub is provided through the pipe concealed by the plate supporting the stopper lever. When installing the stopper assembly, make certain that the overflow is open. Thus, any excess water in the tub will flow out the drain rather than flood the bathroom. Connect the drain to the DWV piping using a P trap or drum trap installed below floor level.

Install the faucets and connect the water supply lines to the hot and cold water piping as described for sinks and lavatories. In cases where two fixtures are installed back to back on either side of a wall, a manifold fitting, Fig. 14-22, simplifies the otherwise complicated water piping. When connecting the water supply lines to the bathtub, make certain that the cold water is controlled by the faucet on the right or connected to the right side of a single control faucet.

## SHOWER STALLS

Shower stalls are available in many shapes, sizes and colors. The walls of shower stalls may be of gel-coated fiber glass, enameled steel, glazed tile or other waterproof materials.

Shower stalls can be either freestanding, Fig. 14-23, or built-in, Fig. 14-24. Shower bases usually are made of terrazzo, fiber glass, cast stone or enameled steel. Shower base dimensions are usually 36 by 36 in. or 36 by 48 in.

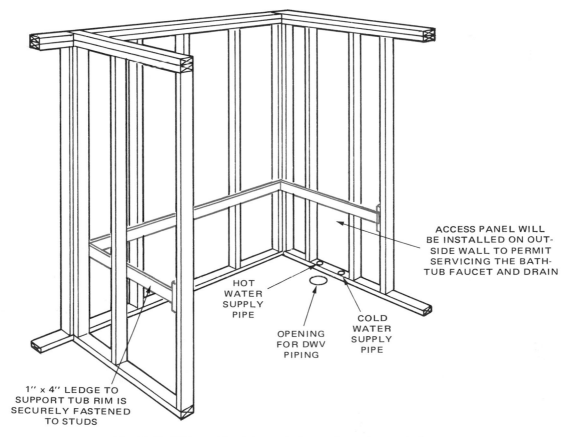

ACCESS PANEL WILL BE INSTALLED ON OUTSIDE WALL TO PERMIT SERVICING THE BATHTUB FAUCET AND DRAIN

HOT WATER SUPPLY PIPE

OPENING FOR DWV PIPING

COLD WATER SUPPLY PIPE

1" x 4" LEDGE TO SUPPORT TUB RIM IS SECURELY FASTENED TO STUDS

Fig. 14-20. Ribbon of 1 x 4s is attached to studs to support tub rim.

A

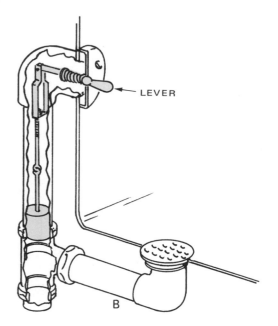

LEVER

B

Fig. 14-21. Overflow piping for bathtubs is practically the same from one manufacturer to the next. But systems for controlling drainage vary. Two styles are shown here. A—Pop-up stopper. B—Plunger type stopper. (Kohler Co.)

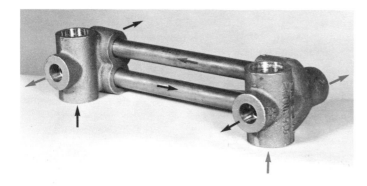

Fig. 14-22. Manifold pipe fitting permits back to back installation of fixtures without complicated crossing of hot and cold water piping. (Precision Plumbing Products, Inc.)

Fig. 14-23. Freestanding shower stall is installed in one piece. (Sears, Roebuck & Co.)

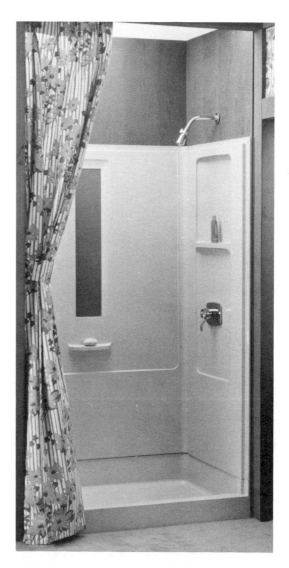

Fig. 14-24. Built-in shower stall is attached to a wall for support. (Owens-Corning Fiberglas Corp.)

Shower stall floors slope 1/4 in. per ft. to a center drain. A chrome, stainless steel or brass strainer covers the drain. The strainer is snapped and twisted into place or is fastened with brass screws.

## INSTALLATION OF SHOWER STALLS

Attach the drain body to the shower base to a 2 in. cast-iron or steel waste pipe and trap. Seal the connection with oakum and lead as in Fig. 14-25. A caulked joint is used when plastic drain pipe is used for the DWV system.

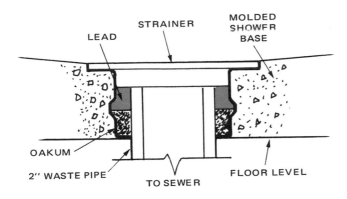

Fig. 14-25. Connection between shower drain and DWV piping must be sealed with lead and oakum or caulking.

Locate hot and cold water lines so that the water valves are within easy reach of the user. Fig. 14-26, illustrates a typical single knob control shower valve. When a shower is installed in a bathtub, a diverter, Fig. 14-27, delivers water to either the spout or the shower head.

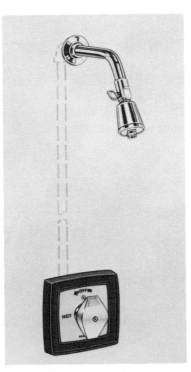

Fig. 14-26. A single handle controls water pressure and temperature on this shower installation. (Kohler Co.)

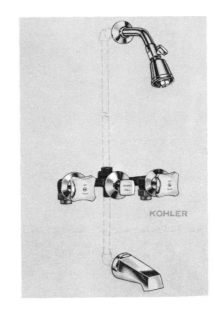

Fig. 14-27. In combination tub and shower faucets a diverter directs the water flow to either the shower head or to the spout.

## WATER CLOSETS

WATER CLOSETS or toilets are manufactured from vitreous china. Enough water is added to the mixture of fine clay, quartz, feldspar and silica to give it the consistency of soft cement. The mixture is poured into plaster of Paris molds where it is allowed to dry. A water closet is cast in about 13 pieces which are skillfully molded into one unit by highly trained workers. In a final step, the completed closet receives a liquid glaze and is fired to a temperature of 2600 F (1426 C). This treatment renders it waterproof.

## TYPES OF WATER CLOSETS

Through the years, four different bowl designs have been developed and are still in use today. They are:
1. Washdown type.
2. Siphon types.
   a. The reverse trap.
   b. The siphon jet.
   c. The siphon action.

## BASIC OPERATING PRINCIPLE

The water closet is designed to carry away solid organic wastes. Water, under pressure or gravity, is directed through passages in the bowl. The water is made to scour the bowl surface and move solid waste up through the trap into the drain.

Removal of the solids depends on a siphoning action. Movement of the water through the downward leg of the trap causes a drop in atmospheric pressure at the trap outlet. This partial vacuum, plus the head of water from the bowl, sweeps the wastes out.

Siphoning action continues to draw the water out of the bowl even after the supply running in, slows to a trickle. But, as the flow through the trap lessens, air is allowed to rush into the trap. This breaks the vacuum and pressure equalizes on both sides of the trap. The flush tank or flush valve delivers enough additional water to seal the trap once more.

While this is the basic operating principle, each type produces the flushing action in a slightly different way. Each has its special uses and advantages.

## WASHDOWN WATER CLOSET

The washdown water closet, Fig. 14-28, is the least expensive, least efficient and noisiest. In many municipal codes, it is no longer acceptable. Staining and contamination are greater because the bowl area is only partly covered by water. The waste passageway is located in the front of the water closet. Usually, this causes the front portion of the closet to protrude. The water passageway is only 1 7/8 in. in diameter.

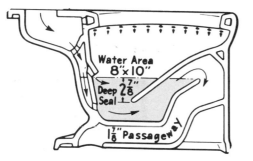

Fig. 14-28. Cutaway of a washdown water closet. This one spills water into the trap from around the rim, and from two larger openings near the trap. (Kohler Co.)

## SIPHON WATER CLOSETS

Siphon type water closets use a jet of water which speeds up the siphon action. In addition, the downstream leg of the siphon is longer than the upstream leg. The result is a partial vacuum which pulls the waste from the bowl.

The reverse trap water closet, Fig. 14-29, is similar to the siphon jet closet. It has a smaller water area, passageway and water seal than the siphon jet or siphon action water closets.

Siphon jet water closets, Fig. 14-30, are designed with large

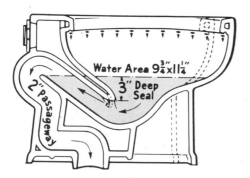

Fig. 14-29. Reverse trap water closet has trap outlet at back of bowl. It can be used with either a flush valve or a flush tank.

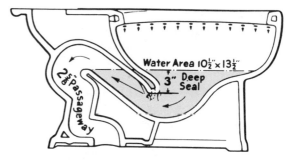

Fig. 14-30. Siphon jet water closet does not need a head of water in the bowl to flush out solids. Stream of water is delivered from closet spud to the outlet of the trap.

passageways and water surface area than the reverse trap closet. This reduces tendency to clog.

The siphon action water closet, Fig. 14-31, is usually a one-piece closet with a low profile. It has quiet flushing action and a larger water surface. The low profile distinguishes this water closet from other designs. The trap passageway diameter is less than the siphon jet closet.

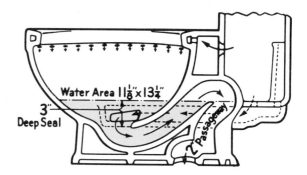

Fig. 14-31. Siphon action water closet usually combines the closet and flush tank in one unit. (Kohler Co.)

## INSTALLATION OF WATER CLOSETS

The connection between a water closet and the DWV piping system is made with a closet flange. It secures the water closet to the floor. The flange, Fig. 14-32, is provided with slots into which closet bolts are placed. The threaded portion extends

Fig. 14-32. Closet flange is the connecting piece between water closet trap and closet bend. It is fastened to the floor with screws.

# Plumbing Fixtures

upward to receive the bowl of the water closet. These bolts secure the bowl to the flange. In addition, they compress the wax or rubber toilet bowl seal to make the joint air and watertight. Fig. 14-33 shows a cutaway of a water closet and tank, along with illustrations of various parts and fittings.

To install the bowl, refer to the exploded view in Fig. 14-34, and proceed as follows:

1. Place the bowl temporarily over the closet flange. Test it for levelness. Use shim, if necessary, to level the bowl. In new construction, floors should be level. However, if the building is older, floors may be uneven.

2. Lift the bowl off and turn it upside down. Place putty around the outer rim of the base to prevent water and dirt from getting under the water closet.

3. Fit a rubber or wax toilet bowl seal on the discharge opening of the bowl.

4. Place the bowl carefully over the closet flange and closet bolts.

5. Check again to make sure it is level and squarely seated; then install and tighten the closet nuts. Fit china covers over the nuts to give a finished appearance to the installation.

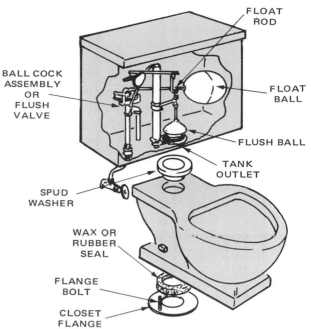

Fig. 14-34. Exploded view shows order and position in which parts and fittings are assembled.

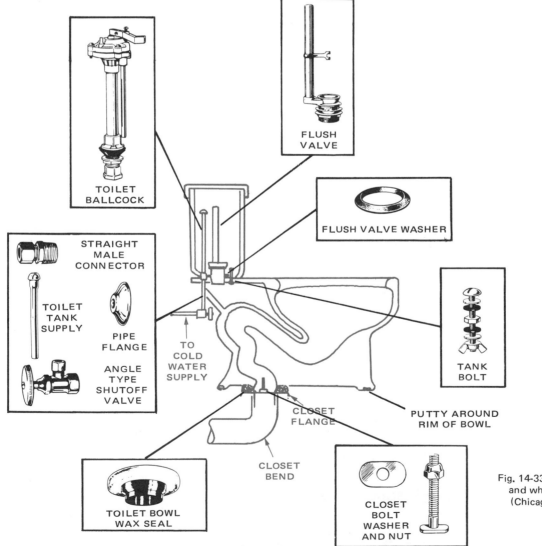

Fig. 14-33. Water closet fittings and where they are attached. (Chicago Specialty Mfg. Co.)

177

To install tank and fittings:

1. Secure ball cock assembly (flush valve) to the bottom of the tank. Make certain that the gaskets provide a complete seal around the opening.
2. Place spud washer over water inlet hole of the bowl.
3. Carefully place the tank into position so that the tank opening fits over the spud washer. Press the tank carefully into place so that the washer is not distorted.
4. Secure with tank bolts.
5. Install the float rod and float ball on the flush valve.
6. Connect the water supply using a flexible chrome-plated copper toilet tank supply tube. Bend it to correct shape and cut off the end which will be fastened to a chromed shutoff valve. Install a compression nut and compression ring. Attach the cut end to the shutoff valve and attach the other end to the inlet of the flush valve.
7. Check completed assembly for leaks.

## BIDET

The BIDET (bih-day′) is a companion fixture to the water closet. It is used for personal hygiene in cleaning the perineal area of the body. The bidet, Fig. 14-35, has long been popular in Europe and South America and it is gaining recognition in America. Being chair height, it is particularly useful for older persons or those who are recovering from an illness which makes bathing difficult.

The user sits astride the bidet facing the faucets which regulate water temperature and rate of flow. The bidet is fitted with a pop-up stopper which permits the accumulation of water in the bowl when desired. Rinsing of the bidet is accomplished through a rim flushing action similar to a water closet.

Fig. 14-35. Bidet provides for additional personal hygiene when used in combination with a water closet. (Kohler Co.)

Installation of the bidet requires the connection of an outlet from the bowl to the DWV piping system through a P trap. This connection is similar to the installation of a lavatory except that it is done below the floor level. Hot and cold water supply line are connected in the same way as for a lavatory.

## URINALS

URINALS, Fig. 14-36, are commonly installed in public restrooms for men. Because of the frequent use which is made of public restrooms, urinals are usually fitted with flush valves. This eliminates the bulk of a water tank.

Fig. 14-36. Urinals are commonly installed in public restrooms. (Eljer Plumbingware Div., Wallace-Murray Corp.)

Installation of the flush valve is not difficult. It is always used in conjunction with a stop valve and a vacuum breaker. The stop valve will shut off the water supply in an emergency as well as regulate the flow of water to the flush valve. Both stop valve and flush valve are attached with slip nuts as shown in Fig. 14-37. Flush valves are discussed in Unit 5.

## WATER FOUNTAINS

In many installations water fountains are used which supply drinking water without cooling it. Water fountains, Fig. 14-38, provide a convenient source of sanitary drinking water. They are available in a wide variety of designs nearly all of which require a 1 1/4 in. drain line and 1/4 or 3/8 in. cold water supply piping.

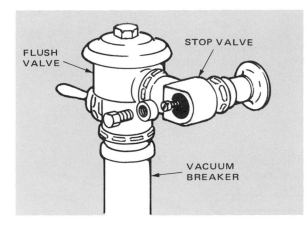

Fig. 14-37. Flush valves are found in commercial buildings. They supply flush water to water closets and urinals.

Fig. 14-38. Water fountains are installed in many public places. (Eljer Plumbingware Div., Wallace-Murray Corp.)

## TEST YOUR KNOWLEDGE – UNIT 14

1. Name the three basic types of lavatories.
2. A lavatory or sink _____ is used to hang a ledge type sink or lavatory.
3. A lavatory or sink strainer is connected to the DWV piping with a P trap. True or False?
4. The function of the P trap is to:
   a. Prevent sewers from backing up.
   b. Prevent sewer gas from entering the building.
   c. Keep water from draining away too rapidly.
5. A _____ wrench is used to tighten the nuts on faucets.
6. Bathtubs are made from the following materials (check all that apply):
   a. Enameled cast iron.
   b. Enameled steel.
   c. Fiber glass reinforced plastic.
   d. All of the above.
7. The standard lengths for bathtubs are _____, _____ and _____ feet.
8. _____ are installed before the wall covering is placed over the framing materials.
9. Hot water is always controlled by the _____ faucet handle.
10. _____ and _____ are used to form a seal between the shower drain and the DWV piping system.
11. Name the four basic types of water closets.
12. The _____ _____ water closet is the quietest.
13. A water closet is secured to the building with closet bolts which are inserted in the _____.
14. To make a water and airtight seal between the closet bowl and the closet flange, a wax or rubber _____ _____ seal is installed.
15. The ball cock is connected to the cold water piping with a _____ and _____ valve.
16. Installation of a bidet is similar to installing a _____.
17. Urinals are generally fitted with _____ valves.

## SUGGESTED ACTIVITIES

1. Study catalogs from several plumbing fixture manufacturers to learn the:
   a. Variety of sink and lavatories available and how the rough-in dimensions may vary.
   b. Variety of water closets available and how their rough-in dimensions may vary.
   c. Different types of bathtubs and showers available and how their installation may be different.
2. Install a lavatory or sink, a water closet and a tub or shower either in a building or in a mock-up building frame. This should be a group project.
3. Invite a sales representative to discuss the merits of the different types of plumbing fixtures.
4. Invite a plumbing contractor to discuss special problems encountered when installing plumbing fixtures.

Fig. 15-1. The water cycle makes it possible for people to repeatedly use water. The water table is the upper limit of earth's crust which is saturated.

# Unit 15

# WATER SUPPLY SYSTEMS

### Objectives

In this unit the basics of well location, types of wells, selection of well equipment and installation of a private water supply system are discussed.

After studying the unit you will be able to:
- List the types of wells and know how they are constructed.
- Describe construction and operation of various types of water pumps.
- Understand and explain the purpose and operation of a pressure tank.
- Describe the process of disinfecting and treating water.

All the fresh water on earth is somewhere in the water cycle, Fig. 15-1. It may be rain or snow; it may be surface water in a lake or pond or ground water from a deep well. As Fig. 15-1 indicates, the cycle is made complete by evaporation of surface water which forms more clouds so that more precipitation (rain or snow) can occur.

Getting water for human use means tapping this water cycle. This is generally done in one of two ways:

1. Collecting surface water from a lake or river or a roof top.
2. Drawing ground water from a well.

Because most cities require tremendous volumes of water, they generally depend upon surface water for their supply. However, this water is likely to contain pollutants injurious to health. It generally requires considerable treatment. Only then is it safe for human consumption.

Homes beyond city water mains use ground water. It offers the advantage of being relatively clean. However, this is true only if the well is properly located and installed.

## LOCATING A WELL

Again, referring to Fig. 15-1, note that the WATER TABLE is a relatively continuous underground surface. Below this table, the sand, sand and gravel or joined rock is filled with water. The depth of the water table below the earth's surface will vary from a few feet to hundreds of feet. It tends to follow roughly the contour of the land.

Accurate methods of finding the depth of the water table before drilling have not been developed. As a general rule, depth is not particularly important in deciding where to locate the well. Other factors, such as relative position of potential sources of contaimination and how near to the place where the water will be used are more important.

Two common sources of contamination are:
1. Septic tank leach fields.
2. Livestock feedlots.

It is recommended that a well be no less than 100 ft. away from such pollution sources. It should also be located on higher ground so that surface water will tend to flow away from the well. Other things being equal, a deeper well is generally preferred because this allows the water to filter through more soil before it is pumped out of the well.

## TYPES OF WELLS

The four most common types of wells, Fig. 15-2, are:
1. The dug well.
2. The driven well.
3. The drilled well.
4. The bored well.

Each has advantages and disadvantages. These must be considered before making a decision about which type to have. The decision about which is most appropriate for a given location must be made with knowledge of the earth's structure where the well is to be located. Otherwise one might begin digging a well only to find that after many days of hard labor the water is too far below the earth's surface to be obtained efficiently by a dug well.

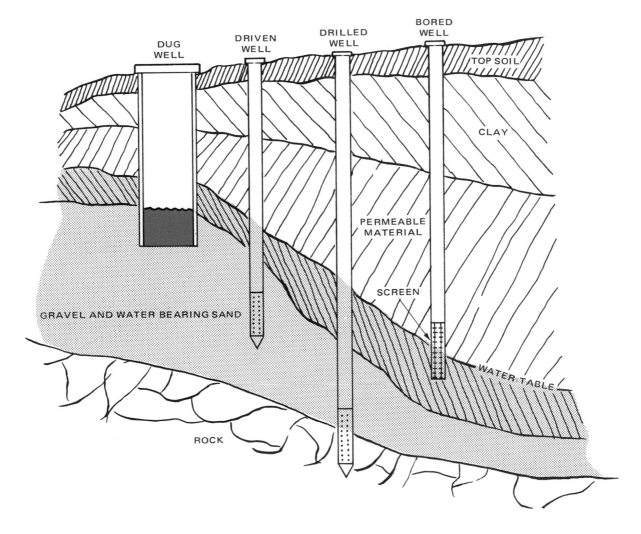

Fig. 15-2. The four most common types of wells. Which is most appropriate, depends on subsoil conditions and depth of the water table.

## DUG WELLS

Dug wells are constructed by digging a circular hole several feet in diameter to a depth somewhere below the water table. Such wells are seldom more than 50 ft. deep because of the difficulty of the excavation. Once the hole is completed, the dug well is lined with stone, brick or concrete pipe to prevent cave-in.

The dug well is very susceptible to contamination by surface water and/or subsurface seepage. Hence, many such wells found in rural areas are polluted.

## DRIVEN WELLS

Driven wells are made by forcing a well point, Fig. 15-3, into the ground. The well point is driven into the earth's subsurface by repeated hammer blows. Lengths of pipe are added as needed.

Driven wells are practical only when the soil is relatively free of rock and the water table is within 50 to 60 ft. of the earth's surface. They are relatively inexpensive. Because no excavation is required, there is little likelihood that surface water or subsurface seepage will contaminate the well.

## DRILLED WELLS

Drilled wells are used:
1. Where greater well depth is required either to reach the water table or to obtain the necessary volume of water.
2. When the subsurface materials are too hard to permit driving a well.

The two most commonly used drilling methods are the percussion and the rotary. The percussion method uses a chisel shaped bit which is raised and lowered by a cable to break up the subsurface material. Fig. 15-4 shows a typical percussion drilling rig.

Fig. 15-4. Percussion drilling rig uses cable to drop heavy drill into hole. This action chisels away subsoil.   (Mobile Drilling Co.)

Rotary drilling requires a drilling rig similar to the one shown in Fig. 15-5. This method is more economical when the water table is relatively far below the surface. The equipment is somewhat more complicated than the percussion drilling rigs. A water slurry is used as a cutting fluid and coolant to speed the drilling operation.

Fig. 15-3. Typical well point uses hard pointed head to drive through subsoil.

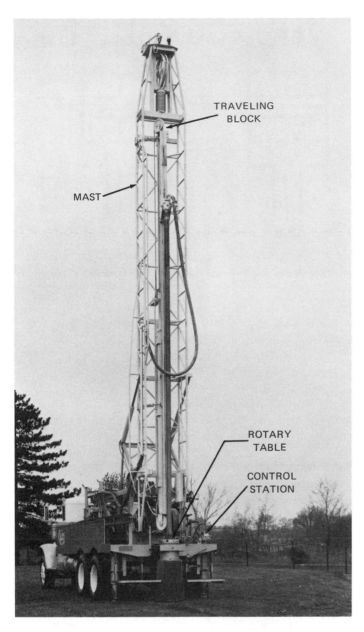

Fig. 15-5. Rotary drilling rig is more efficient than percussion drilling when great depths are involved. (Sanderson Cyclone Drill Co.)

## BORED WELLS

Bored wells are made with an earth auger. It drills a hole which is larger in diameter than the casing. After the water table has been reached, a well casing is inserted to protect the well from contamination.

### TYPES OF PUMPS

The important considerations for selecting pumps are:
1. Well depth.
2. Water demand rate.
3. Ease of maintenance.
4. Initial cost.

Only the basic types of water pumps will be discussed. To make a wise choice, it is necessary to study manufacturers' literature. Determine the capabilities of each pump before making a final choice.

## SHALLOW WELL OR LIFT PUMPS

Lift type pumps can be installed in shallow wells (25 ft. or less). This pump is located at ground level and functions by creating a partial vacuum at the pump inlet. Because the atmosphere exerts a pressure of 14.7 psi on the water surface, the water is pushed up the pipe, Fig. 15-6. Theoretically, this type of pump should be able to lift water 34 ft. at sea level. However, in practical applications, a perfect vacuum is never achieved. In addition, there is some loss due to the friction of the water passing through the pipe.

Several types of lift pumps are available for shallow wells. These include:
1. Reciprocating.
2. Centrifugal.
3. Jet.
4. Rotary.

The basic operating principles of each of these types of pumps will be discussed along with their major advantages and disadvantages.

### Reciprocating pumps

Reciprocating pumps, Fig. 15-7, lift water by moving a piston back and forth in a cylinder. Fig. 15-8 illustrates the

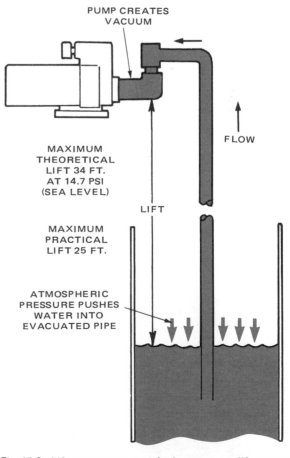

Fig. 15-6. Lift pumps use atmospheric pressure to lift water as high as 25 ft.

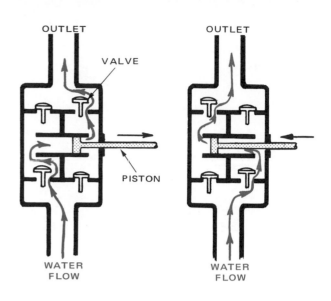

Fig. 15-8. A double piston and four check valves permit a reciprocating pump to supply a steady flow of water.

Fig. 15-7. Reciprocating pumps lift water by moving a piston back and forth in a cylinder.

basic operation of a double acting reciprocating pump.

The advantages of reciprocating pumps include:

1. Ability to pump water containing sand.
2. Adaptable to low capacity water supplies where high lifts are required above the pump.
3. Can operate with a variety of head pressures.
4. Can be hand operated.
5. Will install in very small diameter wells.

The principal disadvantages include:

1. Has pulsating discharge.
2. Vibrates because of the reciprocation of the piston.
3. Has relatively high maintenance costs.
4. Could damage the water piping system if the pump is permitted to operate against closed valves.

## Centrifugal pumps

Centrifugal pumps use a rapidly spinning IMPELLER to lift water. The impeller is a heavy disc mounted in the pump

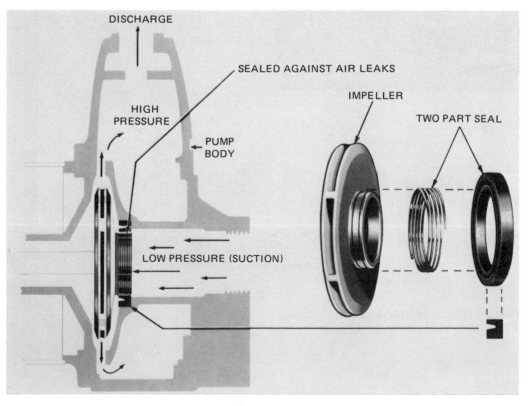

Fig. 15-9. Cutaway showing impeller design of a centrifugal pump.

housing. As it spins, centrifugal force throws water off its outer edge. This creates a vacuum at the center. Atmospheric pressure in the well pushes water into the impeller to fill the vacuum.

Fig. 15-9, shows one type of impeller shape. This type is called an enclosed impeller. Both sides are shrouded. Water enters through the hub and moves to the outer rim along curved ridges or passages called vanes. The water is thrown off the rim of the impeller with some force. This creates a water pressure that pushes the water into the water supply system of the building.

Not all impellers are enclosed. They may take other shapes as shown in Fig. 15-10. Open and semi-open impellers must operate in a close-fitting housing. Water coming out of the impeller goes into a volute. This is a passage shaped like a snail

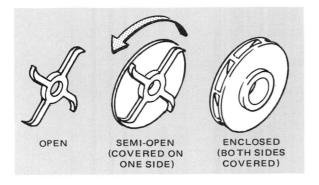

OPEN     SEMI-OPEN (COVERED ON ONE SIDE)     ENCLOSED (BOTH SIDES COVERED)

Fig. 15-10. Impellers, regardless of type, use curved blades to help move water. Vacuum is created when water inside the impeller is thrown out to the discharge side. Water from the well is sucked in to take its place. Impeller must be sealed against leaks. (Peabody Barnes)

shell. It wraps around the impeller. The passage is small at its beginning point but increases in cross-sectional area as it moves to the discharge point. See Fig. 15-11.

This type of pump will not operate unless the housing is filled with water. It must, therefore, be primed by adding

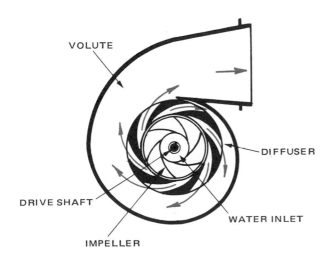

Fig. 15-11. Simplified cross section of an impeller and surrounding volute. Centrifugal force flings water into the volute chambers. Water exits volute chamber under pressure and enters a storage tank or water supply system.

water before it is put into operation.

Since they are not designed for high lifts, centrifugal pumps are located close to the water surface. Foot valves at the inlet end of the suction pipe keep the pump full of water when it is not operating. If this water is lost, the pump must be primed again.

The principal advantages of the centrifugal pump are:
1. Uniform output of water.
2. Pumps water containing sand.
3. Low starting torque.
4. Good service life.
5. Efficient for capacities above 50 gpm.

On the other hand, the centrifugal pump has some major disadvantages:
1. Prime is easily lost.
2. It is efficient only if the proper speed and outlet head pressure are maintained.

## Jet pumps

Like the centrifugal pump, the jet pump uses an impeller. The main difference is a second device, the jet or injector, employed to help lift the water from the well.

Some of the water leaving the impeller under pressure is recirculated through the nozzle and the venturi of the jet. This action creates a vacuum at the nozzle and more water is pushed in from the well.

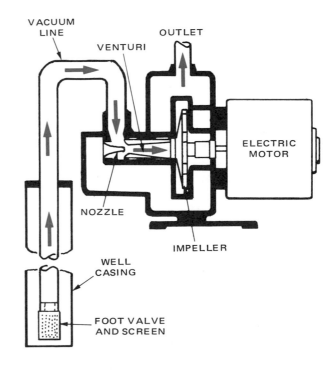

Fig. 15-12. Jet pumps are frequently used in shallow wells because all the moving parts can be located outside of the well casing.

Jet pumps with the ejector (jet) mounted on the pump body itself rather than down in the well are used for shallow well installations, Fig. 15-12. The suction line is placed inside the well casing. It is fitted with a foot valve to prevent the pump from losing its prime.

The chief advantages of the jet pump are:

1. High capacity at low head.
2. Simple operation.
3. No moving parts in the well.
4. It does not have to be installed directly over the well.
   Disadvantages of this type pump are:
1. Air in the suction line will cause pumping to stop.
2. Volume of water pumped decreases as lift increases.

### Rotary pumps

Rotary pumps for shallow wells have a helical rotor driven by the pump motor, Fig. 15-13. The helical rotor rotates inside the molded rubber stator forcing trapped water out the discharge side of the pump.

The advantages of this type of pump are:

1. Constant discharge rate.
2. Efficient operation.

However, sand or silt in the water will rapidly wear the rotor and stator. This reduces the pump's efficiency.

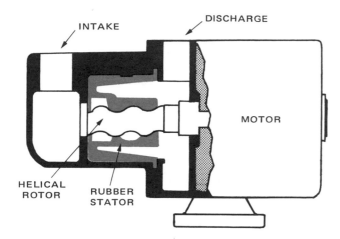

Fig. 15-13. Rotary pumps can be used in place of jet pumps.

## DEEP WELL PUMPS

Wells deeper than 25 ft. require a pumping device installed within the well casing. Although all the parts of the pump do not need to be in the well casing, the primary functioning element of the pump must be installed near the water table. Several of the deep well pumps work exactly like the shallow well lift pumps except that part or all of the pump is installed in the well casing.

The most common types of deep well pumps are the reciprocating, centrifugal, jet and rotary. Each of these types will be explained briefly.

### Reciprocating pumps

Reciprocating pumps for deep wells function by using a cylinder attached to the drop pipe. This cylinder must extend below the water table. To lift water, a plunger (piston) is inserted in the cylinder. The pump rod, as long as 600 ft. connects the plunger to the drive mechanism. The reciprocating pump is well adapted to low capacity, high lift installations. The chief disadvantages are:

1. High maintenance cost.
2. Noise from vibration.
3. The pump must be placed directly over the well.

### Submersible pumps

Centrifugal pumps for deep wells are submersible, Fig. 15-14. The multistage centrifugal pump is designed so that each impeller is mounted on the same shaft rotating at a constant speed. The output of each successive impeller is fed into the next impeller in order to create the necessary pressure to lift the water, Fig. 15-15.

The capacity of the submerged centrifugal pump depends mostly upon the width of the impeller. The water pressure created is determined by the diameter of the impeller, the speed at which the impeller rotates and the number of stages in the pump.

The advantages of this type pump are:

1. Uniform flow.
2. Low starting torque.
3. Self-priming.
4. Frost proof.

Fig. 15-14. Submersible centrifugal pumps permit pumping water from considerable depths. (Peabody Barnes)

Its major disadvantages are:

1. Pump must be pulled to make repairs.
2. Seals to protect electrical equipment from moisture are critical.
3. Abrasion from pumping sand can rapidly reduce efficiency.

### Deep well jets

Jet pumps for deep wells are installed with the ejector body

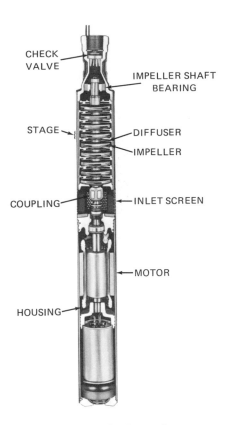

Fig. 15-15. Each stage of a centrifugal pump increases water pressure.

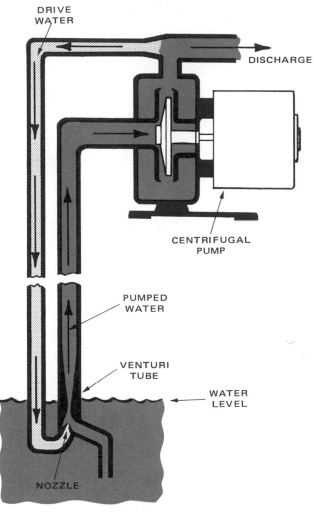

Fig. 15-16. Deep well jet pumps function because the ejector body is positioned inside the well casing at/or below the water table. (Peabody Barnes)

(jet) near or below the water table. Jet pumps function by forcing a relatively small quantity of water under high pressure through a nozzle where water being pumped mixes with the drive water, Fig. 15-16. The drive water and the pumped water then pass through a venturi tube where the speed of the water decreases and the pressure increases. The result is that a relatively small amount of drive water can pump a considerable volume of water.

The advantages of the deep well jet pump are:
1. High volume of water at low pressure.
2. Simple operation.
3. No moving parts in the well.

Principal disadvantages are:
1. Air in either the suction or return line will cause the pump to stop functioning.
2. The volume of water decreases as the depth of the well increases.

### Rotary pumps

The deep well rotary pump is capable of lifting water as high as 500 ft. Its design is similar to the shallow well rotary helical pump except that it is submerged within the well casing. This pump is available with the motor mounted above ground, Fig. 15-17, or in a completely submersible unit, Fig. 15-18. The latter looks like the submersible centrifugal pump.

Advantages of this pump are:
1. Constant discharge rate regardless of head pressure.
2. Efficient operation.
3. Only one moving part in the well (surface mounted motor type only).

The disadvantages of a helical rotary pump are:

1. Rotor and stator are subject to wear.
2. The drive coupling has been the weak point.

## PRESSURE TANKS

Water supply systems must instantly provide water while maintaining constant pressure. To do this, some type of storage tank is required. Municipal water systems generally depend upon water towers to store water and allow gravity to create pressure.

Because of the expense of water tower construction, this method is seldom used in privately owned residential water supply systems. Hydropneumatic tanks are much more economical for a small water system. The word hydropneumatic means that the tank contains both water (hydro) and air (pneumatic) under pressure. Fig. 15-19, illustrates how this tank functions. Water is nearly incompressible; it is necessary to introduce air into the tank. The air compresses providing the force necessary to cause the water to flow through the plumbing system. When the water in the tank goes below a predetermined level, the pump is automatically turned on. As the tank is refilled with water, the air is compressed until the high water level is reached.

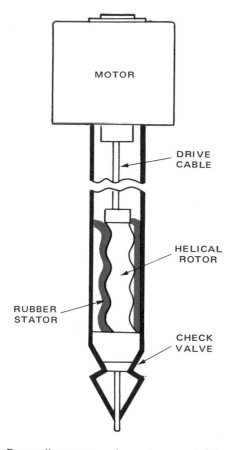

Fig. 15-17. Deep well rotary pump has motor mounted above ground.

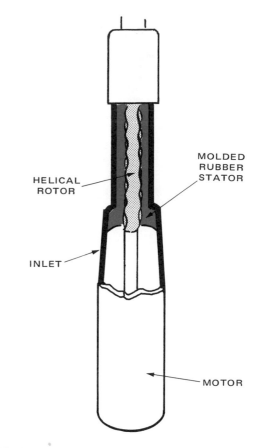

Fig. 15-18. Submersible deep well rotary pump places motor in the well.

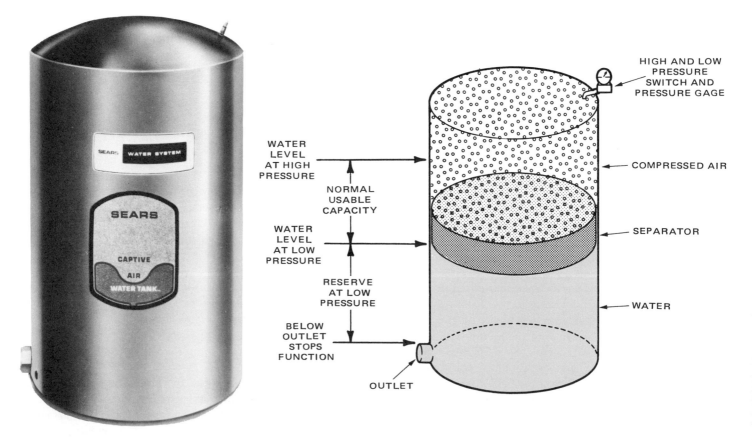

Fig. 15-19. Operation of a hydropneumatic pressure tank. As water volume builds up, so does pressure. Separator prevents tanks from becoming "water logged."

The size of tank required depends upon three factors:

1. Pump capacity vs. maximum demand. If the pump capacity is greater than the maximum demand for water, then the minimum size tank can be selected. If, however, the pump capacity is less than the maximum demand, the tank must be large enough to supply that part of the maximum demand not supplied by the pump.
2. Operating pressure and pressure range. If the system will operate satisfactorily at lower pressure, a greater amount of water can be obtained from a given size tank.
3. Air control and supercharge. Increasing the supercharge on the tank increases the volume of water which can be drawn from the system.

The general rule of thumb for sizing tanks is that the usable capacity should be twice the pump capacity in gallons per minute.

## WELL SANITATION

Preventing the well from becoming contaminated requires three special considerations:
1. Proper location of the well.
2. Proper installation.
3. Proper cleaning and maintenance.

### SANITIZING THE WELL

After drilling is completed, the well driller must pump the drilling residue from the well. This operation not only removes dirt and other contamination, but also reduces the likelihood that the well pump will be damaged by sand or other abrasive particles.

The next step is to chemically disinfect the well. This can be done with a household bleach (sodium hypochlorite) solution. Use two parts of water to one part of bleach. Two quarts of the solution will treat 100 gallons of water. Pour the correct amount of bleach solution into the well casing. The pump is installed and water is pumped from the well until a chlorine odor is obvious. This procedure is repeated several times at one hour intervals to insure that the well is safe. Finally, the solution is poured into the well and permitted to stand 24 hours. The water is then pumped out until it is entirely free of chlorine odor and taste.

### GROUTING THE WELL CASING

The space between the outside of the well casing and the drilled hole must be filled to a depth of at least 10 ft. with a cement grout. This precaution will prevent surface water from entering the well, Fig. 15-20.

### CAPPING

The top of the well casing must be sealed around the drop pipes to keep out contamination. A well seal such as the one shown in Fig. 15-21 should be used.

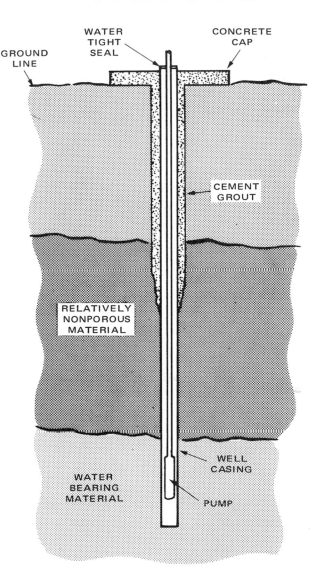

Fig. 15-20. Grout around the well casing prevents contamination of the well by ground water.

## WATER TREATMENT

All water found in nature contains some form of impurity. Even rain drops contain carbon dioxide and dust particles picked up from the atmosphere as they fall to earth.

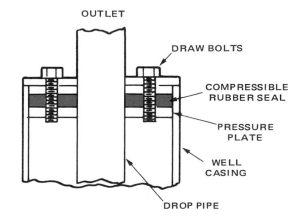

Fig. 15-21. A well seal caps the end of the casing.

189

Impurities found in water can be classified in one of four categories:

1. Gases.
2. Soluble matter.
3. Suspended matter.
4. Living organisms.

Drinking water supplies may have one or more of the four common types of impurities in quantities great enough to require treatment. Most state health departments will test water and recommend a treatment to make it safe for human consumption.

## CHLORINATION

Adding controlled amounts of chlorine to the water supply in the form of sodium hypochlorite (laundry bleach) will act as a disinfectant. Chlorine will not only kill bacteria but it will also remove iron, manganese, hydrogen sulphide (rotten egg gas) and putrid odors from the water by oxidization. The sodium hypochlorite is added automatically with a chlorinator. If the water contains minerals in addition to needing to be disinfected, it may be desirable to use a larger storage tank. Then, the chlorine solution will have sufficient time to oxidize the undesirable minerals before the water is drawn into the piping system.

## REMOVING TURBIDITY

Water requiring considerable chlorination may become "muddy" from oxidized metal, sulfur or organic matter. This turbidity can be removed with a clarifier filter.

## DECHLORINATION

Large treatments of chlorine may give the water an undesirable taste and odor. The chlorine can be removed with an activated charcoal or carbon filter. As the water passes through the filter, the chlorine reacts with the charcoal removing the bad taste and odor. These filters are relatively inexpensive to operate. The activated charcoal or carbon seldom needs to be replaced.

## REMOVING ACID

An acidic water condition can severely corrode the plumbing system. It must be treated if it has a pH value below 5. Treatment consists of a feeder pump which injects soda ash solution into the system. When the water has a pH value above 5 and below 7, it can be treated with a neutralizing filter. This is a conventional clarifier filter filled with an alkaline filter material (usually calcite).

## REMOVING IRON

Iron in water is objectionable. It affects the flavor of foods, stains fixtures and discolors water. Water high in iron content is often clear when freshly pumped but turns yellow or brown as it is exposed to air. This occurs because the iron oxidizes and forms rust.

Iron can be removed by an oxidizing filter charged with manganese processed sand. As the iron contaminated water flows through the filter bed, iron unites with the oxygen in the manganese processed sand. The iron is changed to iron oxide (rust). Maintenance of this filter requires periodic backwash and occasional regeneration of the potassium permanganate oxidizing agent.

## SOFTENING WATER

Excessive amounts of calcium or magnesium in water cause heavy soap consumption. Scale is deposited on containers when the water is heated. Water softeners, Fig. 15-22, remove the dissolved calcium and magnesium by ion exchange. Because this process is reversible, zeolite water softeners are frequently equipped with an automatic regenerative system which allows the zeolite to be recharged indefinitely.

## SELECTING A WATER TREATMENT SYSTEM

Selecting the equipment necessary to properly treat water is a complex task. It requires:

1. Chemical analysis of the water.
2. Determination of the amount of treated water required.
3. Selecting equipment to perform the required functions.

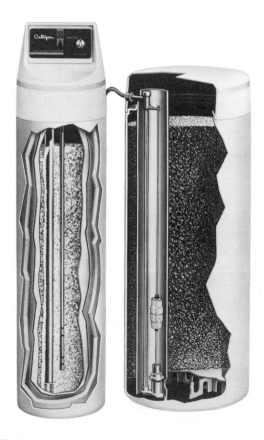

Fig. 15-22. Water softeners remove calcium and magnesium from the water. (Culligan Water Institute)

As can be seen in the chart, Fig. 15-23, one piece of equipment may be able to perform more than one task. With special attachments or carefully selected filter materials, one or two pieces of equipment may give satisfactory treatment. Given the complexity of selecting an efficient water treatment system, it is advisable to consult specialists in this field.

## CONNECTING TO A MUNICIPAL WATER SUPPLY SYSTEM

In most municipalities, water is supplied from a central treatment plant through water mains to each inhabited building. When a new building is constructed, it is necessary to connect to the water main. Using a drilling and tapping machine, it is possible to drill and tap the main without turning off the water. There is no interruption in water supply to other buildings served by the water main. A description of this operation is included in Unit 9.

## TEST YOUR KNOWLEDGE – UNIT 15

1. Water for human consumption is generally taken from the water cycle as (check all that apply):
   a. Ground water.
   b. Ocean water.
   c. Moisture in the atmosphere.
   d. Surface water.
2. Ground water is located below the water _____.
3. Name the four most common types of wells.
4. Lift pumps raise water as much as:
   a. 12 ft.          c. 34 ft.
   b. 25 ft.          d. 50 ft.

5. Deep well pumps have their lifting devices located in the _____ _____; shallow well pumps have them above the _____.
6. Reciprocating pumps make use of a rapidly rotating impeller to lift water. True or False?
7. _____ pumps use an ejector to lift water.
8. Why are pressure supply tanks used in water supply systems?
9. In order to produce a sanitary well, the well must be (check all that apply):
   a. Capped.
   b. Grouted.
   c. Treated with bleach.
   d. At least 100 ft. from a septic tank leach bed.
   e. All of the above.
10. Name the four categories of impurities found in water.
11. The removal of iron from water is accomplished with an _____ filter.
12. Water softeners remove _____ and _____ from the water.

## SUGGESTED ACTIVITIES

1. Disassemble or study cutaways of several different types of water pumps. Identify the principal parts of each.
2. Visit a well drilling site and discuss with the driller, the procedures for locating water, drilling the well, installing the well casing and capping the well.
3. Visit a local distributor of water treatment equipment and discuss the use of each type of equipment.
4. Visit a municipal water treatment plant and consider how this facility compares to a private water treatment system.

FUNCTION

| TYPE OF EQUIPMENT | DISINFECTANT | IRON REMOVAL | MANGANESE REMOVAL | HYDROGEN SULPHIDE REMOVAL | TURBIDITY REMOVAL | CHLORINE REMOVAL | ACID REMOVAL | CALCIUM REMOVAL | MAGNESIUM REMOVAL | IMPROVE TASTE |
|---|---|---|---|---|---|---|---|---|---|---|
| CHLORINATOR | YES | SOME | SOME | YES | | | | | | |
| CLARIFIER FILTER | | | | | YES | | | | | |
| ACTIVATED CHARCOAL FILTER | | | | | | YES | | | | YES |
| FEEDER PUMP | | | | | | | YES | | YES | YES |
| WATER SOFTENER | | SOME | SOME | | | | | YES | YES | |
| PHOSPHATE FEEDER | | SOME | SOME | | | | | | | |
| OXIDIZING FILTER | | SOME | SOME | | | | | | | |

Fig. 15-23. Types and functions of water treatment equipment.

# Unit 16

# PRIVATE WASTE DISPOSAL SYSTEMS

## Objectives

This unit describes the construction and operation of private waste disposal systems as well as systems for moving ground water away from foundations.

After studying the unit you will be able to:
- Explain the operation of a simple septic system.
- List the essential materials and describe methods used in construction of the septic tank and leach field.
- Describe the construction and operation of alternative systems such as the aeration waste treatment system and the closed system.
- Explain how ground water can be removed from around foundations and basements.

The first known sewer system was constructed more than 5000 years ago by the Babylonians. The Romans, too, built extensive sewer systems to remove storm and, later, waste water from their cities. These facts emphasize the importance waste disposal has maintained throughout history. Concern for ecology and the need to conserve our available water supplies has brought renewed interest in waste disposal. Any system employed must be able to render the waste water harmless to the environment before it is discharged. Towns and cities across the country have installed sanitary sewer systems where wastes are collected and treated before being discharged, Fig. 16-1. The connection of the building drain to the sanitary sewer was discussed in Units 9 and 15. This unit will, therefore, only discuss private residential waste disposal systems of the type in use today.

Fig. 16-1. The sewage treatment plant is typical of facilities built by many communities.

## BASICS OF PRIVATE WASTE DISPOSAL SYSTEMS

Local county health officials generally have jurisdiction over the installation of private waste disposal facilities. Their knowledge of the needs in the county and their ability to advise the plumber about specific requirements make them a primary source of information. Any information in this unit should be taken as general minimum guidelines. The requirements established by the local county health department are the final authority.

## HOW SEPTIC SYSTEMS WORK

The essential parts of a private waste disposal system are the SEPTIC TANK and a LEACH BED or leach field. The sketch in 16-2 is a layout of such a system. The distances given should be considered only as general guidelines. Local conditions and requirements will cause variances. The disposal process follows this progression:

1. The waste from the building enters the inlet of the septic tank, Fig. 16-3.
2. In the septic tank the organic solids are permitted to settle to the bottom.
3. The somewhat clarified waste, liquid or water, leaves the septic tank through the outlet and flows through a sealed pipe to the distribution box.
4. The waste water leaves the distribution box, Fig. 16-4, through a number of outlets leading to the leach field.
5. Sealed pipe carries the liquid waste to the leach bed runs.
6. Waste enters the leach bed through perforated pipe or short unjoined sections of pipe buried in gravel.
7. The water, upon entering the leach bed, is absorbed into the ground, thus returning to its source.

## BASIC DESIGN CONSIDERATIONS

The list below presents some considerations about the basic design of a waste disposal system:

1. In no case can a leach bed be expected to function in swampy areas or where flooding is common. The soil would be so saturated with water that absorption of additional water would be impossible. If a system were installed in such a location, the waste would seep through to the surface and produce an unsafe and unpleasant situation.
2. The exact size of the leach bed will depend upon the number of occupants in the residence and the soil characteristics in the leach bed area. A modern home produces about 100 gallons of waste daily for each person living in the house. Because the actual number of people who live in a home may change from time to time, the usual standard is based on the number of bedrooms.
3. The leach bed, being near the surface, must be located where vehicles or other heavy loads will not travel over it. Plantings over leach beds should generally be limited to grass. Larger plants have long roots which are attracted to the water in the leach bed. They could cause extensive clogging of the pipes.
4. The size of trenches and the amount of gravel placed in the leach bed runs, will affect the capacity of the system. Dimensions will generally be specified by county health officials.

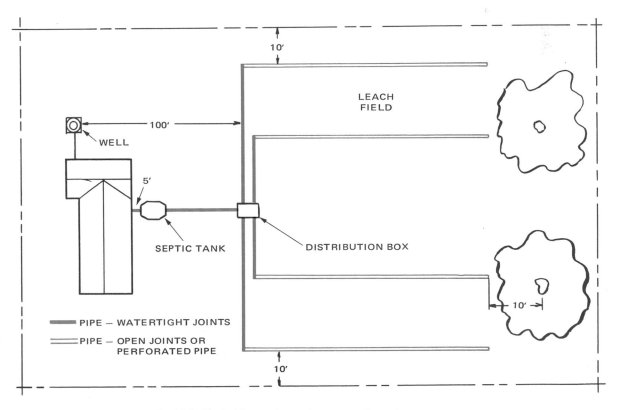

Fig. 16-2. Typical layout for a private waste disposal system.

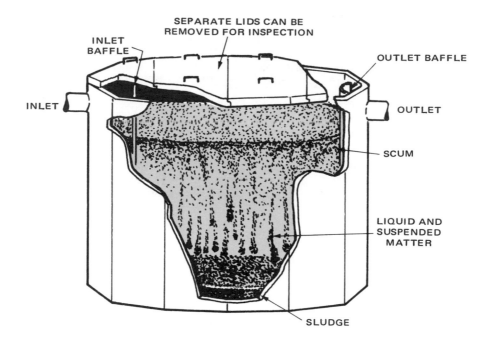

Fig. 16-3. Septic tank permits organic solids to settle out of the waste water.

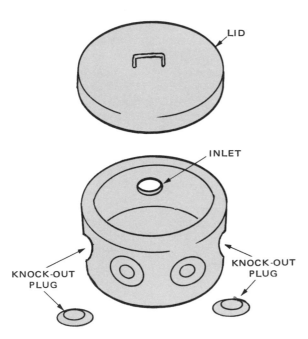

Fig. 16-4. Distribution box distributes the waste water to the many pipes in the leach field.

5. The slope of the pipes in the leach bed runs must be great enough to cause flow to occur but only at a slow rate. This takes careful planning and may require specially designed installations on property where there is considerable slope.

6. Water entering the ground from the leach bed must not pollute wells. This concern is important. It is absolutely essential that the prescribed distance from the well be maintained. This not only is true of the well on the property where the leach field is being installed, but also of any wells on adjacent property.

7. It is generally accepted that leach fields should be installed on ground of relatively low elevation.

## BASIC EQUIPMENT AND PIPE

The septic tank, Fig. 16-3, is only one of many types available. The concrete tank is probably the most commonly used. However, steel and fiber glass reinforced plastic units are also available. The design of the tank is simple. A baffle is placed at the inlet and outlet to reduce the disturbance of the water in the tank when additional waste enters, Fig. 16-5. Some baffles are simply Ts in the inlet and outlet pipes.

The tank must be of sufficient depth to permit the accumulation of solid waste for several years before cleaning is required. The only other requirements are that the tank be watertight and fitted with a lid which can be removed for cleaning.

The DISTRIBUTION BOX is also frequently made from concrete. Its function is to provide a place where the many lines of the leach field connect. Like the septic tank, the distribution box must be watertight. A fitted lid should be easy to remove for inspection.

The piping used in the installation of waste disposal systems is similar to the pipe discussed in Unit 4. Nearly all types of materials used in the manufacture of pipe can be used for septic tank/leach field systems. Clay tile and, more recently, plastic pipe dominate this type of installation. All the sealed runs of pipe would be made with standard pipe and fittings of the chosen type of material.

When plastic pipe is used for the leach bed runs, a pipe similar to regular drainage pipe but with holes in it, is installed. This permits the waste water to enter the gravel beds from the pipe.

When clay pipe is used, short lengths are laid end-to-end, Fig. 16-6, without mortar at the joints. Waste water enters the gravel bed through these open joints.

The most recent development in piping for leach beds and other types of underground drainage installations is a corru-

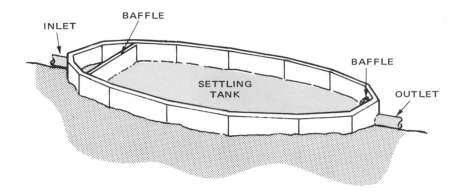

Fig. 16-5. The baffles at the inlet and outlet of the septic tank allow settling to take place.

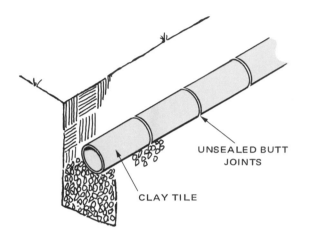

Fig. 16-6. Twelve inch lengths of clay tile butt joined without being sealed work well as leach bed runs.

gated plastic pipe. It is available either perforated for installation in leach bed runs, Fig. 16-7, or solid, Fig. 16-8. The common types of fittings are also available. They are especially designed for rapid assembly, Fig. 16-9.

Fig. 16-7. Perforated corrugated plastic pipe may be used in the leach bed runs. (Hancor, Inc.)

Fig. 16-8. Corrugated plastic pipe is flexible, lightweight and easy to install because it is available in long rolls.

## OTHER TYPES OF WASTE DISPOSAL SYSTEMS

The ever increasing need to reduce all forms of pollution has encouraged the development of new devices to add to or replace the standard septic tank. The aeration type of waste treatment is becoming popular. More recently, a completely closed system has been marketed for use in remote locations. It recycles the flush water.

### AERATION WASTE TREATMENT

The typical aeration waste treatment system replaces the septic tank with a specially designed unit having three compartments, Fig. 16-10. The waste enters the primary treatment chamber where the solid matter settles to the bottom. This solid matter is acted upon by anaerobic bacteria (bacteria that live and grow in the absence of oxygen). After the anaerobic bacteria have acted upon the sludge, it is carried into the aeration chamber. Here, oxygen (air) is passed through it causing aerobic bacteria to further break down the sludge. Untreatable materials, such as sand, remain in the primary compartment.

The final stage of treatment permits any remaining sludge to settle out of the effluent (waste water discharged from the unit) before it is discharged. The sludge which settles out is returned to the aeration chamber where further bacterial action decomposes it.

With an aerated system, it is frequently possible to reduce the size of the leach field because the effluent is clear and not likely to clog the gravel bed. In fact, aeration systems are installed without a leach field in some areas. The effluent of a properly operating unit is clear, odorless and nearly pure enough to drink. This may permit it to be discharged directly into the storm sewer or into other means of draining storm water.

### ORGANIC WASTE TREATMENT SYSTEM

A self-contained waste treatment system called the Clivus Multrum or inclining compost room does away with the need for flush water. Designed in Sweden around 1945, it is now being marketed in the United States. It accepts both bathroom

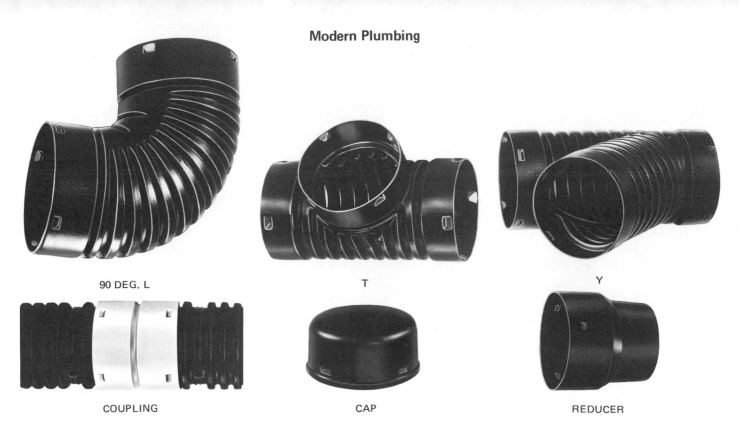

90 DEG. L   T   Y

COUPLING   CAP   REDUCER

Fig. 16-9. Standard fittings are made for corrugated plastic pipe.   (Hancor, Inc.)

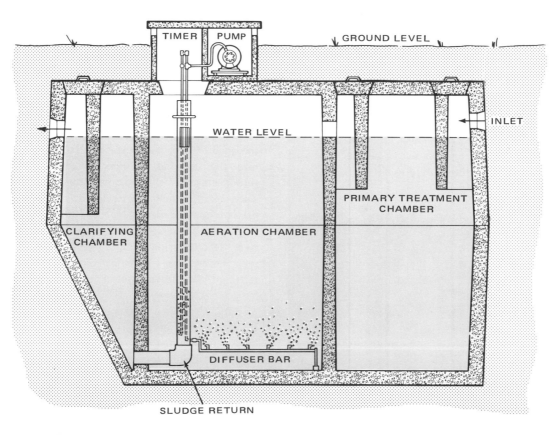

TIMER   PUMP   GROUND LEVEL

WATER LEVEL   INLET

PRIMARY TREATMENT CHAMBER

CLARIFYING CHAMBER   AERATION CHAMBER

DIFFUSER BAR

SLUDGE RETURN

Fig. 16-10. Aeration waste disposal system produces a much cleaner effluent.   (Cast Stone Co.)

and kitchen wastes which are contained in a fiber glass tank, Fig. 16-11. The wastes slide along the inclined bottom as they decompose. Tubes connect the container to a countertop garbage receptacle in the kitchen and to the toilet.

The main by-products are humus (which makes up about 5 to 10 percent of the original volume), carbon dioxide and water vapor. Gases and vapor are exhausted through the roof by way of a draft tube. After two to four years, the humus is safe for use in gardens since disease producing organisms have been destroyed by soil bacteria.

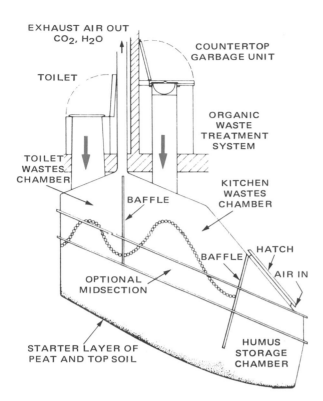

Fig. 16-11. Diagram of a waterless organic waste disposal system which converts kitchen and bathroom wastes into humus. Unit is charged with a thin layer of peat and top soil to promote bacterial action. (Clivus Multrum USA, Inc.)

The unit is designed to be installed in one and two story homes, in public buildings or in vacation homes. The container may be placed in basements, crawl spaces or outdoors under a specially designed structure.

Use of this type of disposal system depends upon local approval. To date, installations have been made in 39 states and eight Canadian provinces.

## CLOSED SYSTEM OF WASTE TREATMENT

A system developed primarily for portable toilets (commonly seen on construction sites and for use in travel vehicles) is finding some use in homes where the installation of other systems is hampered by distance, terrain and soil conditions. In this system, water is stored in a tank. The same tank serves as a settling basin for solid waste, Fig. 16-12. From the tank, the water passes through a filter into a chamber where metered amounts of chemicals are added to remove odors and help purify the water. The water is then ready to be reused to flush the water closet. Depending upon the capacity of the holding tank, this unit will need to be pumped clean and recharged with water and chemicals every 160 to 1000 times it is used.

## DISPOSING OF EXCESS GROUND WATER

In many areas of the country, basements would be wet and foundations damaged if the natural rain water were allowed to collect around the foundation and under the basement floor. To prevent this problem, drain pipe or tile are buried in gravel around the perimeter of the foundation. These pipes function exactly the opposite of those in a leach field, they collect water from the soil and channel it to the storm sewer. Generally, the water is directed to the curb along the street where it can flow naturally into the storm sewer.

In some cities, it has been the practice to connect these drains to the sanitary sewer or private waste disposal systems. This practice unnecessarily overloads the treatment system

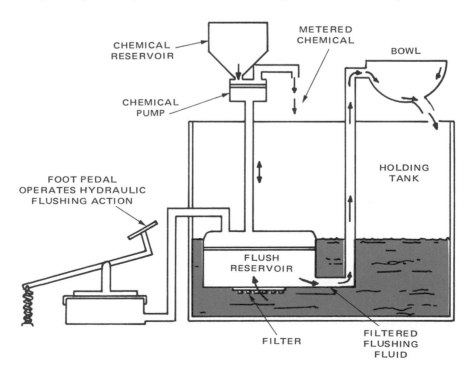

Fig. 16-12. Completely self-contained waste disposal systems recycle the flushing liquid. (Monogram Industries, Inc.)

with water that needs no treatment. For homes which have basements below the grade of the storm sewer it is necessary to install a sump which will collect the ground water. A sump pump lifts the water to street level where it can run off naturally. The sump, Fig. 16-13, is simply a closed container. It will collect water until a predetermined depth is reached. Then the sump pump, Fig. 16-14, is turned on by the float controlled switch and the water is pumped out of the sump.

Fig. 16-13. This sump is made from injection molded plastic.

## TEST YOUR KNOWLEDGE – UNIT 16

1. Before designing a private waste disposal system, which of the following government officials must be contacted?
   a. City mayor.
   b. Water department.
   c. County health department.
   d. Building inspector.
2. The primary purpose of a septic tank is to provide a place where organic solids can _____.
3. The pipe which is buried in the runs of a leach field is _____.
4. Swampy ground is an excellent place to install a leach field. True or False?
5. Why is the distance from a well to a leach field considered very important?
6. To prevent the water in a septic tank from being disturbed when additional waste enters, a _____ is required at the inlet.
7. Which of the following types of pipe can be used in constructing a leach field.
   a. Plastic.
   b. Clay tile.
   c. Corrugated plastic.

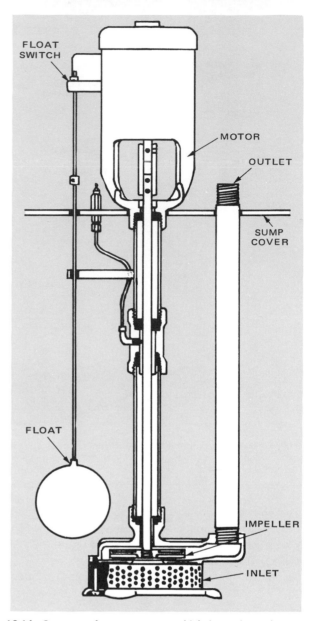

Fig. 16-14. Cutaway of a sump pump which is used to raise ground water from below the basement floor to the street level where it is carried away by the storm sewer. Check valves are often installed to keep water from backing up into the sump.

8. An aeration system provides three stages of treatment. List the three stages and briefly describe what occurs during each stage.
9. Ground water which would otherwise cause the basement to be moist can be collected in a _____ if drain tiles are installed around the foundation.
10. Ground water which is collected in the sump can be removed by a _____ pump which raises the water to the street level where it flows to the _____ sewer.

## SUGGESTED ACTIVITIES

1. Invite a representative of the county health department to discuss the advantages and limitations of various types of private waste treatment systems.
2. Visit a site where a private waste disposal system is being installed. Discuss the installation and maintenance procedures with the contractor.

# Unit 17
# HYDRAULICS AND PNEUMATICS

### Objectives

This unit will explain the nature of liquids and gases and how they act in plumbing systems. Their properties are illustrated with practical examples.

After studying the unit you will be able to:
- Explain the characteristics of fluids under pressure in water systems.
- Apply the characteristics of gases in plumbing applications to develop better plumbing systems.
- Apply practical methods of computing useful pressures, given type and size of piping and head pressures of water supplies.
- Demonstrate undesirable characteristics of gases in incorrectly designed drainage systems.

The principles of hydraulics and pneumatics explain many characteristics of water supply and drainage systems. A basic knowledge of these principles will help plumbers understand how plumbing fixtures and piping arrangements function. They will also come to understand what causes some of the problems encountered on the job.

## HYDRAULICS

HYDRAULICS is the study of the characteristics of fluids. The first principle of hydraulics explains why water pressure at the bottom of a tank increases as the depth of the water in the tank increases.

### WATER PRESSURE

A cubic foot of water weighs 62.4 lb. and exerts 62.4 lb. of pressure per square foot (psf) on the bottom of its container, Fig. 17-1. If a tank 1 ft. wide, 1 ft. across and 2 ft. high is filled with water, the pressure on the bottom of the tank is 124.8 pounds per square foot, Fig. 17-2.

The general formula for finding the pressure exerted by water of varying depths is given in Fig. 17-3. Note that the pressure was computed in pounds per square foot (psf). It is common practice to compute pressure in pounds per square inch (psi). Because there are 144 square inches of surface area in a square foot, (12 in. x 12 in. = 144 sq. in.) the formula is modified as shown in Fig. 17-4 to compute pressure in pounds per square inch.

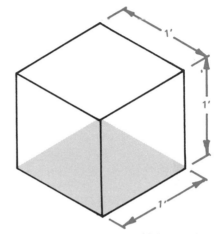

Fig. 17-1. A cubic foot of water weighs 62.4 pounds and exerts a pressure of 62.4 pounds per square foot on the base of its container.

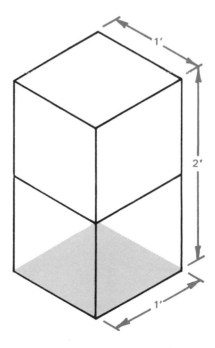

Fig. 17-2. As the depth of the water increases the pressure exerted on the bottom of a tank is increased.

This same basic principle accounts for the increase in water pressure after installing a water tower as part of a municipal water supply system. If water is stored in a tower which is 100 ft. tall, the water pressure at the base of the tower can be

P = 62.4 x D
P = PRESSURE PER SQUARE FOOT
D = DEPTH OF THE WATER IN FEET
EXAMPLE: WHAT IS THE PRESSURE PER SQUARE FOOT ON THE BOTTOM OF A TANK WHICH IS FILLED TO A DEPTH OF 12'-6" (12.5 FT.)?
P = 62.4 x D = 62.4 x 12.5 = 780 POUND PER SQUARE FOOT (PSF).

Fig. 17-3. Formula for computing the pressure exerted by water of varying depth in pounds per square foot.

$P = \dfrac{62.4}{144} \times D$
P = PRESSURE PER SQUARE INCH
D = DEPTH OF THE WATER IN FEET
EXAMPLE: WHAT IS THE PRESSURE PER SQUARE INCH TO THE BOTTOM OF A TANK WHICH IS FILLED TO A DEPTH OF 20'-9" (20.75)?
$P = \dfrac{62.4}{144} \times D = \dfrac{62.4}{144} \times 20.75 = 8.00\ PSI$

Fig. 17-4. Formula for computing the pressure exerted by water of varying depth in pounds per square inch.

computed as shown in Fig. 17-5. From the previous illustration, it can be seen that the water pressure available at buildings on the same water supply system will vary depending on their elevation with respect to the water tower.

Pressure in a water supply system is not just a downward force. If a pressure of 65 psi is available at a faucet, the 65 psi is exerted in all directions equally. When a valve or faucet is opened, the water will flow with equal force regardless of what direction the valve is pointing.

## PRESSURE HEAD

PRESSURE HEAD is the pressure available at some point in the water system. It is measured by the depth of a column of water above the point where the measurement is taken. For example, the pressure head at the base of the water tower in Fig. 17-5 is 100 ft.

From scientific studies, it is known that a column of water 1 ft. high produces a pressure of 0.43 psi. Another way of thinking about the relationship between pressure head and pressure measured in pounds per square inch is that a column of water 2.31 ft. tall will produce a pressure of 1 psi, Fig. 17-6.

For some calculations, pressure head or head loss is used rather than pressure or pressure loss. It is important that these two concepts be understood. Otherwise, calculations dependent on these measurements cannot be made correctly.

In some cases, it will be necessary to convert pressure head to pounds per square inch or to convert pounds per square inch to pressure head. The formulas and examples in Fig. 17-7, describe the appropriate procedures for making these conversions.

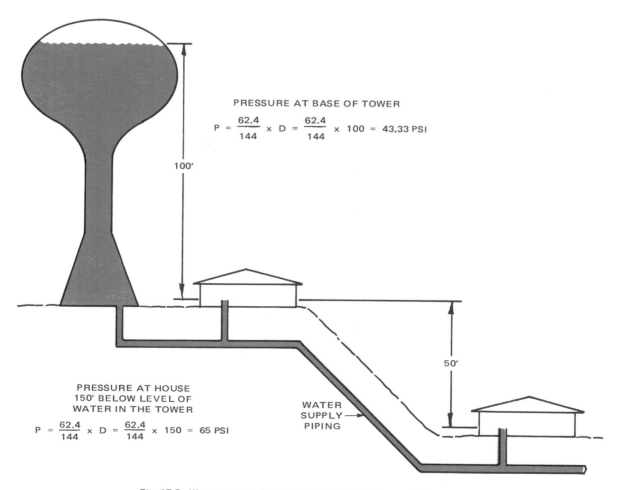

PRESSURE AT BASE OF TOWER
$P = \dfrac{62.4}{144} \times D = \dfrac{62.4}{144} \times 100 = 43.33\ PSI$

100'

50'

WATER SUPPLY PIPING

PRESSURE AT HOUSE
150' BELOW LEVEL OF WATER IN THE TOWER
$P = \dfrac{62.4}{144} \times D = \dfrac{62.4}{144} \times 150 = 65\ PSI$

Fig. 17-5. Water pressure increases as the depth of the water increases.

## FRICTION LOSS

You have just learned that water pressure is affected by elevation. The plumber must also consider the effects of friction on water pressure and flow rates when designing piping systems.

Water running through a relatively large, smooth pipe at low pressure has a smooth, streamlined flow. At the other extreme, when the pipe is rough, small in diameter and the pressure on the water high, the flow is turbulent. Streamlined flow produces little friction while turbulent flow can produce considerable friction.

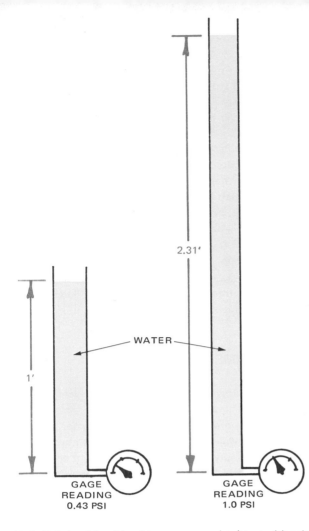

Fig. 17-6. Relationship of head to pressure can be shown with columns of water and pressure gauges.

TO CONVERT PRESSURE HEAD OF WATER TO PRESSURE MEASURED IN POUND PER SQUARE INCH:

$P = H \times .43$

P = PRESSURE (MEASURED IN PSI)

H = PRESSURE HEAD (MEASURED IN FEET OF WATER)

EXAMPLE: HOW MUCH PRESSURE (PSI) IS EXERTED BY A PRESSURE HEAD OF 45 FT. OF WATER?

$P = H \times .43$

$P = 45 \times .43$

$P = 19.35$ PSI

TO CONVERT WATER PRESSURE MEASURED IN POUND PER SQUARE INCH TO PRESSURE HEAD:

$H = 2.31\ P$

EXAMPLE: HOW MUCH PRESSURE HEAD IS REQUIRED TO CREATE A PRESSURE OF 80 PSI?

$H = 2.31\ P$

$H = 2.31 \times 80$

$H = 184.8$ FT.

Fig. 17-7. Changing pressure head to psi and psi to pressure head.

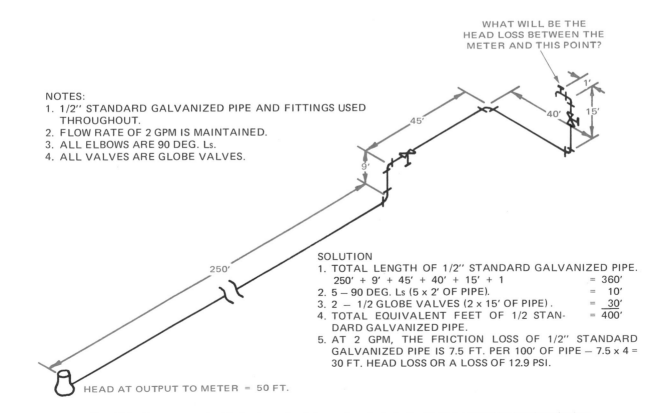

NOTES:
1. 1/2" STANDARD GALVANIZED PIPE AND FITTINGS USED THROUGHOUT.
2. FLOW RATE OF 2 GPM IS MAINTAINED.
3. ALL ELBOWS ARE 90 DEG. Ls.
4. ALL VALVES ARE GLOBE VALVES.

WHAT WILL BE THE HEAD LOSS BETWEEN THE METER AND THIS POINT?

HEAD AT OUTPUT TO METER = 50 FT.

SOLUTION
1. TOTAL LENGTH OF 1/2" STANDARD GALVANIZED PIPE.
   250' + 9' + 45' + 40' + 15' + 1 = 360'
2. 5 — 90 DEG. Ls (5 x 2' OF PIPE). = 10'
3. 2 — 1/2 GLOBE VALVES (2 x 15' OF PIPE). = 30'
4. TOTAL EQUIVALENT FEET OF 1/2 STANDARD GALVANIZED PIPE. = 400'
5. AT 2 GPM, THE FRICTION LOSS OF 1/2" STANDARD GALVANIZED PIPE IS 7.5 FT. PER 100' OF PIPE — 7.5 x 4 = 30 FT. HEAD LOSS OR A LOSS OF 12.9 PSI.

Fig. 17-8. To compute the friction loss of a piping system, include the friction created by bends and valves.

Actually, the friction comes from two sources.
1. The water rubbing against the rough walls of the pipe.
2. Particles of water striking one another.

Water pressure or pressure head is lost. It is used up in overcoming the friction.

The amount of friction loss has been determined for various kinds of piping materials. This information is given in the graphs and charts on pages 260 through 267 in the Useful Information Section. As you will see from studying them, friction losses vary for different kinds of piping materials even when diameters are the same. For example, consider the pressure head losses for the following types of 1/2 in. pipe when the flow rate is 5 gallons per minute:
1. Schedule 40 plastic pipe: head loss of 23.44 ft. per 100 ft. of pipe.
2. L grade copper pipe: head loss of 11 ft. per 100 ft. of pipe.
3. Standard galvanized pipe: head loss of 40 ft. per 100 ft. of pipe.

This head loss can be translated into a pressure drop by multiplying the head loss by the amount of pressure exerted by each foot of water depth. Thus, for every 100 ft. of 1/2 in. Schedule 40 piping that the water passes through, pressure will be reduced by 10.07 psi (23.44 x 0.43).

The loss of 1/2 in. copper would be only 4.73 psi (11 x

0.43). Galvanized piping, having a rougher wall surface, would show a loss of 40 x 0.43 = 71.2 psi.

Friction created by fittings and valves must also be considered. As will be noted from the example which follows, the effect of fittings and valves becomes extremely important when small diameter water piping is being installed.

Tables of friction loss due to fittings and valves are given in the Useful Information Section, pages 266 and 267. These losses are expressed in an equivalent number of feet of pipe of the same material. For example:
1. A 1/2 in. galvanized steel 90 deg. elbow produces a friction loss equal to 2 ft. of 1/2 in. standard galvanized pipe.
2. A 1/2 in. copper 90 deg. elbow produces a friction loss equivalent to 1 ft. of 1/2 in. copper pipe. Using the tables in the Useful Information Section, page 264 as a reference, the loss of a piping system can be computed as described in Fig. 17-8.

## WATER HAMMER

When water flowing through a pipe is suddenly stopped by closing a valve, considerable force must be absorbed by the piping system in order to stop the flow of water. Being nearly incompressible, water is like a battering ram moving toward a fixed object (rapidly closing valve). The weight and velocity of the water creates a large amount of force. So great is the force that it changes the shape of the object which is hit. Fig. 17-9 illustrates how water hammer occurs.

Water hammer frequently causes pipes to vibrate creating considerable noise in the piping system. It is not uncommon for pipes to burst as a result of water hammer. The principles of pneumatics come into play in solving this problem.

## PNEUMATICS

The study of compressible gases (usually air) is known as PNEUMATICS. Air has common uses in plumbing systems:
1. The shallow well pump depends upon air pressure to force water from a well.
2. The hydropneumatic water storage tank is commonly used to provide a reliable supply of water at a constant pressure.
3. Antihammer devices use trapped air to cushion and absorb the energy of water hammers.

All of these uses were discussed in Unit 15, Water Supply Systems.

Pneumatics can also create some undesirable effects when air is trapped in water supply or drainage systems.

## AIR LOCKS

An upward bend in a low-pressure piping system (DWV) which extends above the regular flow line of the pipe run, Fig. 17-10, is likely to allow air to accumulate in the bend. The air acts as a blockage in the pipe and either reduces or completely stops the flow through the pipe. Because of this potential problem, it is necessary to eliminate upward bends on horizontal runs of DWV piping.

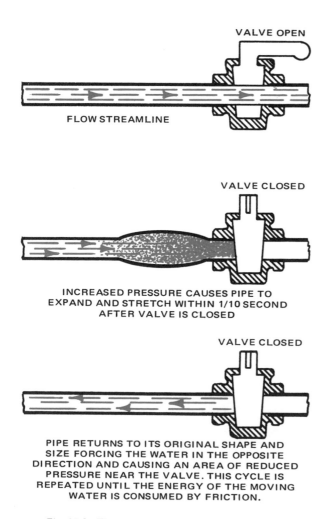

VALVE OPEN

FLOW STREAMLINE

VALVE CLOSED

INCREASED PRESSURE CAUSES PIPE TO EXPAND AND STRETCH WITHIN 1/10 SECOND AFTER VALVE IS CLOSED

VALVE CLOSED

PIPE RETURNS TO ITS ORIGINAL SHAPE AND SIZE FORCING THE WATER IN THE OPPOSITE DIRECTION AND CAUSING AN AREA OF REDUCED PRESSURE NEAR THE VALVE. THIS CYCLE IS REPEATED UNTIL THE ENERGY OF THE MOVING WATER IS CONSUMED BY FRICTION.

Fig. 17-9. The causes and effects of water hammer.

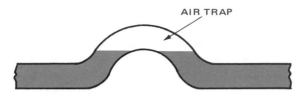

Fig. 17-10. An upward bend in piping can cause air locks which will block the flow through the pipe.

## PNEUMATICS OF DWV STACKS

The DWV piping system functions because gravity carries the waste water downhill. Under normal conditions both the sanitary and storm sewer are only partially filled with water. The remaining space is occupied by air. To understand why it is necessary to select the proper size DWV piping and install vents correctly, examine Fig. 17-11. It is a simple example of what happens when water is discharged through the sewer system. The drawing illustrates two DWV piping systems.

Building A has a correctly sized and installed DWV system. The DWV stack is large enough to prevent the discharge from a fixture from completely filling the stack.

The smaller stack in Building B causes water to travel through the DWV piping as a slug much like a piston moves in a cylinder. The water traveling through the stack compresses the air ahead of it and may force the water out of traps at lower levels in the building. If this were to occur, sewer gas could enter the building through the unfilled traps. The building drain for Building B is incorrectly installed because it is below the normal flow line of the sewer main. This means that pressure increases in the building drain ahead of the slug of water. The trapped air must escape either by blowing through the water in the main or by blowing back through the onrushing slug of water. In either case, the flow of the sewage through the building drain will be slowed. Eventually, this reduced flow will certainly cause trouble. Water which is stopped or which flows too slowly allows solids to settle. Eventually, a blocked building drain will result.

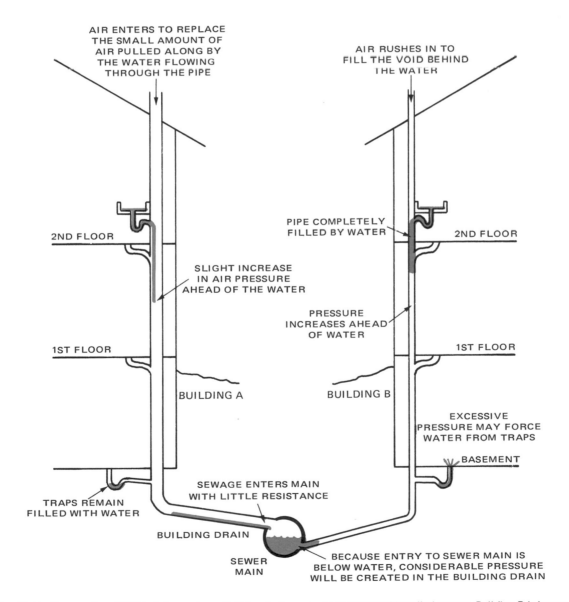

Fig. 17-11. Pneumatics of DWV piping systems. Building A shows properly sized and installed system. Building B is incorrect.

## TEST YOUR KNOWLEDGE — UNIT 17

1. A column of water 14 ft. tall will exert a pressure of _____ pounds per square foot on the bottom of its container.
2. A column of water 22 ft. tall will exert a pressure of _____ pounds per square inch on the bottom of its container.
3. The pressure head available at the base of a 50 ft. water tower is 50 ft. True or False?
4. The pressure in pounds per square inch available at the base of the tower in the previous question will be:
   a. 34 psi.
   b. 21.5 psi.
   c. 18 psi.

5. The friction loss per 100 ft. of 3/4 in. type M copper pipe is _____ psi when the flow rate is 5 gallons per minute (see Useful Information, page 264).
6. A 1 in. plastic 90 deg. L has a friction loss equal to _____ ft. of 1 in. plastic pipe (see Useful Information, page 260).

## SUGGESTED ACTIVITIES

1. Assume that the pipe and fitting in Fig. 17-8 are all 1/2 in. type M copper. Otherwise no change has been made in the piping. Compute the head loss due to friction using the tables in the Useful Information Section, page 264.
2. Repeat Activity 1 assuming that all pipe and fittings are 1/2 in. Schedule 40 plastic.

# Unit 18
# INSTALLING AIR CONDITIONING SYSTEMS

### Objectives

This unit explains systems for conditioning air and describes the plumber's tasks in installing them.

After studying the unit you will be able to:
- Show the basic operation of each system and its parts.
- Explain the jobs plumbers are required to do while installing the systems.

Air conditioning provides a comfortable climate within a structure. This means more than having enough heat to keep warm in winter and enough refrigeration to be cool in summer. A good air conditioning system controls all the principal factors that affect human comfort. This means the right temperature, the right amount of moisture and a controlled supply of fresh, clean, odorless air. The term air conditioning is being used in its technical sense: any system which controls three or more of the environmental factors which contribute to human comfort. Therefore, cooling is only one of the major elements of climate control.

## BASIC COMPONENTS OF AN AIR CONDITIONING SYSTEM

The basic units of an air conditioning system are listed in Fig. 18-1. A heating unit and a refrigeration unit give full temperature control. Moisture is controlled within a HUMID-IFIER and a DEHUMIDIFIER. The air is cleaned by a filter. In some systems, fans are used to move the air.

We will talk about each part of the system briefly. But procedures will cover only the installation of pipe and fittings which connect the air conditioning components to the water, gas and sewer piping. The design and total installation of such systems is beyond the scope of this book. Yet, the plumber needs to know about the basic parts of the systems and how they are combined. Only then can the plumbing portions of the installation be done correctly.

## TEMPERATURE CONTROL

Temperatures in inhabited buildings are controlled with heating and refrigeration devices. A heating system consists of a complex of mechanical and electrical devices. These devices:
1. Convert some form of energy into heat.
2. Circulate and distribute the heat.
3. Automatically control heating through an automatic thermostat.

Many different systems can be used to heat a structure. Some systems combine several air conditioning functions.

### Heat transfer

Heating and cooling systems both depend on heat transfer for their operation. Heat is only transferred in one direction. It moves from a warmer object to a cooler object. In heating, the heating system transfers heat from the warmer heat source to

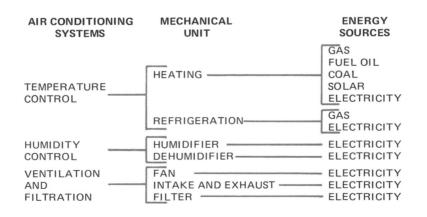

Fig. 18-1. Air conditioning systems control three or more of the environmental factors which contribute to human comfort within a structure.

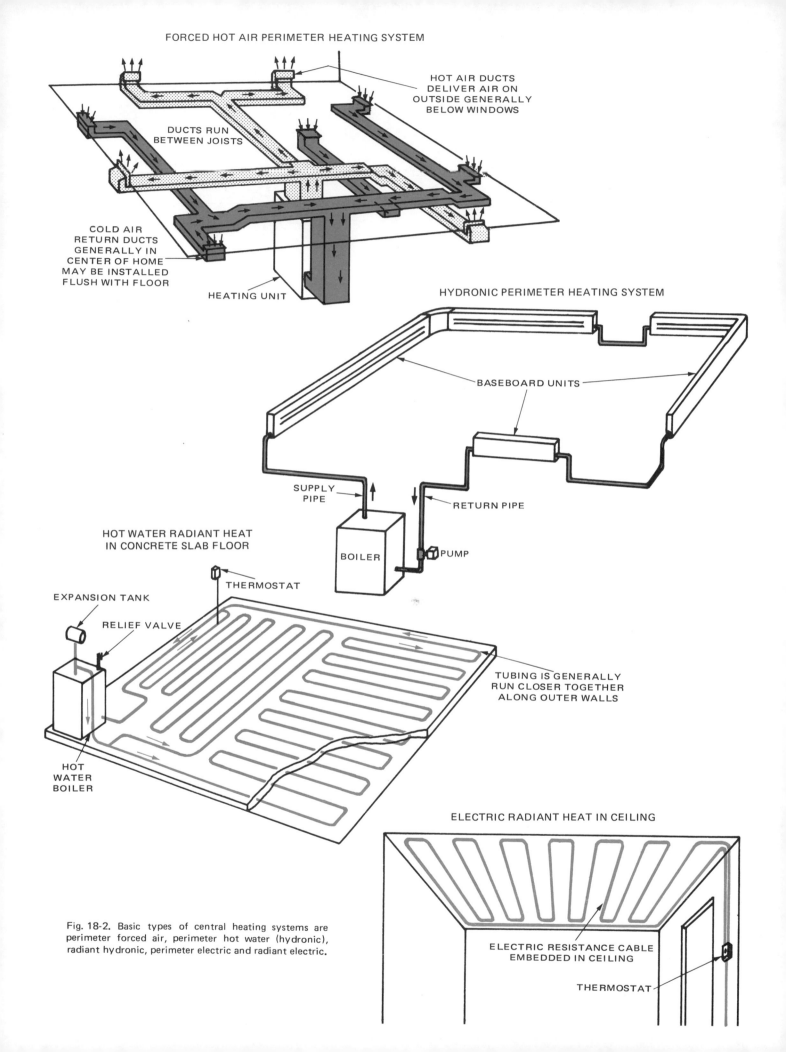

FORCED HOT AIR PERIMETER HEATING SYSTEM

HOT AIR DUCTS DELIVER AIR ON OUTSIDE GENERALLY BELOW WINDOWS

DUCTS RUN BETWEEN JOISTS

COLD AIR RETURN DUCTS GENERALLY IN CENTER OF HOME MAY BE INSTALLED FLUSH WITH FLOOR

HEATING UNIT

HYDRONIC PERIMETER HEATING SYSTEM

BASEBOARD UNITS

SUPPLY PIPE

RETURN PIPE

BOILER

PUMP

HOT WATER RADIANT HEAT IN CONCRETE SLAB FLOOR

EXPANSION TANK

RELIEF VALVE

THERMOSTAT

TUBING IS GENERALLY RUN CLOSER TOGETHER ALONG OUTER WALLS

HOT WATER BOILER

ELECTRIC RADIANT HEAT IN CEILING

ELECTRIC RESISTANCE CABLE EMBEDDED IN CEILING

THERMOSTAT

Fig. 18-2. Basic types of central heating systems are perimeter forced air, perimeter hot water (hydronic), radiant hydronic, perimeter electric and radiant electric.

the cooler house. Conversely, the cooling system absorbs the heat from the house and moves it to the outside air.

Residential heating systems are available in endless varieties. However, most differences relate to:

1. The available energy sources.
2. The particular heating devices in use.
3. The heat distribution design used.

Energy to produce the heat for residential structures comes from several sources:

1. Gas.
2. Fuel oil.
3. Coal.
4. Electricity.
5. Solar.

The choice of which to use depends on factors such as:

1. Availability of the energy source.
2. Reliability of the source.
3. Compatibility with the total system design.
4. Initial cost.
5. Operating cost.

Residential heating is commonly provided either by several individual room/area heaters or by one centrally located heating plant distributing heat to the entire structure. Individual room/area heaters use different methods of heat distribution and use a variety of fuels. Heat units include fireplaces, stoves, floor furnaces, in-the-wall heaters or portable heaters. These individual room/area heaters are adequate for heating single enclosed areas but one heater cannot be expected to heat an entire house uniformly.

### Central heating

Central heating has almost completely replaced individual area heaters in residential construction. In central heating, a larger single heating plant produces heat at a centrally located point. The heat is circulated and distributed throughout the structure by perimeter heating or radiant heating, Fig. 18-2.

In perimeter heating, the heat outlets are placed on the outside walls of the structure since the main loss of heat is through windows and outside walls. The heat outlets, usually on or near the floor, allow the heat to rise and warm these cold areas. Consequently, the entire house is warmed. Heat is moved from the heating plant through the building by:

1. Warm air circulating systems.
2. Hot water radiators/convectors.
3. Electrical resistance heating radiators/convectors.

Radiant heating is generally placed in the floors or ceilings of the structure. Hot water piping or electrical resistance wiring is "buried" under the surface. The heat from the warmed surfaces is transferred by radiation, convection and conduction. Electric resistance wiring systems eliminate the need for a central heating plant.

### SOLAR HEAT

Up to now, the sun has been used mostly to help the heating system. Dwellings have been designed to take advantage of the sun's heat energy:

1. By properly placing the house on the site, it is exposed to

the winter sun which is lower on the southern horizon. This provides heat gain so the heating unit consumes less fuel.

2. This heat gain is further increased by having large windows on the south side of the dwelling.

Because of the need to conserve fuel, the use of solar collectors should become more popular. These collectors can eventually provide the primary heat supply for a structure.

The most flexible heating system in use today appears to be a central heating plant combined with a perimeter warm air distribution system. The system is readily adaptable to cooling, humidity control, air filtration and ventilation. From the point of view of the plumber, the hydronic heating systems are the most interesting because their installation requires extensive piping.

### HYDRONIC HEATING SYSTEMS

The simplest form of a hydronic heating system is the single pipe forced circulation system illustrated in Fig. 18-2. In this system, the water is heated in the boiler. Then it is pumped through the pipe to the radiators where heat is given off. From the radiators, the water flows back to the boiler where it is reheated.

The expansion tank serves as a reservoir for the increased volume of water after heating. The relief valve is a safety device which will release excess water pressure in case the system malfunctions. The pressure reducing valve limits the pressure of the incoming fresh water supply. The air vents allow any air which may enter the system to escape at high points in the piping.

A major disadvantage of the one-pipe system is its inability to heat buildings evenly. Only one control activates the pump which circulates the hot water. When the control calls for heat, all radiators receive it whether the room is in need of it or not. In some one-pipe systems, the water must flow from one radiator to the next. The water gives up more and more heat as it goes from one radiator to the next radiator in line. As a result, rooms which receive the heat first can tend to overheat while those at the end of the line tend to be cooler.

These objectional features can be overcome by installing a system of risers off the main piping and several zone valves. See Fig. 18-3.

Fig. 18-4 shows a hydronic heating system of this type. Valves have been installed to control the flow of hot water through two parts of the system which receive heat independently of each other. In installations of this type, for example, one thermostat and valve will control heating of bedrooms. A second thermostat and valve will control heat to the rest of the house.

By installing thermostats and valves (zone valves) at each radiator, Fig. 18-5, the temperature of each room can be controlled individually. This type of hydronic heating system is known as a multizone hydronic system.

Two-pipe forced circulation systems, Fig. 18-6, provide more uniform heating than the simple one-pipe system because the water entering each radiator is heated closer to the same temperature. This more even temperature is a result of the hot

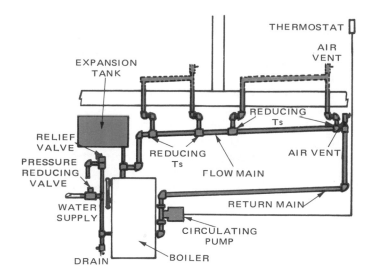

Fig. 18-3. One-pipe system is more efficient if risers connect radiators to the flow main.

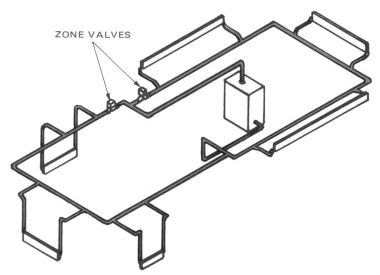

Fig. 18-4. Zone valves can be installed to control separate sections of a one-piece hydronic heating system.

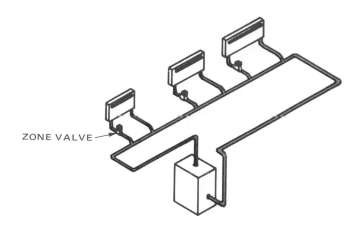

Fig. 18-5. Installation of a zone valve at each radiator permits individual control of the temperature in each room.

water passing through fewer pipe and radiators. The two-pipe system can be designed as a zone or multizone system by installing the zone valves at the right locations in the piping.

## COOLING SYSTEMS

Residential construction normally uses the same air distribution system for heating and cooling. The cooling system must remove heat from the house. This is usually done through a perimeter air distribution system.

The two basic cooling systems in common use in residential construction are the remote system and the integral (self-contained) system. Both use the same basic mechanical refrigeration device.

A central air cooling unit for a house consists of a condensing unit located outdoors and a heat exchange coil or evaporator which is located in the bonnet of the furnace.

The furnace fan circulates warm air from the rooms through the heat exchange coil. The coil removes heat and moisture from the air. The cooled air is redistributed through the house.

At the same time, the moisture in the air is removed. It simply condenses on the cool surface of the heat exchange coil and is drained away. The heat from the air is absorbed by a liquid refrigerant, such as freon, circulating inside the heat exchange coil. The refrigerant, while absorbing the heat, changes from a liquid to a gas. It is then pumped to a compressor and compressed to a high pressure. The absorbed heat is forced out of the refrigerant by:

1. Being subjected to high pressure.
2. Being cooled in a condenser.

The heat given off during the refrigerant's condensation is released to the outside air. This refrigerating cycle continues absorbing heat inside the house and releasing it to the outside until the desired inside temperature is reached, Fig. 18-7.

Some air cooling units use refrigerants to cool a solution of water and glycol (antifreeze). This is circulated through the heat exchange coil in the furnace bonnet.

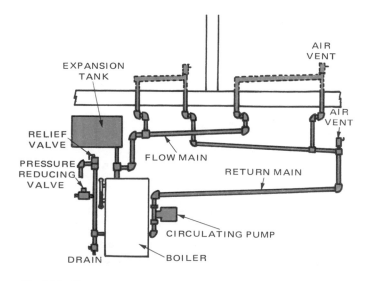

Fig. 18-6. Two-pipe hyrdronic system provides more uniform heating than the simple one-pipe system.

# Installing Air Conditioning Systems

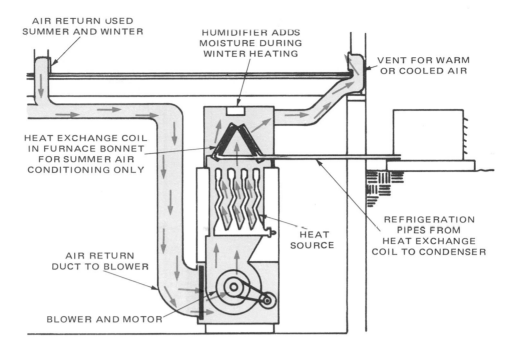

AIR RETURN USED
SUMMER AND WINTER

HUMIDIFIER ADDS
MOISTURE DURING
WINTER HEATING

VENT FOR WARM
OR COOLED AIR

HEAT EXCHANGE COIL
IN FURNACE BONNET
FOR SUMMER AIR
CONDITIONING ONLY

HEAT
SOURCE

REFRIGERATION
PIPES FROM
HEAT EXCHANGE
COIL TO CONDENSER

AIR RETURN
DUCT TO BLOWER

BLOWER AND MOTOR

Fig. 18-7. Basic parts of the refrigeration unit of a central air conditioning system include heat exchange coil and condenser. They are attached to an existing forced air furnace.

## HEAT PUMP

The heat pump used for residential heating and cooling is like a mechanical refrigeration device. It operates on the principle just described. With a reversing valve, it produces heated air on the heating cycle and cooled air on the cooling cycle. Changing the valve causes the condenser and the evaporator (heat exchange coil) to reverse roles. The condenser becomes the evaporator and the evaporator becomes a condenser.

The remote central air conditioning system has the compressor and condenser outside the house. The refrigerant is piped to the heat exchange coil in the furnace bonnet or air duct.

It is common practice to set the remote condensing unit on a concrete base in the yard. This system is becoming increasingly popular because it places the major source of noise and vibration where it will least affect the house.

The self-contained unit has the heat exchange coil in the same cabinet as the compressor and condenser. Placed either inside or outside the house, warm air passes through the cabinet where it is cooled and returned to the house's air distribution system.

## HUMIDITY CONTROL

Humidity control is the process by which moisture is added to or removed from a given quantity of air. The comfort range for most people is 30 to 70 percent relative humidity.

*Relative humidity is the amount of moisture (water vapor) in the air compared to the amount it could actually hold at a given pressure and temperature.* The comparison is expressed in percentages. For example, a relative humidity of 50 percent means there is half as much water vapor in the air as it could

hold at that temperature and pressure.

Excess moisture in the home comes from many sources, such as cooking, cleaning, washing and from outside air. To remove this moisture from the air, adequate ventilation and a dehumidification system are necessary. A dehumidification system removes the moisture from damp air by passing it over cold coils where it condenses.

If the air is too dry, a humidification system adds moisture to the air by:
1. Exposing it to a large surface of water.
2. Spraying atomized water into it.
3. Introducing heated water vapor into the air (steam).

## VENTILATION AND CLEANING THE AIR

*Ventilation is the process of changing air within an enclosed space by supplying and distributing fresh air and exhausting used air.* Closely allied to ventilation is air filtration. Adequate ventilation and air filtration alter the properties of the air in an enclosed area in the following ways:
1. Replaces oxygen used in human metabolism.
2. Removes carbon dioxide, a waste product of metabolism.
3. Reduces the intensity of odors.
4. Removes dust particles.
5. Reduces the bacteria content of the air.
6. Reduces the pollen content of the air.
7. Helps control the moisture in the air.

The simplest type of ventilation system is cross ventilation through open windows. However, exhaust fans and forced air filtration and ventilation are generally provided in residential construction. Electrostatic air cleaning, by attracting dirt to a charged grid, further improves the properties of air in an enclosed area.

### CAPACITY OF PIPE TO CARRY GAS IN CU. FT./HR.

| LENGTH IN FEET | NOMINAL IRON PIPE SIZE, INCHES | | | | | | | | |
|---|---|---|---|---|---|---|---|---|---|
| | 1/2 | 3/4 | 1 | 1 1/4 | 1 1/2 | 2 | 2 1/2 | 3 | 4 |
| 10 | 132 | 278 | 520 | 1050 | 1600 | 3050 | 4800 | 8500 | 17,500 |
| 20 | 92 | 190 | 350 | 730 | 1100 | 2100 | 3300 | 5900 | 12,000 |
| 30 | 73 | 152 | 285 | 590 | 890 | 1650 | 2700 | 4700 | 9700 |
| 40 | 63 | 130 | 245 | 500 | 760 | 1450 | 2300 | 4100 | 8300 |
| 50 | 56 | 115 | 215 | 440 | 670 | 1270 | 2000 | 3600 | 7400 |
| 60 | 50 | 105 | 195 | 400 | 610 | 1150 | 1850 | 3250 | 6800 |
| 70 | 46 | 96 | 180 | 370 | 560 | 1050 | 1700 | 3000 | 6200 |
| 80 | 43 | 90 | 170 | 350 | 530 | 990 | 1600 | 2800 | 5800 |
| 90 | 40 | 84 | 160 | 320 | 490 | 930 | 1500 | 2600 | 5400 |
| 100 | 38 | 79 | 150 | 305 | 460 | 870 | 1400 | 2500 | 5100 |
| 125 | 34 | 72 | 130 | 275 | 410 | 780 | 1250 | 2200 | 4500 |
| 150 | 31 | 64 | 120 | 250 | 380 | 710 | 1130 | 2000 | 4100 |
| 175 | 28 | 59 | 110 | 225 | 350 | 650 | 1050 | 1850 | 3800 |
| 200 | 26 | 55 | 100 | 210 | 320 | 610 | 980 | 1700 | 3500 |

Fig. 18-8. The correct size of black iron pipe required to install a gas-fired heating unit can be selected by following the requirements given in this chart. (North American Heating and Air Conditioning Wholesalers Assoc.)

## INSTALLATION OF AIR CONDITIONING SYSTEMS

Generally, the plumber's work on the installation of an air conditioning system is limited. It involves the installation of piping which connects the various components to the water, gas, oil supply piping and possibly to the sanitary sewer. In addition, hot water and steam (hydronic) heating units require extensive piping to circulate the heated water or steam. The special practices employed by the plumber to install an air conditioning system will be discussed briefly in the remainder of this unit. The basic skills of cutting and fitting pipe were discussed in Unit 9 and need not be repeated. Only where a particular type of air conditioning component requires special piping will these requirements be considered.

### INSTALLING HEATING UNITS

Gas-fired heating units require piping of natural gas from the meter to the heating unit. The first concern is to select pipe and fittings large enough to deliver the proper amount of gas to the heating unit. See Fig. 18-8. The pipe size must be selected by matching the requirements of the heating unit with the capacity of the various sizes of pipe. Black iron pipe is generally used instead of the more expensive galvanized iron

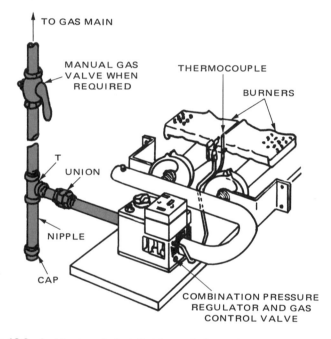

Fig. 18-9. A drip trap is installed in vertical gas piping immediately before the piping connects to the heating unit.

pipe. This is satisfactory because gas is dry and has little tendency to rust pipe and fittings.

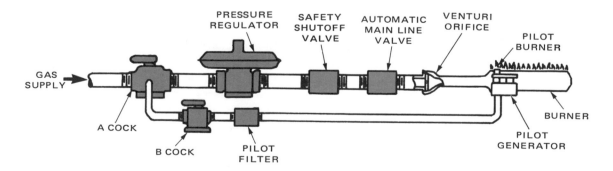

Fig. 18-10. A typical burner unit of a gas-fired heating unit. (North American Heating and Air Conditioning Wholesalers Assoc.)

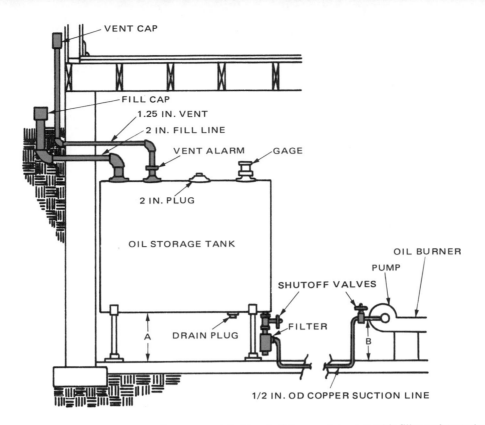

Fig. 18-11. Installation of a fuel oil storage tank inside a building requires an outside filler and vent pipe.

The gas supply piping should be assembled as shown in Fig. 18-9. The drip trap below the T connection is designed to catch any foreign particles in the gas piping which might damage the heating unit regulator and valves, Fig. 18-10. The manual shutoff valve is installed as a safety device and to permit disconnecting of the heating unit without turning off all other gas-fired equipment.

Oil-fired heating units require installation of a storage tank for fuel oil and piping to carry the fuel to the heating unit. Fig. 18-11, illustrates the correct installation of a fuel oil storage tank inside a building. Note that a 2 in. filler pipe and a 1 1/4 in. vent pipe extend to the outside of the structure. A shutoff valve and filter are installed at the outlet of the tank to permit control of the fuel supply and trap any particles which

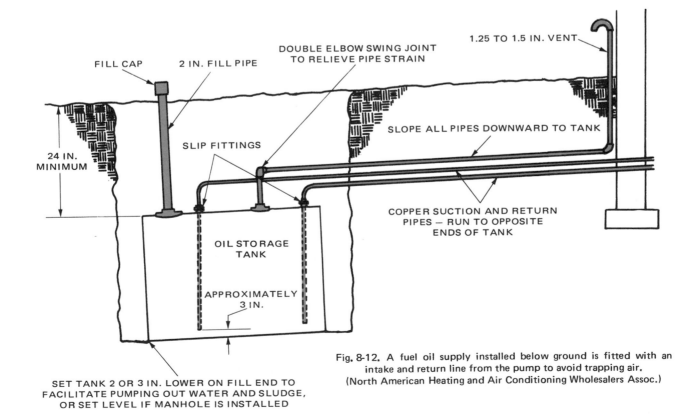

Fig. 8-12. A fuel oil supply installed below ground is fitted with an intake and return line from the pump to avoid trapping air.
(North American Heating and Air Conditioning Wholesalers Assoc.)

may have gotten into the tank. Fig. 18-12 describes the installation of a fuel oil tank underground. This type of installation requires a pump to deliver the fuel oil to the burner. Because all the oil delivered by the pump may not be burned, a return line is installed so the excess will flow back to the tank. Some codes require the installation of safety valves near the heating unit which will automatically turn off the fuel oil supply in case of fire.

Solar heating units are receiving considerable attention now that fuel supplies are scarce and expensive. The installation of the collector panels, Fig. 18-13, will require piping to a heat exchanger. Neither the panels nor the piping are standardized. The plumber will find it necessary to follow the plans carefully. These are supplied by the architect or engineer.

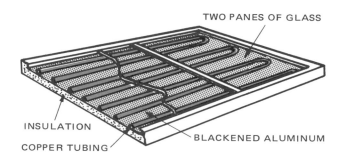

Fig. 18-13. Solar heating panels are now being marketed which collect heat by circulating water through tubing attached to a blackened plate. The plate absorbs heat which is then conducted to the tubing.

Installation of piping for hydronic or radiant heating requires quality work in cutting and fitting the pipe. Because of the pressure exerted by the hot water and steam, special precautions should be observed to make certain that all pipe and fittings are carefully installed. In the case of the radiant system, the pipe is assembled and embedded in the flooring material. *Such piping must be tested before the floor is placed. Leaks are expensive to repair once the flooring is installed.*

Piping connections in hydronic heating systems should be made according to the manufacturers' directions unless local codes require a different piping arrangement. The local codes must always be followed.

In forced circulation hot water systems, it is not necessary to pitch horizontal sections of piping so that all water will drain toward the boiler. However, piping must be installed in a manner which will prevent air pockets in the pipes. Trapped air will reduce the efficiency of the system and may cause it to be noisy during operation. When it is impossible to eliminate potential air traps, vents must be installed as in Fig. 18-14.

Install drain valves at all low points to permit all water to be drained from the system. Water trapped in the piping would freeze causing considerable damage should the system be shut down during cold weather, Fig. 18-15.

To reduce vibration and prevent sagging of pipes, pipe hangers are installed as in Fig. 18-16. The spacing of the hangers can be determined from the chart in Fig. 18-17.

Since metal pipe expands when it is heated, hangers should allow some movement in the pipe. Pipe should not bind on any part of the building. Expansion of iron and copper pipe

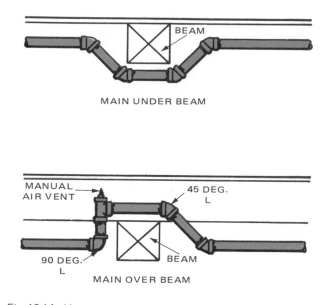

Fig. 18-14. Vent potential air pockets on hydronic heating systems. (North American Heating and Air Conditioning Wholesalers Assoc.)

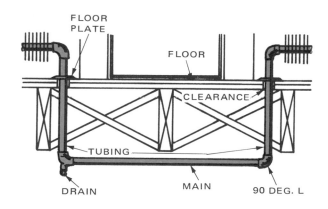

Fig. 18-15. Drains are installed at low points in hydronic heating systems to permit the system to be completely drained.

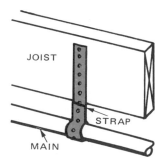

Fig. 18-16. Providing support for piping prevents noise and sagging. (North American Heating and Air Conditioning Wholesalers Assoc.)

can be calculated from the chart in Fig. 18-18. By making the necessary allowances, noise and leak causing pressure on fittings can be eliminated.

The installation of radiators is illustrated in Fig. 18-19. *Note that some radiator/convectors should be installed with a slight pitch.* Consult the manufacturers' directions before attaching the radiator/convectors.

Fig. 18-20, shows the connection of a unit heater. Valves

| NOMINAL PIPE OR TUBE SIZE INCHES | MAXIMUM SPAN FEET |
|---|---|
| 1/2 | 5 |
| 1 | 7 |
| 1 1/2 | 9 |
| 2 | 10 |
| 3 | 12 |
| 3 1/2 | 13 |
| 4 | 14 |

Fig. 18-17. Correct spacing of pipe hangers on hydronic piping can be estimated from this table.

| AVERAGE WATER TEMPERATURE F | EXPANSION PER IRON INCHES | 100 FT. OF PIPE COPPER INCHES |
|---|---|---|
| 160 | 0.7 | 1.0 |
| 180 | 0.9 | 1.3 |
| 200 | 1.0 | 1.5 |
| 220 | 1.2 | 1.7 |

Fig. 18-18. Expansion of heated metal piping requires allowances for small amounts of movement.

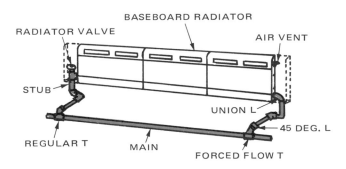

STANDARD PIPING ARRANGEMENT

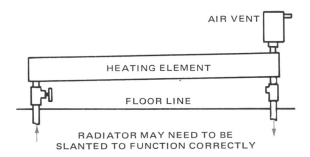

Fig. 18-19. Connecting radiators/convectors to the hydronic system will be made using Ts and elbows.

control the heat supply to the heater, the return to the boiler and drainage of the unit.

## INSTALLING REFRIGERATION UNITS

Some refrigeration units require a drainage connection to the DWV piping system. As air passes through the refrigeration unit, moisture condenses on the coils and must be drained

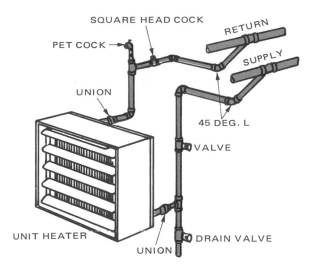

Fig. 18-20. Standard piping for a hydronic unit heater includes valves

away. This is usually a simple process. The plumber connects a small diameter pipe to the DWV piping or runs a pipe to a location near a floor drain.

## INSTALLING HUMIDIFIERS AND DEHUMIDIFIERS

Humidifiers require that water be piped to them. Generally, soft copper tubing is used. In many cases, it is connected to the cold water supply piping by a saddle valve, Fig. 18-21. The saddle valve can be installed as follows:

1. Remove dirt and corrosion from the surface of the pipe at the point where the tap is to be made.
2. Install saddle strap and valve housing around the pipe.
3. If water supply system is in operation, shut off the water.
4. Some saddle valves have guides for drilling. See Fig. 18-22. Attach a drill guide to the housing. The guide has threads which match those on the valve housing.
5. Using a portable electrical drill and a metal cutting bit, (usually 1/4 in.) drill a hole in the wall of the pipe.
6. Remove drill guide and clean out metal chips.
7. Install the faucet or the stem and packing nut (depending on the type of saddle valve).

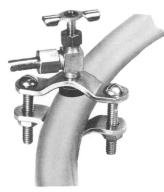

Fig. 18-21. A saddle valve is frequently used to connect humidifiers to the water supply piping. (Parker — Hannifen Corp.)

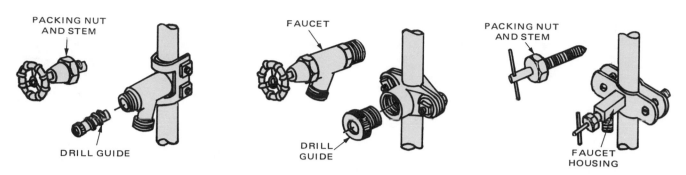

Fig. 18-22. Three types of saddle valves as they are attached to pipe, ready for drilling opening in the pipe.

8. Turn on the water supply valve.
9. Test the saddle valve for proper operation and to flush out remaining chips of metal left behind by the drilling operation.

Plumbing for dehumidifiers on central air conditioning systems is installed as described for the refrigeration unit. Sometimes separate dehumidifiers are installed. Most of these units are portable and are not connected to the plumbing.

## TEST YOUR KNOWLEDGE – UNIT 18

1. The term air conditioning means any system which controls three or more of the environmental factors which contribute to human comfort within a structure. True or False?
2. The basic components of an air conditioning system which provides temperature control are the _____ unit and the _____ unit.
3. The energy sources most commonly used to produce heat include gas, fuel oil, electricity and coal. True or False?
4. Dehumidifiers add moisture to the air. True or False?
5. Ventilation and air filtration alter the properties of the air in a building in seven ways. List four of the seven.
6. A gas-fired heating unit which requires 30 ft. of gas pipe to be connected to the main and consumes gas at a rate of 130 cu. ft. per hour should be connected with pipe having a diameter of:

a. 1/2 in.
b. 3/4 in.
c. 1 in.
d. 2 in.

7. To prevent solid particles from passing through the gas piping where they could damage the regulator and valves in the heating unit, a _____ is installed.
8. The efficiency of hydronic heating systems is reduced by air trapped in the system. _____ are installed at any high point in the piping to eliminate trapped air.
9. Pipe hangers on 3/4 in. diameter pipe used to install a hydronic heating system should not be spaced more than _____ feet apart.

## SUGGESTED ACTIVITIES

1. Have students study the air conditioning system in their own homes and, if possible, several other houses to:
   a. Find out what type of components are in the system.
   b. Learn how the heating unit is connected to the energy supply.
   c. Determine what special requirements of other components have been satisfied through some type of piping.
2. Investigate the air conditioning system in your school and compare its components and their installation to those usually found in a residential structure.

# Unit 19
# MAINTAINING AND REPAIRING PLUMBING SYSTEMS

## Objectives

The hardest part of maintaining plumbing systems is in recognizing where the trouble lies. This unit deals with the procedure for troubleshooting and repair by observing the symptoms.

After studying this unit you will be able to:
- Recognize plumbing problems through the symptoms the system displays.
- Describe an orderly method of checking and testing which will confirm actual problem.
- Demonstrate or explain procedures for making proper repair when the problem is located.

Proper maintenance and repair of plumbing systems is the only way to insure that these systems will continue to give healthful, sanitary service. A good plumber, like a good doctor, learns to observe the "patient" carefully before deciding what must be done. First the symptoms are noted. Then the causes are probed step-by-step. Finally, the remedy is decided upon.

This unit, then, will stress symptoms. In fact, it is organized around them.

Plumbing troubles fall into three major groups:
1. Those which relate to the water supply.
   a. Faucets and valves.
   b. Water closets.
   c. Water supply piping.
2. Those which relate to the drainage system.
   a. Water closets.
   b. Lavatories, tubs, showers or sinks.
   c. DWV piping.
3. Those which relate to hot water.
   a. Water not being heated.
   b. Too little hot water.
   c. Water too hot.
   d. Leaks in the tank.
   e. Noise in the tank.

Within each of these major groups specific symptoms can be identified. In this unit we will first describe these symptoms. Then we will suggest the possible causes. Finally,

the proper repair will be described.

No repair should be made without first isolating the problem. This unit of Modern Plumbing is based on this advice. Follow these three basic steps:
1. Identify the symptoms by observing what is happening to the water.
2. When more than one problem could cause the symptom, test to see which is causing it.
3. When the cause is found, use proper repair procedures.

Using this approach, the plumber will always be able to find the answer to the basic question: Is the problem with the water supply system, with the drainage system or with the water heater?

## PROBLEMS IN THE WATER SUPPLY SYSTEM

Problems with the water supply system fall into four subgroups: faucets and valves, water closets, pipes and hot water. Each of these will be discussed systematically. As the experienced plumber knows, many different problems produce the same symptoms. It is the plumber's task to track down the true cause.

### FAUCETS AND GLOBE VALVES

Because they receive such heavy use, faucets are often a trouble spot and need repairs. Globe valves suffer from the same malfunctions, though not as often. The importance of maintaining them against even minor problems is shown in Fig. 19-1.

Though often neglected, their repair is quite simple once their design is understood. The cutaway drawings in Fig. 19-2 are typical. They illustrate the relationship of parts and identify them. As you study the symptoms in the rest of the unit, refer to them. They will help you understand why certain symptoms occur and how certain repairs should be made.

#### Dripping valves and spouts

When a valve or faucet does not completely stop water flow, there are three possible causes:
1. A defective handle or stem prevents the washer from being

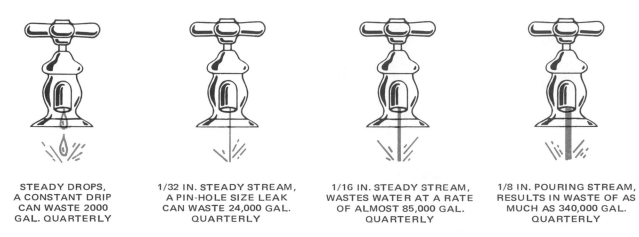

STEADY DROPS,
A CONSTANT DRIP
CAN WASTE 2000
GAL. QUARTERLY

1/32 IN. STEADY STREAM,
A PIN-HOLE SIZE LEAK
CAN WASTE 24,000 GAL.
QUARTERLY

1/16 IN. STEADY STREAM,
WASTES WATER AT A RATE
OF ALMOST 85,000 GAL.
QUARTERLY

1/8 IN. POURING STREAM,
RESULTS IN WASTE OF AS
MUCH AS 340,000 GAL.
QUARTERLY

Fig. 19-1. Even small leaking faucets can cause enormous waste of water if they go unrepaired. (J.A. Sexauer Mfg. Co., Inc.)

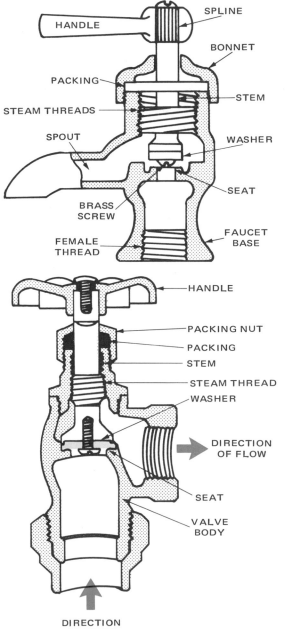

Fig. 19-2. Cutaway view of a typical faucet and globe valve. They are very much alike.

brought up against the seat.

2. The washer has deteriorated or broken.

3. The seat is worn or pitted.

The first possibility is the easiest to check. Remove the screw which secures the handle to the stem, Fig. 19-3. Then remove the handle. Lift or carefully pry it off. Inspect the splines on the inside of the handle and on the end of the stem, Fig. 19-4. If one or both are badly worn, they will need to be replaced. In the case of the handle, this is simple. Obtain a new one of the correct size and shape. Secure it to the stem with the right screw.

If the problem is with the stem, turn off the water supply. Remove the stem by loosening the bonnet and turning the stem, Fig. 19-5, until it can be lifted out of the valve body.

This same procedure must be used to check out other possible causes of the problem. While the stem is off, check the washer and seat for signs of wear. Replace the washer if it is split, deteriorated or has lost its pliability. Washers are made in many sizes as shown in Fig. 19-6. Select the correct size. Install it with a new brass machine screw. Make sure that any mineral deposits around the base of the stem are removed. The washer must seat properly on the end of the stem.

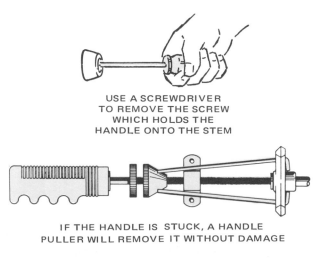

USE A SCREWDRIVER
TO REMOVE THE SCREW
WHICH HOLDS THE
HANDLE ONTO THE STEM

IF THE HANDLE IS STUCK, A HANDLE
PULLER WILL REMOVE IT WITHOUT DAMAGE

Fig. 19-3. Remove the screw which secures the handle to the stem. A puller, such as the one pictured, will lift stubborn handles from the stem without damage to the faucet. (J.A. Sexauer Mfg. Co., Inc.)

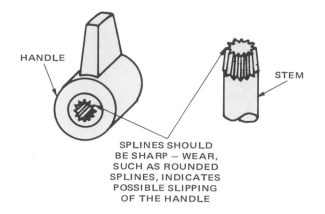

Fig. 19-4. Inspect the splines on the inside of the handle and on the end of the stem. If worn, replace the parts.

Inspect the faucet or valve seat for wear. If it is pitted or worn so that its surface is uneven, it must be refaced. Use the tool shown in Fig. 19-7.

With the cutter in contact with the valve seat, rotate it several times in each direction. Remove the tool and inspect the seat again. If the surface is smooth and shiny, install the repaired stem and carefully secure it with the bonnet.

To prevent damage to the new washer, hold the stem as the

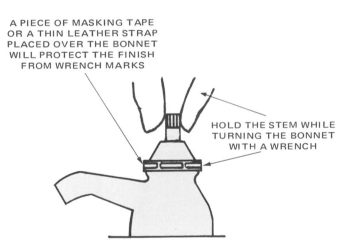

Fig. 19-5. To remove the stem, loosen the bonnet carefully to avoid damage to the chromed surfaces.

bonnet is secured. Otherwise, the stem may be forced against the seat with enough force to ruin the washer.

If the resurfacing of the seat is not satisfactory, the seat must be removed and replaced with a new one. See Fig. 19-8. If the seat is not removable, the faucet must be replaced. Procedure for removal is explained under: *Faucet or globe valve vibrates and is noisy when water is running.*

## Faucet or globe valve vibrates and is noisy when water is running

There are four possible causes for this problem:

1. Washer is loose.
2. Packing has deteriorated.
3. Stem is worn.
4. Faucet or valve base is worn.

To isolate the problem, turn off the water supply and remove the handle and stem as described under *Dripping valves*

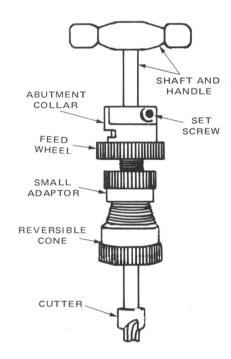

Fig. 19-7. Faucet seat reforming tool puts new face on seat.
(J.A. Sexauer Mfg. Co., Inc.)

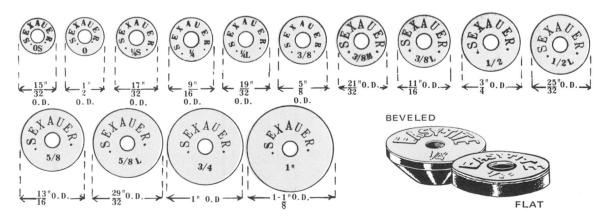

Fig. 19-6. Washers are produced in standard sizes. Any size is available either beveled or flat.

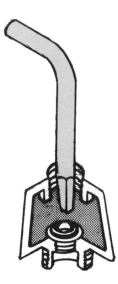

Fig. 19-8. Some seats are threaded into the valve body. They are removed either with an Allen wrench or with a four-sided tool as shown. Turn counterclockwise to remove.

*and spouts.* Inspect the washer to make certain the screw holds the washer firmly in the seat. Replace the washer if it shows signs of deterioration or wear.

Check the packing in the bonnet. It should be pliable and snugly fitted around the stem. If it does not, replace the packing. Also examine the stem for signs of heavy wear in the threads which engage the female threads in the faucet base. If these threads are worn, the stem must be replaced. On older valves and faucets the threads in the body of the faucet into which the stem threads, may also be worn. In this case, the entire faucet or valve should be replaced.

To replace a faucet, disconnect the water supply from the underside of the faucet, Fig. 19-9. Remove the locknut from the base of the faucet under the sink and lift out the faucet.

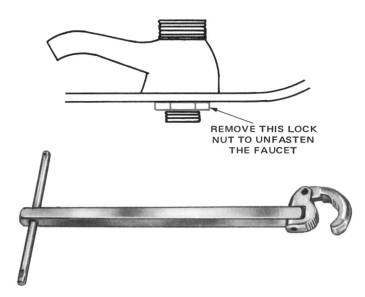

Fig. 19-9. Removing a faucet from sink or lavatory may require a special tool. The basin wrench jaws swivel 180 deg. and will either tighten or loosen the locknut on the base of the faucet. This is one of the few instances when a tooth-jawed tool is used on a chromed nut. An open end wrench or adjustable wrench should be used where the nut can be reached.

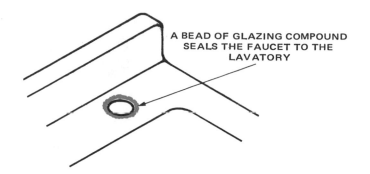

Fig. 19-10. Apply glazing compound around opening in the sink where the faucet will be installed.

Globe valves can be removed by taking out the piping which connects to the inlet and outlet.

Installing a new faucet is essentially the reverse of removing the old one. Thoroughly clean the sink or lavatory at the point where the faucet will be installed. Apply glazing compound, white lead putty or place the washer provided with the new faucet around the opening in the sink, Fig. 19-10. This makes a watertight seal at the base of the faucet. It prevents water from running under the faucet and into the cabinet or onto the floor below.

Next, position the new faucet, install the washer and finger tighten the locknut below the sink. Check the alignment of the faucet and tighten the locknut with a wrench. Attach the water supply piping.

### Handle rotates without changing water flow

One of two defects are responsible for this problem:
1. Defective splines on the handle.
2. Defective splines on the stem.

Either one prevents the handle from turning the stem. For the appropriate repair procedure, refer to the symptom: *Dripping valves and spouts.*

### Water leaks around stem

Water leakage around the stem when the faucet or valve is open is caused by either:
1. Deteriorated packing.
2. A worn stem.

The proper procedure for disassembly and reassembly of faucets and valves is described under the symptom: *Dripping valves and spouts.* Carefully check the stem for wear where it passes through the bonnet. If heavy wear is noted, replace the stem.

If the stem is not worn, the packing is at fault. It should be replaced.

### Slow flow of water from spout

Reduced water flow has several possible causes:
1. The cutoff valve in the water supply piping below the fixture is partially closed.
2. Aerator screen at the end of the faucet is clogged.
3. Pressure on the branch line is reduced.

First, check the shutoff valve in the water supply piping. If it has been partially closed, it will restrict flow through the pipe leading to the faucet. This valve should be in the water supply piping below the sink. If not, check the pipe below the floor until a valve is located. Open the valve completely and check the faucet.

If flow is still not good, check the aerator screen, Fig. 19-11. Remove it and if it is clogged, thoroughly clean and replace. Turn on the faucet to see if the problem has been solved.

If the flow has still not improved, check for heavy flow of water to another outlet on the same branch. Generally, this type of problem is intermittent (off and on). For example, if the kitchen sink and the washing machine are connected to the same branch, water pressure may drop markedly while the washer is filling. The only solution is to install larger piping in the branch serving the washer and kitchen sink.

But, if such a condition does not exist, then the reduced pressure condition will need to be investigated further. Check both the hot and cold water flow. If the cold water is flowing faster than the hot, it is possible that mineral deposits have built-up on the inside of the pipe and are restricting the water flow. This condition occurs more quickly in the hot water piping because the higher temperature causes the minerals to deposit faster. This is more likely to happen in galvanized iron pipe than in copper or plastic.

The only remedy for mineral deposits is to remove and replace the affected pipes. Horizontal runs of pipe are more likely to be severely clogged than risers. These can frequently be replaced from below the floor without replacing the risers which extend to the fixture from the horizontal piping. Procedures for replacing pipe are discussed later.

Another reason for reduced pressure at a faucet is a crushed water supply pipe. This type of damage can occur in an area where the pipe is exposed. Copper pipe is more likely to be

crushed than galvanized because copper is softer. To repair this type of damage, the crushed portion of pipe must be replaced.

## Mixing faucet leaks at joint between spout and faucet base

If water leaks around the joint of the spout and faucet, Fig. 19-12, the problem is one of the following:
1. O-ring washers are deteriorated.
2. Spout is worn.
3. Swing spout post is worn.

To check these conditions, turn off the water supply and remove the nut or set screw which secures the spout to the faucet base. Carefully lift the spout from the faucet and remove the swing spout post. Inspect the O-ring washers to see if they are worn, broken or no longer pliable. If any of these defects are present, the rings must be replaced.

Inspect the bearing surfaces of the swing spout post for signs of wear. If wear exists, the O-ring washers will not seat properly. The swing spout post should be replaced.

The inside of the swing spout may be so badly worn that the O-ring washers will not seal properly. If this is the case, the swing spout must be replaced.

## Single-handle faucet leaks

Faucets having a single handle to control both volume and temperature of the water are widely used. In these faucets the standard compression washer has been replaced by another means of controlling the water flow.

Many different types of single-control faucets are marketed. Common mechanisms are shown in Fig. 19-13.

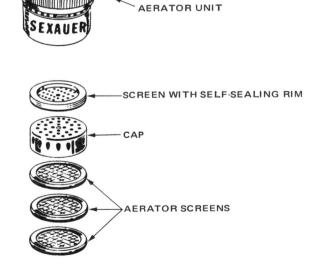

Fig. 19-11. Exploded view of aerator for kitchen faucet shows typical set of screens.

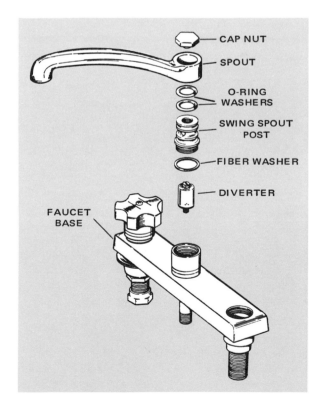

Fig. 19-12. Leakage at the joint between the swing spout and the base of the faucet generally requires replacement of the O-ring washers on the swing spout post. (J.A. Sexauer Mfg. Co., Inc.)

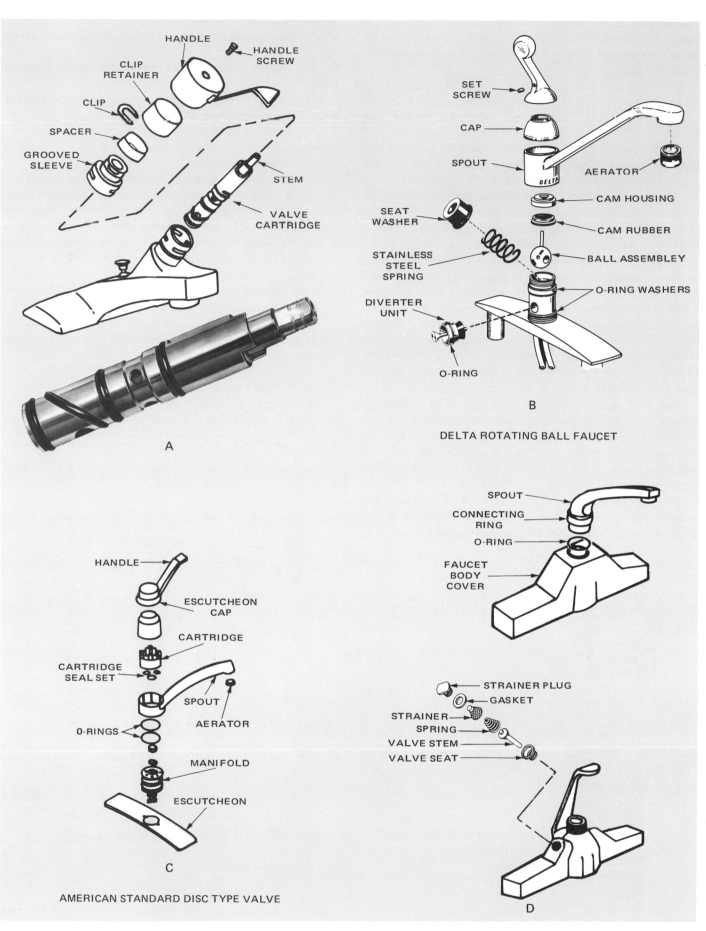

HANDLE

HANDLE
SCREW

CLIP
RETAINER

CLIP

SPACER

GROOVED
SLEEVE

STEM

VALVE
CARTRIDGE

A

SET
SCREW

CAP

SPOUT

AERATOR

CAM HOUSING

CAM RUBBER

SEAT
WASHER

STAINLESS
STEEL
SPRING

BALL ASSEMBLEY

O-RING WASHERS

DIVERTER
UNIT

O-RING

B

DELTA ROTATING BALL FAUCET

SPOUT

CONNECTING
RING

O-RING

FAUCET
BODY
COVER

HANDLE

ESCUTCHEON
CAP

CARTRIDGE

CARTRIDGE
SEAL SET

O-RINGS

SPOUT

AERATOR

MANIFOLD

ESCUTCHEON

C

AMERICAN STANDARD DISC TYPE VALVE

STRAINER PLUG

GASKET

STRAINER

SPRING

VALVE STEM

VALVE SEAT

D

Fig. 19-13. Exploded views of the most common types of single control, noncompression type faucets.

The following procedures for repair are based on common units for each type and should be used only as a general guide. Steps may vary slightly with each manufacturer. Instructions are usually packaged with repair parts. These should always be read carefully.

The cartridge faucet, also called a rotating cylinder faucet, controls water through ports in the cylinder. Fig. 19-13, view A, shows a single cylinder design. Fig. 5-30 in Unit 5 shows a cutaway of a cartridge faucet with two cylinders. One cylinder controls flow of hot and cold water while the other maintains even pressure between hot and cold water.

### TO DISASSEMBLE AND REPAIR A CARTRIDGE FAUCET:

1. Shut off the water. Turn on the faucet to drain the water.
2. Refer to view A, Fig. 19-13, and remove the handle. This is usually secured with a screw. Some screws are concealed under screw-in or snap-on caps.
3. Remove the clip retainer and the clip. Some clips are external and must be removed before lifting off the handle.
4. Remove remaining parts as shown in the exploded view.
5. Remove dirt and deposits. Check parts for cause of defect. Replace worn cartridge or O-rings as needed. Kits are available to replace worn parts.
6. Reassemble and turn on the water. Test faucet. Consult manufacturer's instructions for additional information on repair.

The rotating ball faucet uses a ball with ports to control water. The handle moves the ball back and forth and from side to side. When a port in the ball is aligned with one of the inlet ports in the faucet body, water is allowed to flow through the spout. This is illustrated in Unit 5. Refer to Fig. 5-29.

Another style of faucet uses a pair of ceramic discs to control water flow. One disc is stationary while the other is moved by the handle. When holes in the discs line up water will flow.

### TO DISASSEMBLE AND REPAIR THE ROTATING BALL FAUCET:

1. Shut off the water and drain the faucet.
2. Loosen the setscrew located in the handle. (See view B of Fig. 19-13.)
3. Unscrew the cap.
4. Grasp the protruding stem to remove cam housing, cam rubber and ball assembly.
5. Lift off spout if unit has swinging spout.
6. Remove the two seat washers and their springs.
7. Remove O-ring washers from faucet body.
8. Inspect cam housing, cam rubber and ball assembly. Replace if worn or corroded.
9. Reassemble according to manufacturer's instructions. *Caution: In replacing the ball assembly, be sure oblong slot in the ball is placed over the metal peg projecting from one side of the cavity. O-ring washers should receive a light coating of heat resistant grease.*

The noncompression valve faucet, popular in kitchens is relatively simple in operation. Little effort is required to disassemble and reassemble. The single handle rotates an eccentric left or right. (The eccentric is a cylinder-like piece with knobs on it.) The knobs push on the valve stems to open the valve allowing water to flow into the spout.

### TO DISASSEMBLE AND REPAIR THE DISC TYPE CARTRIDGE FAUCET:

1. Shut off the water and drain the faucet.
2. Grasp the handle and lift it as high as it will go.
3. Loosen setscrew recessed under the handle. Lift the handle off. (Some units have retaining screw through the top.)
4. Remove the escutcheon cap to expose the cartridge.
5. Remove the screws and lift off the cartridge.
6. If wear, not dirt, is the cause of the leak, replace the cartridge with a new one. Align the ports on the cartridge bottom with the three holes in the faucet body and screw the new cartridge to the body.

### TO DISASSEMBLE AND REPAIR NONCOMPRESSION VALVE FAUCETS:

1. Shut off water and drain faucet.
2. Loosen connecting ring and remove swing spout.
3. Lift off faucet body cover to expose faucet body.
4. Unscrew strainer plug. Remove entire assembly down to and including the valve seat.
5. Remove valve seat with Allen wrench or valve seat removal tool.
6. Clean the strainer.
7. Check valve stem, strainer, gasket and valve seat. Replace if worn or corroded.
8. Replace O-ring on the spout. Lubricate with heat resistant grease.
9. Reassemble following reverse order of disassembly.

## WATER CLOSETS

This section identifies four common symptoms of problems in the supply of fresh water to the water closet. Repair procedure will be suggested. Fig. 19-14 identifies the basic parts of a water closet and can be used as a reference for this section.

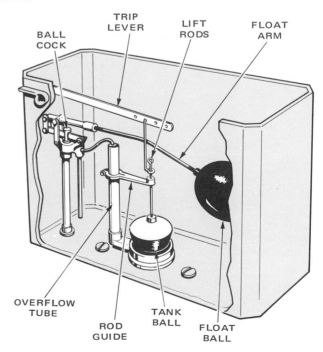

Fig. 19-14. Cutaway view shows principle parts of a water closet tank. Some parts are known by several different names. (Fluidmaster Inc.)

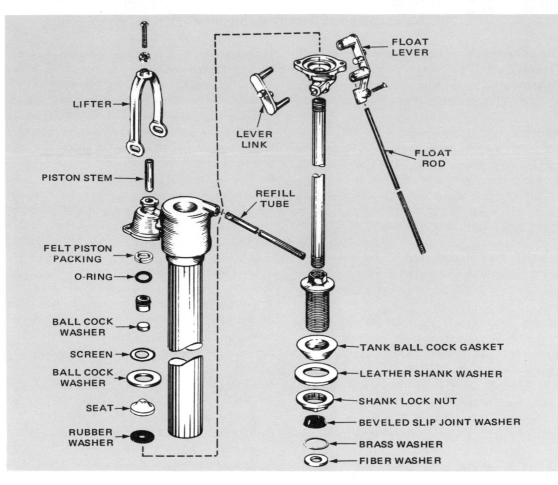

LIFTER

PISTON STEM

FELT PISTON PACKING

O-RING

BALL COCK WASHER

SCREEN

BALL COCK WASHER

SEAT

RUBBER WASHER

LEVER LINK

REFILL TUBE

FLOAT LEVER

FLOAT ROD

TANK BALL COCK GASKET

LEATHER SHANK WASHER

SHANK LOCK NUT

BEVELED SLIP JOINT WASHER

BRASS WASHER

FIBER WASHER

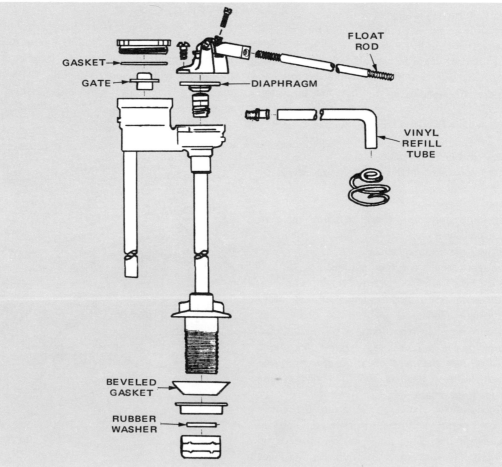

GASKET

GATE

DIAPHRAGM

FLOAT ROD

VINYL REFILL TUBE

BEVELED GASKET

RUBBER WASHER

Fig. 19-15. Common types of float valves (ball cocks). Parts most likely to need replacing are the ball cock washers and the seats. (Continued)

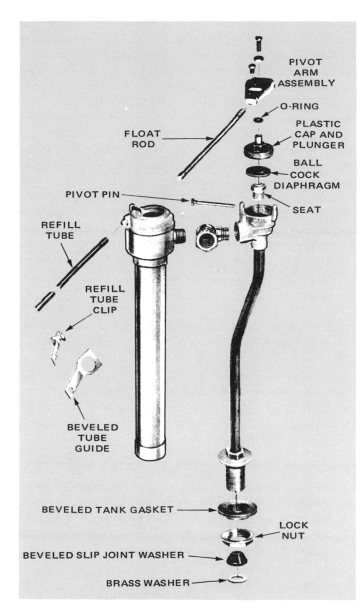

Fig. 19-15. (Continued) Common types of float valves (ball cocks). Parts most likely to need replacing are the ball cock washers and the seats. (J.A. Sexauer Mfg. Co., Inc.)

## Water flows continuously

When this condition occurs, one of the following problems exists:

1. Float valve (ball cock) is faulty.
2. Float ball is improperly adjusted.
3. Float ball is faulty.
4. Overflow tube has deteriorated.
5. Flush valve (tank ball) does not align properly with its base (seat).
6. Flush valve is worn or damaged.
7. Pressure flush valve is worn or malfunctioning because of dirt or mineral particles.

To determine which part of the mechanism is at fault, the plumber will make a series of checks. Defective parts or whole assemblies of parts may need repair or replacement.

CHECKING THE BALL COCK. With the tank cover removed, lift the float as high as it will go. If the water does not stop running, the ball cock is the cause. Fig. 19-15 shows three different types of ball cocks.

If the ball cock housing or body is a sealed type, or if the housing is deteriorated, replace the entire unit. The replacement unit may be a duplicate of the old unit or a universal type, Fig. 19-16.

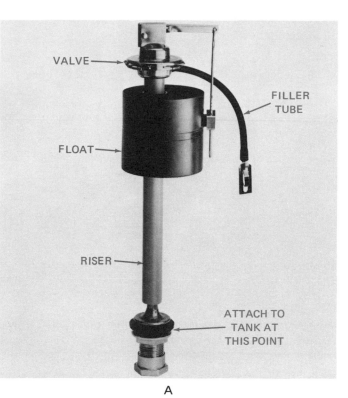

A

B

Fig. 19-16. New type of ball cock assembly combines the float with the riser of the valve mechanism. A—Valve mechanism is at top of assembly. B—Valve mechanism located beneath the float. (Fluidmaster Inc.)

To install a new ball cock, turn off the water supply. Drain the tank by flushing. Disconnect the water supply piping below the tank. Remove the locknut which secures the unit in the tank. See Fig. 19-17.

When installing the new unit, carefully position the rubber washer over the hole inside of the tank so the tank will not leak. Attach the metal washer and locknut. Tighten down the locknut until the rubber seal bulges slightly around the collar on the inside of the tank.

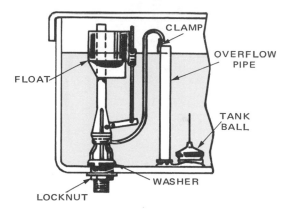

Fig. 19-17. Cutaway of tank shows new type ball cock assembled properly in tank.

Attach the float mechanism and the water supply piping. Turn on the water supply. Watch the tank fill to the water level line marked on the inside of the tank. Adjust the float so that the float valve closes when the water has reached that level. Flush the water closet to make sure the unit is working properly. Make adjustments as needed.

If the ball cock is not a sealed unit, the washers on the inlet valve plunger, Fig. 19-18, can be replaced. To release the plunger, remove the two pins that fasten the lever and float arm to the body of the ball cock. Pull the plunger up and out.

Remove the washers. Clean recesses behind the washers of all dirt or mineral deposits. Do not mar the plunger. Install new washers of the same type and size. Reassemble and test.

CHECKING FLOAT BALL ADJUSTMENT. If the water stops running when the float is lifted, the float may be at fault. Note the water level in the tank. If it is above the water level line and water is flowing into the overflow tube, the float may need adjusting. Bend the float arm downward until the water shuts off at the proper level.

CHECKING FOR FAULTY FLOAT BALL. Observe the position of the float ball in the water. If it does not rise to the surface it is leaking, replace it.

CHECKING THE OVERFLOW TUBE. If the tank does not fill with water, several problems may exist. First, examine the overflow tube for leaks which permit the water to leak down into the bowl.

If this tube is leaking, turn off the water supply. Flush the tank to drain most of the water. Remove and replace the overflow tube, Fig. 19-19. Refill the tank and check the float adjustment.

CHECKING THE FLUSH VALVE FOR ALIGNMENT. A flush valve which is misaligned will also prevent the tank from filling. It allows water to flow continuously from the tank into the bowl. This problem can generally be corrected by adjusting the guide rod on the flush valve.

CHECKING FOR WORN OR DAMAGED FLUSH VALVE. If the ball on the flush valve is worn, it will not seal properly. It must be replaced. Fig. 19-20 shows a satisfactory replacement. Occasionally, the seat of the flush valve will deteriorate to the point where the entire flush valve must be replaced. Procedures for this installation are discussed later under the symptom: *Water closet tank leaks.*

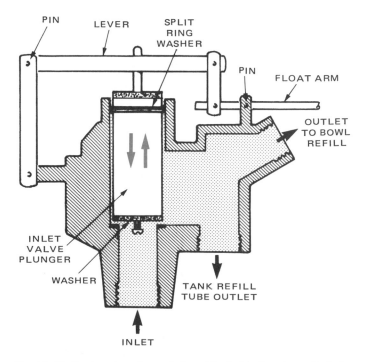

Fig. 19-18. Section view shows design of typical ball cock, particularly the plunger and the valve seat.

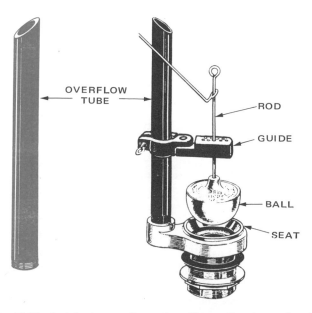

Fig. 19-19. A defective overflow tube will permit water to flow into the bowl. It will be impossible for the tank to fill. To replace, detach guide and unscrew pipe from the housing. (J.A. Sexauer Mfg. Co., Inc.)

Fig. 19-20. A new ball will frequently repair a leaky flush valve.

To determine if small quantities of water are actually leaking through the flush valve, it may be helpful to pour food coloring into the water in the tank. Allow it to sit for several hours. If the water in the bowl becomes the same color as the water in the tank, the flush valve leaks.

CHECKING PRESSURE FLUSH VALVES. Pressure flush valves are of two types: the diaphragm and the piston. See Figs. 19-21 and 19-22. Basically, the interior of both these valves consists of two chambers. The chambers are separated by a relief valve mounted on a rubber diaphragm. When the relief valve is seated and held in place by water pressure in the upper chamber, the valve permits no water to flow. But when the handle is moved, the plunger upsets the relief valve and causes water to flow to the fixture past the diaphragm.

When these units fail, it is usually the result of:
1. Dirt or mineral deposits which lodge in the passages and prevent the diaphragm from reseating.
2. Wearing of parts.

SERVICING PRESSURE FLUSH VALVES. To service the

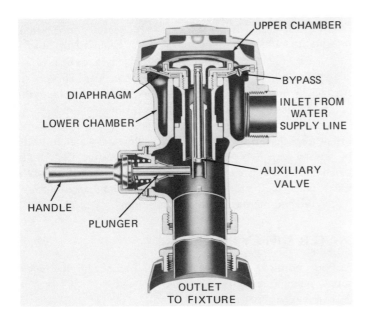

Fig. 19-21. Cutaway of a diaphragm type pressure valve shows the unit in "off" or neutral position. Handle at left moves the plunger to upset the relief valve when the unit is operated.

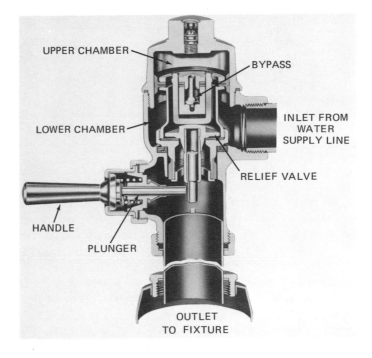

Fig. 19-22. Piston type pressure flush valve is similar in operation to the diaphragm type. (Sloan Valve Co.)

diaphragm pressure valve, first shut off the water by turning the setscrew of the stop valve all the way in. (This valve is always located alongside the pressure valve.) Then, unscrew the outer cover and lift off the inner cover. Pull out the diaphragm assembly. Lift out the relief valve. If there is "gravel" or other foreign matter lodged in the orifices of the diaphragm, wash it out, reassemble the valve and test.

If unit still does not work properly, disassemble as described. Then break down the diaphragm assembly. Hold the lower portion of the assembly with a wrench and remove the rubber valve seat which is the top of the assembly. Carefully replace the valve seat and the diaphragm. Reassemble and test.

Servicing of the piston pressure valve is similar to that described for the diaphragm pressure valve. Since several different designs are manufactured, the plumber should carefully read the manufacturer's instructions packaged with repair kits.

### Water closet will not flush

When the water closet will not flush, proceed as follows:
1. Check for a broken or disconnected flush lever. If the arm which extends from the handle to the flush valve linkage is broken, the ball on the flush valve will not be lifted off its seat. Replace the flush lever, Fig. 19-23. Make sure that the linkage to the flush valve is properly connected and aligned.
2. Check for lack of water in the flush tank. If there is no water in the supply tank, check all valves in the supply line leading to the bathroom. If they are all open, the problem is in the float valve. It must be removed, cleaned and repaired. Follow procedures outlined in the preceding subsection: *Water flow continuously.*

### Water closet tank leaks

When water is leaking from the water closet tank, the most likely causes are:

Fig. 19-23. Flush tank lever consists of two parts: a handle and a long arm which is linked to the flush ball by two lift rods. (J.A. Sexauer Mfg. Co., Inc.)

1. Condensation on the outside of the tank. Dry the wet portions of the tank. Watch where the water reappears. If the water is dripping as a result of condensation on the outside of the tank, install a tank liner. This measure will insulate the tank so that its outside surface does not become too cool. An alternative is to reduce the humidity in the bathroom with a dehumidifier or ventilating fan.

2. Loose inlet water supply piping. If the water supply piping is loose either at the valve or where it connects to the float valve, tighten the compression nuts. If this procedure fails, it will be necessary to shut off the water supply and disassemble the leaking joint to determine if the fitting or the compression ring are worn to the point of needing replacement.

3. Cracked or broken tank.

4. Loose or deteriorated washer at joint between the tank and the bowl. If the source of the problem is either No. 3 or No. 4, the tank will have to be removed. Disconnect the water supply and completely drain the tank. A sponge will help remove water which does not run out of the tank when it is flushed.

Tanks are generally attached to the bowl with two or three bolts. The bolts compress a rubber spud washer to seal the joint around the flush valve, Fig. 19-24. Carefully remove these bolts. Lift the tank from the bowl.

Remove the flush valve by taking off the locknut on the underside of the tank. This allows the entire assembly to be pulled from the tank. Before installing a new flush valve, all remains of the washer at the base of the tank should be removed so that the new washer will seal properly.

When installing the new flush valve, be careful to align the valve so the small pipe from the float valve will direct water into the overflow tube. Tighten the locknut only enough to insure that the gaskets are compressed tightly against the tank. Too much pressure will break the tank.

If a new tank is being installed, remove the float valve and the flush lever from the old tank and install them as described in the previous section. Then install the tank on the bowl of the water closet.

A new closet spud gasket, Fig. 19-25, must be installed between the bowl and the tank. The gasket compresses to form the seal when the closet bolts are tightened. Again, use care to prevent breaking either the tank or the closet bowl during this process.

## WATER SAVING DEVICES FOR WATER CLOSETS

Because of the increasing cost of water and the ever increasing demand being made on water treatment facilities, many devices are being developed which helps to conserve water. In a typical residential building the flushing of toilets

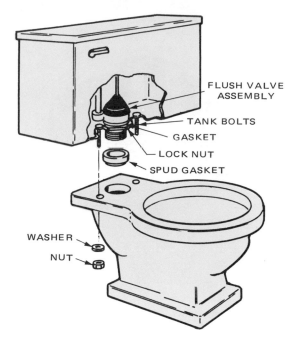

Fig. 19-24. Exploded view of tank and bowl shows position of spud washer and underside of flush valve assembly.

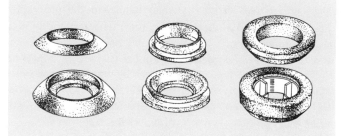

Fig. 19-25. Closet spud gaskets come in many shapes. (J.A. Sexauer Mfg. Co., Inc.)

consumes about half of the water used. Therefore, the potential for savings are great if effective devices can be developed.

One such device is a flush tank which uses normal water pressure to reduce the amount of water required to flush a toilet. See Fig. 19-26. Fig. 19-27 pictures a device which allows air to bleed from the tank ball or flapper. This causes the ball to stop the flush action sooner. However, holding the control handle down will cause normal flushing.

A second means of conserving water in a water closet is to install a special flush valve which will allow the water closet to be flushed with two different quantities of water, Fig. 19-28. When the handle is depressed, the fixture flushes as a normal toilet. When the handle is raised, a smaller amount of water is discharged and the fixture flushes as a urinal.

## WATER SUPPLY PIPING

Four problems that are likely to occur with the water supply piping follow:

1. Leaks at fittings.
2. Leaks in the pipe.
3. Broken pipes.
4. Restricted flow through the pipe.

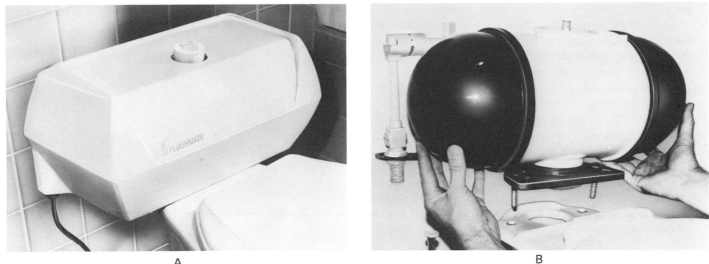

A

B

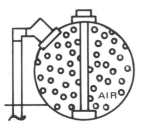

1. Empty tank containing air.

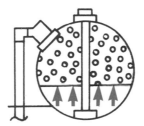

2. Water entering tank compresses this air.

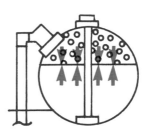

3. When the forces of air pressure and water pressure are equal, water flow stops.

4. Depression of the push button lifts main valve inside tank. This allows tank water to "escape" into the bowl, being pushed by the force of compressed air as well as pulled by gravity.

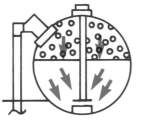

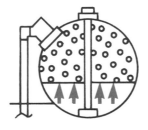

C

5. The drop in pressure inside the tank automatically closes the main valve; Flushmate immediately refills for next use in approximately 60 seconds. (Refill cycle of conventional tank-type toilets is between one and two minutes.)

Fig. 19-26. This flush tank unit is designed to reduce water usage. A—Unit with cover installed. B—Unit with cover removed during installation. C—How flush tank operates. (Water Control Products/N.A., Inc.)

Fig. 19-28. Flush valve is designed to permit flushing of water closet as a urinal or as a toilet. This device can save as much as 25 percent in water consumption. (3-M Mfg. Co.)

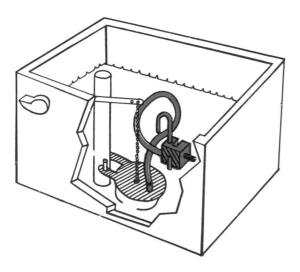

Fig. 19-27. Attachment allows tank ball to sink and seat faster thus saving water. (Savway Dual Flush Co.)

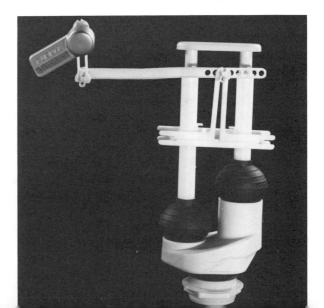

### Leaks at fittings

Leaks sometime occur at fittings as a result of other work done on the piping system. Twisting pipes may loosen a joint and start a leak. But occasionally a joint will begin to leak for no apparent reason. Sometimes a leak at a fitting in galvanized piping is a danger sign that the piping system is deteriorating. The rest of the system should be inspected.

Procedures for repairing leaks at fittings are different for each type of material. Copper, galvanized steel and plastic repairs will be discussed separately.

INSPECTING AND REPAIRING COPPER FITTINGS. A leak at the joint of copper fittings can generally be repaired by resoldering the joint. Sometimes this can be done without disassembling the joint.

Before attempting to resolder the joint, turn off the water and drain the pipe completely. Any water in the pipe near the joint will cool the material and the solder will not melt. Clean the area around the joint with abrasive paper. Apply a coat of flux to the cleaned area. Direct the heat on the pipe. When the pipe is heated, apply solder. Immediately after the solder has flowed into the joint, wipe away the excess. Refer to Unit 10 for more detailed instructions on soldering.

If this procedure fails to work, the joint is probably dirty inside the fitting. To correct this problem, use a torch to heat the joint while pulling the pipe out of the fitting with a pair of pliers.

When the pipe and fitting are separated, wipe off the excess solder immediately with a damp cloth. Clean the fitting and pipe with abrasive. Apply a coat of flux to the inside of the fitting and to the end of the pipe. Reassemble and solder.

If the shutoff valve leaks slightly or if water slowly enters the pipe from some other source and prevents the solder from melting properly, try inserting a ball of bread into the pipe to act as a sponge. The bread will absorb the water and permit the soldering job to be completed. When the water supply is turned on, the bread will break up and flow out through a faucet.

REPAIRING LEAKS IN GALVANIZED (THREADED) FITTINGS. One of the disadvantages with threaded pipe is that tightening one joint will loosen the joint on the other end of the pipe. Still, it is frequently possible to fix a leaking joint by simply tightening it with two pipe wrenches as shown in Fig. 19-29.

If this procedure fails to stop the leak in the pipe or opens a leak in the joint at the opposite end, there are two alternatives:

1. Disassemble the joint. Inspect the pipe and fittings for rusting and corrosion. If present, install new pipe and fittings. If parts are not deteriorated, apply pipe compound or teflon joint tape and reassemble. This should stop the leak.

2. If it is not practical to disassemble the joint, cut the pipe as shown in Fig. 19-30 and install a union. In making allowance for the union, subtract its length; then add the thread depth in the union. Refer to the chart in Unit 9, Fig. 9-52 for thread depth allowances.

With a union in the pipe, it is possible to tighten each joint independently of the other.

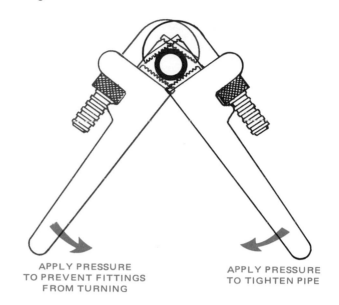

APPLY PRESSURE TO PREVENT FITTINGS FROM TURNING

APPLY PRESSURE TO TIGHTEN PIPE

Fig. 19-29. How to tighten a joint in galvanized pipe.

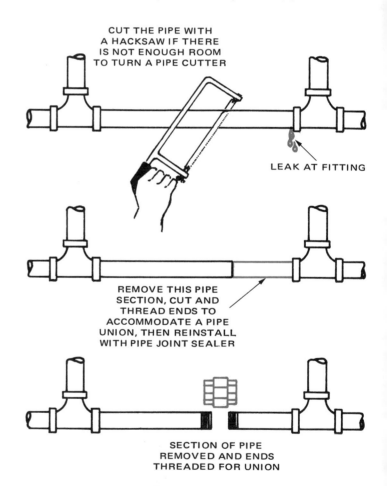

CUT THE PIPE WITH A HACKSAW IF THERE IS NOT ENOUGH ROOM TO TURN A PIPE CUTTER

LEAK AT FITTING

REMOVE THIS PIPE SECTION, CUT AND THREAD ENDS TO ACCOMMODATE A PIPE UNION, THEN REINSTALL WITH PIPE JOINT SEALER

SECTION OF PIPE REMOVED AND ENDS THREADED FOR UNION

Fig. 19-30. Preparing to install a union in a length of galvanized pipe.

REPAIRING LEAKS IN PLASTIC FITTINGS. Plastic fittings which develop leaks present their own set of problems. Most plastic water supply piping is permanently cemented together. It is impossible to separate the pipe and fittings without destroying one or both. The procedure outlined in Fig. 19-31 shows how to replace entire fitting and part of pipe so that new, completely cemented joints can be formed.

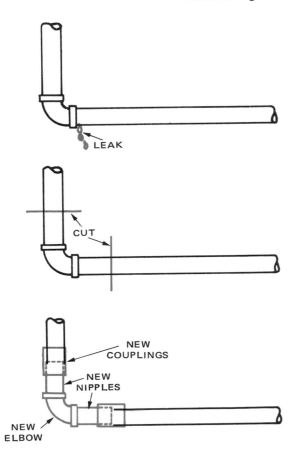

Fig. 19-31. To repair a leak in a plastic pipe fitting, the entire fitting, plus part of the pipe, must be replaced.

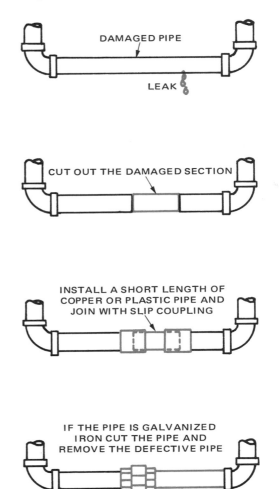

Fig. 19-32. Damaged section of pipe is repaired by removing portions and installing new pipe.

## Leaks in pipe

The principal causes of leaks in pipes are:

1. Deterioration.
2. Damage due to impact or puncture.
3. Damage due to freezing.

REPAIRING LEAKS CAUSED BY DETERIORATION. Of the three major kinds water supply piping materials, galvanized pipe is the most likely to leak as a result of deterioration. At any point where the galvanized coating has been removed or was not properly adhered to the steel originally, the pipe is likely to rust. The ends of the pipe where threads have been cut, the galvanized coating was removed and the pipe wall reduced in thickness, are most likely to deteriorate.

If deteriorated pipe is the problem, remove the defective pipe and replace it. See Unit 9 for a complete discussion of measuring, cutting, threading and installing pipe.

REPAIRING DAMAGE FROM IMPACT OR PUNCTURE. Any pipe damaged by impact or puncture can be repaired by cutting out the damaged section and replacing it, Fig. 19-32. An alternative method is to install a specially designed clamp, Fig. 19-33. This type of repair is temporary. It should be used only until it is possible to replace the defective pipe.

REPAIRING FREEZE DAMAGE TO PIPES. Pipe will split when freezing water expands in it. Repairing this kind of damage requires the same procedure outlined in Fig. 19-33.

Precautions must be taken to prevent the pipe from freezing again. One of the easiest ways to solve this problem is to attach electric heat tape, as shown in Fig. 19-34.

## Restricted flow of water

Reduced water pressure can result from defects in a valve or faucet. Consult the section on faucet problems. If that source is eliminated, it is time to inspect the pipes for mineral deposits.

Minerals in the water have more of a tendency to be deposited on the inside of galvanized pipes than on other piping materials. As these deposits accumulate, the inside diameter of the pipe gets smaller and the flow of water is reduced.

To check for this condition, disassemble a horizontal run of pipe at a convenient location and inspect the inside of the pipe. If this pipe and others are clogged, they will need to be replaced before normal water pressure can be restored. Unit 9 should be consulted for a pipe and fitting installation.

## DRAIN, WASTE, VENT (DWV)

A chart of symptoms for troubleshooting the DWV system is shown in Fig. 19-35. Lavatory, sink, tub and shower drains are considered together because the same troubleshooting and repair procedures apply to all four.

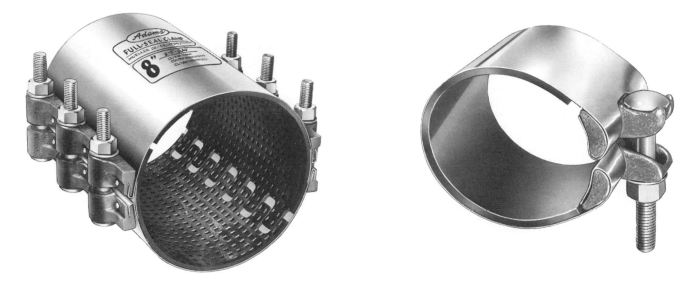

Fig. 19-33. Pipe clamps can be installed to stop a small leak in a pipe. Two styles are shown.   (Mueller Co.)

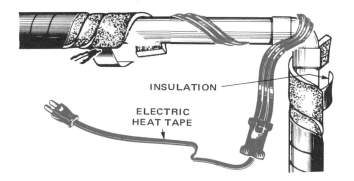

INSULATION

ELECTRIC
HEAT TAPE

Fig. 19-34. Wrapping with electric heat tape is a simple way to prevent pipes from freezing.   (Wrap-on Co., Inc.)

## WATER CLOSET DRAINS

The common drainage malfunctions of a water closet can be divided into two categories:
1. Failure of the fixture to drain.
2. Leaks near the base of the closet.

The following discussion analyzes these problems and suggests correct repair procedures.

### Water closet will not drain

Fig. 19-35 indicates that stoppage in either the stack or the building drain will generally cause more than one fixture or floor drain to back up. If this condition does not exist, it is safe to assume that the problem is located either in the fixture trap or in the branch piping connecting the fixture to the stack.

The simplest way to clear a blocked water closet is to use a "plumbers' friend," (force cup) Fig. 19-36. This tool forces water through a drain under pressure. In many cases this is enough to dislodge the obstruction and permit the drain to return to normal operation. The water ram, Fig. 19-37, is a more recent development. It uses air to apply even more pressure to force the obstruction on through the pipe.

This device should be used with care. It is possible to blow the water out of nearby traps rather than force the blockage through the pipe. This problem can be reduced by closing the stopper mechanism at each of the fixtures on the branch.

DRAIN, WASTE, VENT — TROUBLESHOOTING AND REPAIR

**WATER CLOSET
WILL NOT DRAIN**
1. STOPPAGE IN TOILET BOWL TRAP.
2. STOPPAGE IN DWV BRANCH PIPING.
3. STOPPAGE IN STACK.
4. STOPPAGE IN BUILDING DRAIN.

MORE THAN ONE
FIXTURE WILL GENERALLY
MALFUNCTION WHEN THESE
CONDITIONS EXIST.

**TUB, LAVATORY, SHOWER
OR SINK DRAIN CLOGGED**
1. STOPPAGE IN STOPPER MECHANISM.
2. STOPPAGE IN FIXTURE TRAP.
3. STOPPAGE IN DWV BRANCH PIPING.
4. STOPPAGE IN STACK.
5. STOPPAGE IN BUILDING DRAIN.

**WATER CLOSET
LEAKS AT BASE**
1. CRACKED OR BROKEN BOWL.
2. LOOSE BOWL.
3. WAX CLOSET BOWL GASKET DAMAGED.

Fig. 19-35. Problems with the DWV system can be grouped into three categories.

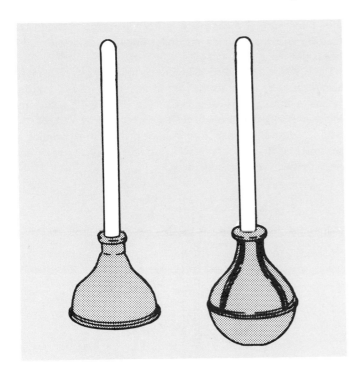

Fig. 19-36. Force cups are manufactured in two basic styles.
(J.A. Sexauer Mfg. Co., Inc.)

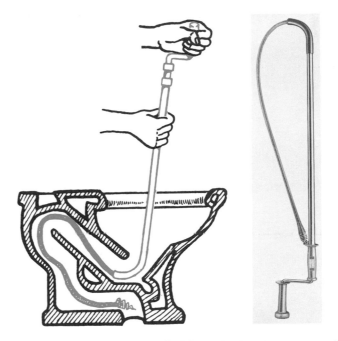

Fig. 19-38. Closet auger has a flexible steel cable which can usually reach and remove blockage from a water closet trap.
(The Ridge Tool Co. & J.A. Sexauer Mfg. Co., Inc.)

the branch DWV piping can be cleaned. The plumber should attempt this only when all other methods fail.

REMOVING THE WATER CLOSET. Turn off the water supply and attempt to force as much of the water out of the toilet bowl as possible with a force cup before the water closet is removed.

Disconnect the flexible supply from the base of the ball cock, Fig. 19-39. Remove the nuts from the closet bolts and

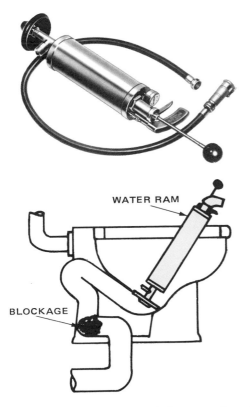

Fig. 19-37. Water ram uses compressed air to open clogged drains.
(General Wire Spring Co.)

If the force cup and water ram fail to remove the obstruction, use a closet auger, Fig. 19-38.

If these simple procedures fail to open the drain and if there is reason to believe that the blockage is not in the stack or building drain, drain and remove the water closet so that

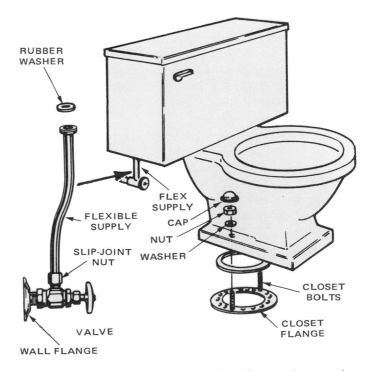

Fig. 19-39. Removing a water closet involves disconnecting a number of parts. It generally creates a mess which the plumber must clean up when the job is completed.

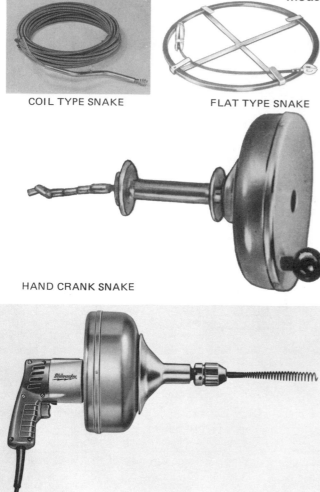

COIL TYPE SNAKE          FLAT TYPE SNAKE

HAND CRANK SNAKE

ELECTRICALLY DRIVEN SNAKE

Fig. 19-40. Any of these snakes are suitable for cleaning drains.
(Sears, Roebuck & Co.)

lift the entire water closet off of the closet flange. Place the bowl on some boards or blocks so it will not tip over.

With the water closet removed, it is easy to insert a snake into the closet bend. Fig. 19-40 shows several types that are satisfactory. Move the snake backward and forward and rotate it until the blockage is cleared.

Replace the water closet carefully to prevent damage. There must be a good seal between the base of the bowl and the DWV piping. It is recommended that the wax closet bowl gasket, Fig. 19-41, be replaced. The old one may not reseal.

Clean the closet flange carefully and place the new wax closet bowl gasket in position. Also, clean the underside of the

Fig. 19-41. Wax closet bowl gasket is used to seal the joint between the closet bowl and the closet flange.   (J.A. Sexauer Mfg. Co., Inc.)

toilet bowl so that a complete seal can be made between the wax ring and the base of the toilet bowl. Installation of the bowl is described in Unit 14.

BLOCKAGE IN THE STACK. A blocked stack can be cleared easily by going to the roof of the structure and running the snake through the vent stack. See Fig. 19-42. When the blockage has been broken, water should flow out of the fixture. Additional water should be run through the drains as the snake is moved. This will produce a clean drain and prevent reoccurrence of the problem.

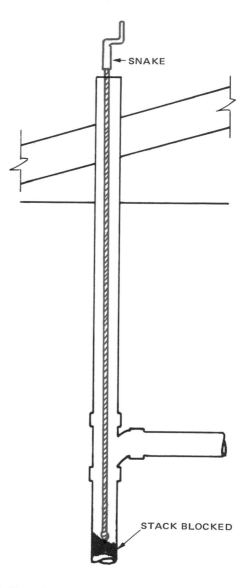

Fig. 19-42. Clear blockage in the stack by running the snake through the vent stack.

UNCLOGGING A BUILDING DRAIN. If the stack is clear and the water has still not drained out of the fixture, the problem is probably in the building drain. To unclog a building drain, remove the cleanout cover at the base of the stack, Fig. 19-43. Run a snake or a drain cleaning machine, Fig. 19-44, through the building drain. When the pipe is clear, replace the cleanout cover. The system should be in working order. There is one possible exception. If the building drain is broken, it will have to be replaced.

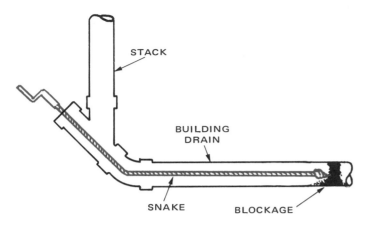

Fig. 19-43. A blockage in the building drain can be removed with a snake routed through the cleanout at the base of the stack.

## Water closet leaks at the base

Water closet leaks near the base of the bowl may be caused by:

1. A cracked or broken bowl.
2. Loose closet bolts.
3. Damaged wax closet bowl gasket.
4. Damaged or poorly fitted spud gasket between bowl and water supply tank.

CRACKED OR BROKEN TOILET BOWL. Wipe the bowl dry near the base and watch where the water comes from. If the water comes through the bowl either when the water closet is flushed or when the bowl is filled, it is certain that the bowl is cracked and must be replaced. Follow procedures given earlier and in Unit 14 for removing and reinstalling a water closet.

LOOSE CLOSET BOLTS. If the closet bowl bolts are loose, the entire closet will move. Frequently, when this occurs, the seal at the wax closet bowl gasket is broken, permitting the leak. Tightening the closet bowl nuts may solve the problem.

If not, the wax closet bowl gasket will need to be replaced.

Damage to the toilet bowl gasket will cause the toilet bowl to leak when the water closet is flushed. To repair this defect, follow the procedures for removing and reinstalling a water closet as described under the symptom: *Water closet will not drain.*

## Tub, lavatory, shower or sink drain blocked

The failure of a tub, lavatory, shower or sink drain to flow properly can be traced to one of five causes:

1. Blockage at the stopper mechanism.
2. Blockage in the fixture trap.
3. Blockage in the DWV branch piping.
4. Blockage in the stack.
5. Blockage in the building drain.

A tub and lavatory are likely to have a pop-up stopper built into the fixture drain, Fig. 19-45. The stopper, Fig. 19-46, can generally be removed by turning and lifting. Clean the stopper and inspect the drain for other foreign matter. If this does not solve the problem, check the fixture trap.

BLOCKAGE IN THE FIXTURE TRAP. Foreign matter is easily cleared by removing the P trap. Two different types are shown in Fig. 19-47. If the trap is fitted with a cleanout, it will only be necessary to remove the plug to inspect the condition of the trap.

If there is no cleanout, loosen the compression nuts and remove at least one section of the trap, Fig. 19-48. Assuming that the P trap was clogged and that it was thoroughly cleaned before being replaced, the problem should be solved.

However, if the drain still does not flow freely, it will be necessary to use a snake to clean the DWV branch piping as described under the heading: *Water closet will not drain.* If the problem persists, the procedure for cleaning the stack and the building sewer described in the same section should be followed. When the drain has been cleared, the trap and all other removed parts should be reinstalled and tested for leaks.

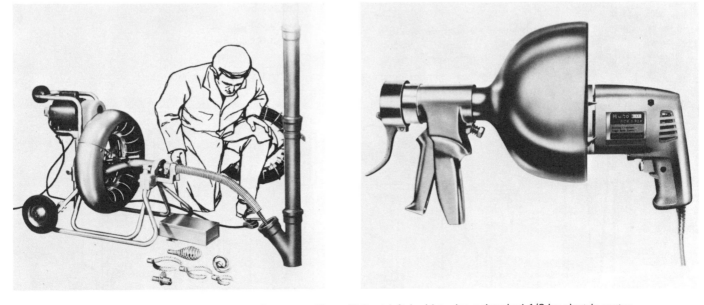

Fig. 19-44. Drain cleaning machines may be power driven. Unit at left is driven by a detached 1/2 hp electric motor. Operator controls motor with foot switch. Detachable heads are designed to remove any type of obstruction. Hand held unit, right, will handle up to 50 ft. of snake. (Marco Products Co.)

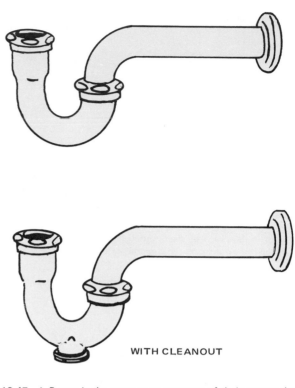

WITH CLEANOUT

Fig. 19-47. A P trap is the most common type of drain connecting the lavatory, sink or tub with the DWV piping system.

## MAINTAINING AND REPAIRING WATER HEATERS

Automatic water heaters function so well that they are often neglected until they do not work. A few simple maintenance procedures can prolong the life of the heater and increase its operating safety. In addition to presenting suggestions for maintenance of water heaters this section will deal with the following typical repair problems:

1. No water being heated.
2. Insufficient amount of hot water.
3. Water too hot.
4. Leaking tank.
5. Noise in the tank.

Fig. 19-45. Typical pop-up mechanisms for lavatories and tubs. (Kohler Co.)

Fig. 19-46. Pop-up stoppers are made in a variety of styles. (J.A. Sexauer Mfg. Co., Inc.)

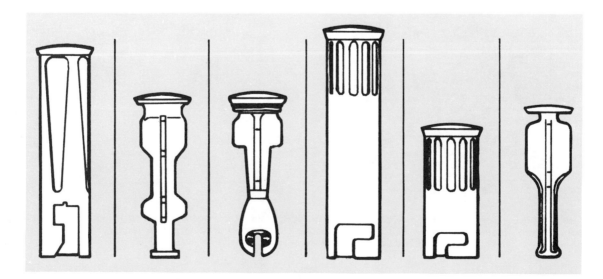

## MAINTAINING WATER HEATERS

Sediment in the bottom of the tank comes from settling of particles in the water. These can be easily removed by draining a few gallons of water from the tank every six months.

To insure that the pressure/temperature relief valve is working, lift the lever and allow a small quantity of water to

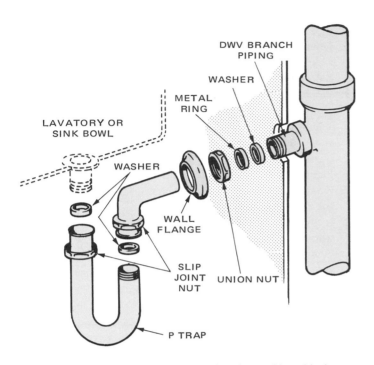

Fig. 19-48. P trap must be disassembled to clear stubborn blockage.

escape. Check to be certain that the valve has closed completely when it is released. If the valve does not operate properly, it should be replaced immediately.

Gas water heaters should receive two additional items of maintenance service:

1. The burner and pilot should be cleaned annually.
2. The adjustment of the flame should be checked. It should be clean and blue.

The vent pipe to a gas water heater should be inspected for

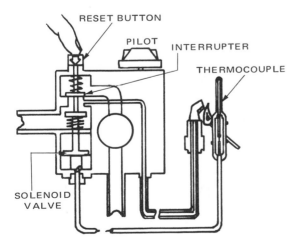

Fig. 19-49. Depressing the reset button allows the pilot to be ignited.

blockage or leaks. Unless the burned gases can escape through the vent, there is danger of carbon monoxide poisoning.

Before attempting to reignite the pilot, clean it and the burner with a brush and vacuum cleaner. The presence of an excessive draft can result from a fan being placed too close to the heater. Simply remove or redirect the fan to eliminate the draft.

To reignite the pilot:

1. Set the thermostat valve on pilot, Fig. 19-49.
2. Depress and hold the reset button.
3. Hold a lighted match in front of the pilot.
4. Continue to hold the reset button down a minute or two until the thermocouple has an opportunity to heat.
5. Release the reset button.
6. Replace the cover and turn the thermostat valve to the desired temperature. The burner should ignite.

### No hot water

When an electric water heater does not produce hot water, first check the circuit breaker or fuse in the main electrical panel. If the circuit breaker is tripped to the off position or the fuse is blown it probably means that one of the heating elements has shorted or the controls have malfunctioned causing an overload. Unless you are an experienced electrician and have the proper equipment to test electrical circuits, it is probably best to seek help rather than attempt to work with the 240 volt electrical current serving the water heater.

### Insufficient amount of hot water

The most obvious cause for shortage of hot water is a heater that is too small. Generally, this condition will be noted only when heavy demands are made on the water heater. The only cure for this problem is to install a larger heater.

If the water is continually lukewarm regardless of the amount of water used, check the thermostat. Sometimes the setting is accidentally moved. Simply reset the thermostat and allow the water temperature to rise to the desired level.

A third possible cause is leaking of the dip tube near the top of the heater, Fig. 19-50. In such cases, incoming cold water mixes with the hot water at the top of the tank. This causes the lukewarm temperature at the outlet. In many heaters, it is possible to remove and replace the dip tube. However, this may not be advisable if the tank is old — 15 years or more. It is likely that the tank will not last much longer under any circumstances.

### Water too hot

The most likely reason for the water being too hot is a thermostat set too high. Readjust the thermostat and check the water after an hour or two.

If the high temperature persists, the thermostat may not be functioning properly. If this condition is allowed to continue, the high limit protector on an electric water heater or safety cutoff thermostat on a gas water heater should cut off supply of energy before steam is created in tank. See Unit 6. Even if this safety device fails, the tank should not become a safety hazard. The pressure/temperature relief valve will allow water to escape if a very high temperature is reached.

To repair or replace a defective thermostat:

1. Turn off the energy supply.
2. Turn off the water supply.
3. Drain the tank.
4. Remove the defective part.

Once the new thermostat is installed, completely fill the tank before turning on the electricity to an electric water heater. Otherwise, the element will burn out.

### Leaking tank

If the tank begins to leak, it will be necessary to replace the water heater. The water supply and energy source must be turned off. Drain the tank completely. Disconnect the water

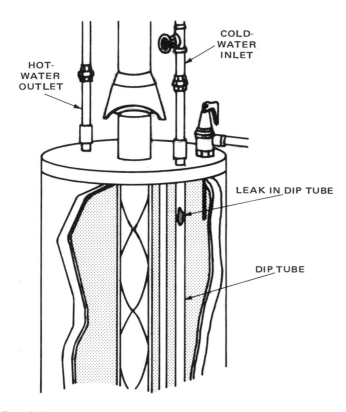

HOT-WATER OUTLET

COLD-WATER INLET

LEAK IN DIP TUBE

DIP TUBE

Fig. 19-50. A leak in the dip tube can allow cold water to mix with outgoing hot water.

pipes noting which is the incoming cold water. Disconnect the gas or electric supply and remove the tank. For information about the installation of water heaters, see Unit 6.

### Noise in the tank

A rumbling sound in the tank as the water is heating is probably caused by sediment in the tank. Periodic draining of several gallons of water should eliminate the problem.

### TEST YOUR KNOWLEDGE – UNIT 19

1. Describe the basic steps in the complete overhaul of a leaking faucet.
2. Continual leaking of water around the stem of the faucet when the faucet is turned on is caused by what two things?

3. A loose washer can cause a faucet to vibrate and make noise. True or False?
4. Which of the following can also cause a faucet to be noisy?
   a. Deteriorated packing.
   b. Sand in the water.
   c. Water hammer.
   d. Worn seat.
5. The aerator screen on a kitchen faucet may _____ the flow of water if it is clogged.
6. What conditions should be checked when a faucet delivers very little water?
7. Before removing a faucet stem or disconnecting the water supply piping from the base of the faucet, what must the plumber do?
8. The inspection procedures for a water closet in which the water flows continually are carried out in a particular order because it is an efficient way to isolate the problem. List the six steps in the correct order.
9. What should the plumber do before removing any part of the float valve?
10. If the water in the tank of a water closet is above the water level line and running into the overflow tube, the _____ adjustment is incorrect.
11. If the water level in the water closet tank never reaches the water level line, the possible causes are:
    a. A deteriorated _____ tube.
    b. A misaligned _____ valve.
    c. A damaged or deteriorated _____ valve.
12. When a new flush valve is installed, care must be taken so that the _____ will not be broken when the locknut is tightened.
13. A closet spud gasket is installed in the joint between:
    a. The flush tank and the toilet bowl.
    b. The ball cock and bottom of tank.
    c. The closet flange and the toilet bowl.
14. Repair of a leaky fitting in a galvanized piping system may require the pipe to be _____ and a _____ installed so that the leaking joint can be repaired.
15. Repairing a leaky fitting in a plastic piping system requires that the entire _____ plus a small amount of each _____ be cut out and a new one installed.
16. List the three principal causes of leaks in pipes.
17. To prevent pipes from freezing an electrical _____ _____ can be wrapped around the pipe.
18. When a water closet does not drain, the simplest maintenance procedure is to use a _____ _____.
19. If more than one drain is blocked, it is likely that the problem is in either the _____ or the _____ drain.
20. Using a snake to clean a stack can often be accomplished most easily by going to the _____ of the house and inserting the snake in the vent.
21. It is recommended that the joint between the base of a toilet bowl and the closet flange be sealed with:
    a. Caulking.
    b. Oakum.
    c. Putty.
    d. Wax ring.

22. A blocked tub, lavatory, shower or sink drain can be caused by five different problems. List the five problems.

## SUGGESTED ACTIVITIES

1. Practice plumbing maintenance and repair procedures by working on actual or simulated plumbing installations which have symptoms such as:
   a. Faucet dripping at end of spout.
   b. Water closet will not flush.
   c. Faucet leaking at base of swing spout.
   d. Sink drain clogged.
   e. Building drain clogged.
   f. Water closet leaking at base.

2. Study the common repair parts available for faucets, valves and water closets:
   a. How are they sized or otherwise distinguished from one another?
   b. What advantages and disadvantages do different products have which are designed to serve the same function?

3. Examine the special tools needed to perform plumbing repairs:
   a. Which are most useful?
   b. Which are the best designed and likely to give the best service?
   c. What tools should be in the tool box of a plumber specializing in residential plumbing repair?

# Unit 20
# SPRINKLER SYSTEMS

## Objectives

This unit describes the components and materials used in lawn and garden sprinkler systems. It outlines basic principles for installation of such systems.

After studying this unit you will be able to:
- List four basic considerations for satisfactory operation of sprinkler systems.
- Explain the importance of water pressure in the operation of sprinkler heads.
- List the factors which can cause pressure loss.
- Name and describe the operation of three principle types of sprinkler heads.
- Describe the processes of designing, laying out and installing a lawn or garden sprinkling system.

Lawn or garden sprinkling systems consist of an underground network of piping and sprinkler heads. They have become increasingly popular for homes, commercial buildings and golf courses. Part of this growing popularity is the result of cheaper installation costs stemming from introduction of durable plastic materials. A second factor is new technology which has cut down installation time. Moreover, the production of a wide variety of sprinkler heads makes it possible to custom design an effective sprinkler system for any lawn or garden.

## SPRINKLER SYSTEM DESIGN

Equipment used in the sprinkler system will determine, to a large extent, where the heads are located. However, no matter what components are employed, there are basic considerations which must be satisfied. Otherwise, the system will not function effectively. The considerations are:
1. The design must provide for controlled coverage of the area to be watered.
2. The sprinkler system must make the most of existing water pressure.
3. Cost of the sprinkler system should include maintenance and repair.
4. In cold climates, the system must be designed to prevent freeze damage.

## CONTROLLED PRECIPITATION COVERAGE

From a detailed plot plan of the property, the sprinkler system designer will determine the type and location of sprinkler heads. Such a plan is shown in Fig. 20-1. The factors which must be considered when making these decisions include:
1. The type of plant life to be watered.
2. The slope of the ground.
3. The porosity of the soil.
4. The amount of water a certain type of sprinkler head will deliver.
5. The area that a given type of sprinkler head will water.
Study Fig. 20-1 again. Note the designer has placed the sprinkler heads. No part of the lawn is left unwatered.

## EFFECT OF WATER PRESSURE

Water pressure is an important consideration in the design of sprinkling systems because it has such a great affect on sprinkler head operation. The heads require considerable water pressure. The designer, then, must be aware of the pressure available as well as the pressure losses that can be expected from friction.

Friction is the drag or resistance to water flow exerted by the walls of the pipe and fittings. Several factors must be combined to accurately estimate total friction loss:
1. Diameter of pipe.
2. Length of pipe.
3. Number of fittings.
4. Type of fittings used.
The designer can minimize pressure drop in several ways:
1. By keeping pipe runs as short as possible.
2. By keeping fittings to a minimum.
3. By dividing a system into several units which sprinkle at different times. This method is often used when the system will require more water flow than can be delivered by the normal water supply piping in a residential structure.

## PROTECTION AGAINST FREEZING

Where freezing temperatures are likely, the sprinkler system must be designed so that it can be drained. Water expands

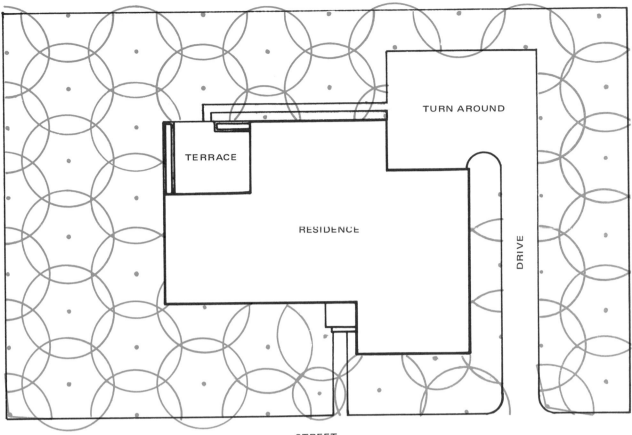

Fig. 20-1. Using a plot plan, the designer will place sprinkler heads so that all lawn or garden areas are adequately watered.

when it freezes. If allowed to stand in the system, in cold weather, it will result in broken pipe, valves and sprinkler heads. *Drains should be placed at low points in the system. Carefully slope all piping toward the drains.*

## TYPES OF SPRINKLER HEADS

Sprinkler heads fall into three principal types:
1. The spray type.
2. The rotary type.
3. The wave type.

Each type is available in several models designed to meet specific requirements. The wide selection of heads allows the designer to fit the system to whatever the need.

The type or model selected for any single installation will depend upon:
1. The location of the head in the sprinkler system.
2. The type of plant life to be watered.
3. The distance from other sprinkler heads.

Many of the heads are adjustable. They can be made to limit either the amount of water which passes through them or the area they will cover.

## SPRAY TYPE HEADS

Probably the most common for residential lawn watering is the pop-up spray sprinkler head, Fig. 20-2. This head is installed flush with the top of the ground. When the water is

Fig. 20-2. Pop-up spray valves are generally installed in sprinkler systems for residential lawn areas. Two makes are shown. (Weather-matic Div., Telsco Ind. and Rain Bird)

turned on, water pressure causes the spray nozzle to rise above the grass. This action permits the spray head to deliver water without interference, Fig. 20-3. When the water is turned off,

Fig. 20-3. The nozzles of pop-up valves rise above the grass when the water is turned on.

the spray nozzle drops back into the spray head. This feature permits unobstructed use of the lawn while preventing damage to the sprinkler heads.

A variation of the spray type sprinkler head is installed on the end of a short length of pipe which extends above the foliage, Fig. 20-4. This fixed spray head does not have the pop-up feature and is generally installed in flower beds or areas planted in some type of ground cover which is of uniform height. Because traffic in flower beds is very limited, these spray nozzles are not likely to be damaged.

## ROTARY TYPE

Rotary type sprinkler heads will cover a larger area than the spray type. However, they require higher water pressure. The pop-up rotary sprinkler is widely used for residential installations, Fig. 20-5. Many pop-up rotary sprinkler heads contain two nozzles. One provides coverage of the ground at the greatest distance from the nozzle. The second covers the ground near the nozzle, Fig. 20-6.

Rotary sprinkler heads are also made for installation above ground, Fig. 20-7. These nozzles function effectively when installed in locations where they will not be damaged.

Fig. 20-4. The fixed spray head is generally installed in flower beds. (Weather-matic Div., Telsco Ind.)

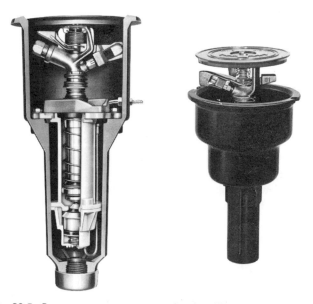

Fig. 20-5. Pop-up rotary type spray heads will cover large areas of lawn. Cutaway shows spring arrangement which retracts nozzle after water is shut off. (Rain Bird)

Fig. 20-6. The two nozzles of a rotary head provide uniform sprinkling. (Weather-matic Div., Telsco Ind.)

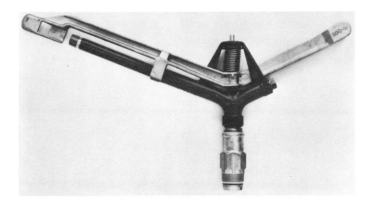

Fig. 20-7. This rotary head can be permanently mounted above ground level. (Rain Bird)

## WAVE TYPE SPRINKLER HEADS

Wave type sprinkler heads are designed to cover a large rectangular area, Fig. 20-8. Fewer sprinkler heads are needed then with rotary and pop-up units. The nozzle is adjustable to rectangles of various sizes.

## SPRINKLER SYSTEM INSTALLATION

Each type of sprinkler system has its own special installation requirements. Therefore, only general installation practices common to most systems will be discussed. Manufacturers' instructions and local codes will provide the detailed information required.

## LAYOUT OF THE SYSTEM

Working from the designer's plot plan, the installer will need to:
1. Locate the water sources.

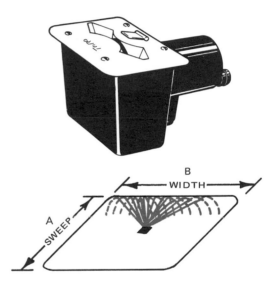

Fig. 20-8. Wave sprinkler head distributes water over rectangular area. Sweep times width equals total area of coverage. (Toro Mfg. Corp.)

2. Find a location for the controls.
3. Lay out the sprinkler heads.

Once these tasks have been accomplished, the pipe runs can be established. Because of the necessity to balance the water supply and the nozzles on a given pipe run, the installer should follow the plan carefully. No changes should be made without approval of the designer. Unless the sprinkler heads are carefully placed, they may not provide complete, uniform watering of the area. This requires careful measurements and frequent checks for accuracy.

## PIPE LAYING

New pipe laying equipment has simplified sprinkler installation. Hand digging of trenches, Fig. 20-9, has been replaced by a small trencher, Fig. 20-10, and the automatic pipe laying

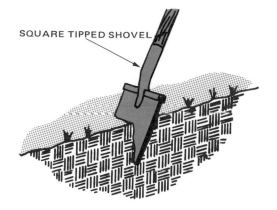

SQUARE TIPPED SHOVEL

Fig. 20-9. Trenching with a shovel. This method has largely been replaced by faster methods.

Fig. 20-11. The automatic pipe laying machine cuts through the ground and buries the pipe all in a single operation. (Davis Mfg. Co.)

machine, Fig. 20-11. The trencher cuts a narrow ditch about 7 in. deep and places the excavated dirt alongside the trench where it can easily be used for backfill.

The automatic pipe laying machine cuts through the ground and forces the pipe into the cut. The dirt is pushed back together and the pipe laying operation is completed with little damage to the lawn.

Because they are easier to install and cheaper, flexible plastic pipe and fittings are used for most sprinkler systems. Polyethylene is preferred because of its resistance to expansion and water hammer. However, polyvinyl-chloride is nearly as good. Pipe is purchased in long rolls and is installed in

Fig. 20-10. The trencher cuts a narrow ditch and places the excavated earth alongside the trench. (Charles Machine Works, Inc.)

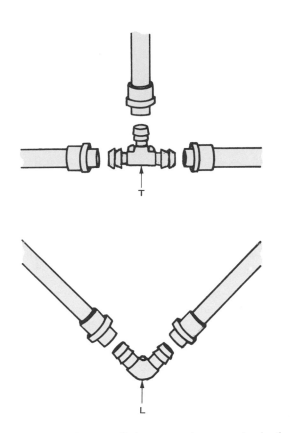

Fig. 20-12. Compression type fittings are used to connect polyethylene pipe and fittings.

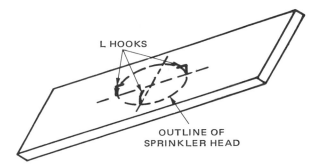

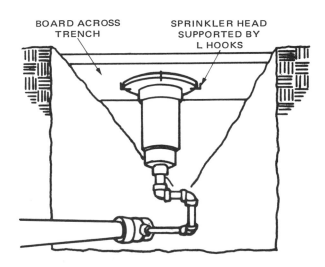

Fig. 20-13. Sprinkler heads can be supported with a board across the ditch during backfilling. This assures proper alignment of the sprinkler head with ground level.

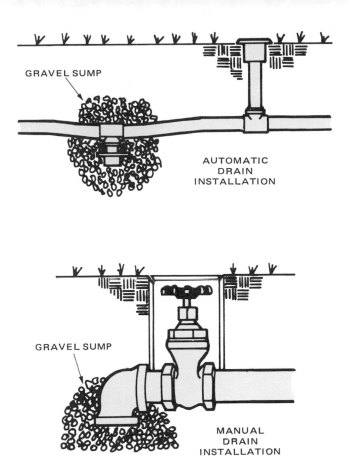

Fig. 20-14. Drain valves are installed to remove water from the piping system and prevent freeze damage.

continuous pieces so that the number of fittings is minimized. Where fittings are required, the compression type, Fig. 20-12, are used on polyethylene. Polyvinyl-chloride pipe uses the standard solvent-cemented fittings.

The installation of pop-up sprinkler heads is made easier by attaching the head to a board as shown in Fig. 20-13. Three L hooks will support the grass shield of the sprinkler head. When the backfilling is completed, hooks can be twisted aside to release the board.

Drain valves must be installed at low points in the piping. Proper installation is shown in Fig. 20-14. The automatic drain valves open each time the water is turned off and permit the water in the pipe to drain into a gravel bed below the valve. Manual valves are hand operated to drain the system at the end of the watering season. *All pipes must slope uniformly toward one of the drain valves.* This prevents the trapping of water which could freeze and burst the pipe, Fig. 20-15.

The sprinkler system is generally connected to existing hose bibs by a faucet adapter. A series of valves are then attached to the adapter. One valve is required for each set of sprinkler heads, Fig. 20-16. The valves are controlled by a central control panel, Fig. 20-17, located at some convenient place. The control valves work on a timer which automatically turns each set of sprinkler heads on and off at a prescribed time.

## TEST YOUR KNOWLEDGE — UNIT 20

1. Protection against freezing is a basic consideration in the design and installation of sprinkler systems. True or False?
2. List the three basic types of sprinkler heads.

243

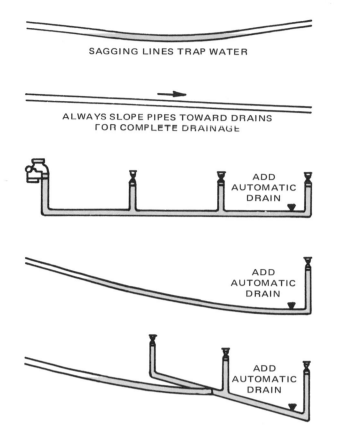

SAGGING LINES TRAP WATER

ALWAYS SLOPE PIPES TOWARD DRAINS FOR COMPLETE DRAINAGE

ADD AUTOMATIC DRAIN

ADD AUTOMATIC DRAIN

ADD AUTOMATIC DRAIN

Fig. 20-15. Drain valves must be installed at all low points in the piping.

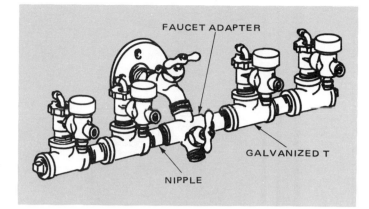

Fig. 20-16. Faucet adapter is used to attach valve units to hose bib.

3. Which of the following factors contribute to reduction of water pressure in a sprinkler system?
   a. Pipe diameter.          d. Lack of drain valves.
   b. Pipe length.            e. Type of fittings.
   c. Number of fittings.     f. All of the above.

4. The pop-up design for spray and rotary sprinkler heads has the advantage of being less likely to be _____ by lawn mowers and people walking through the yard.
5. The _____ type of sprinkler waters a rectangular area.
6. Describe the three steps prior to establishing the pipe runs in a sprinkler system.
7. Automatic or manual drain valves are located at low points in the piping system. True or False?
8. The sprinkler system is generally connected to the water supply piping at a hose _____. To make this connection a faucet _____ is installed.

## SUGGESTED ACTIVITIES

1. Practice installing a single run of pipe and spray heads for a sprinkler system.
2. Visit a job site where a sprinkler system is being installed.

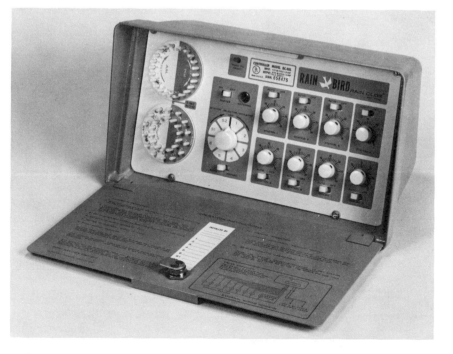

Fig. 20-17. The master control turns each section of the sprinkler system on and off automatically.

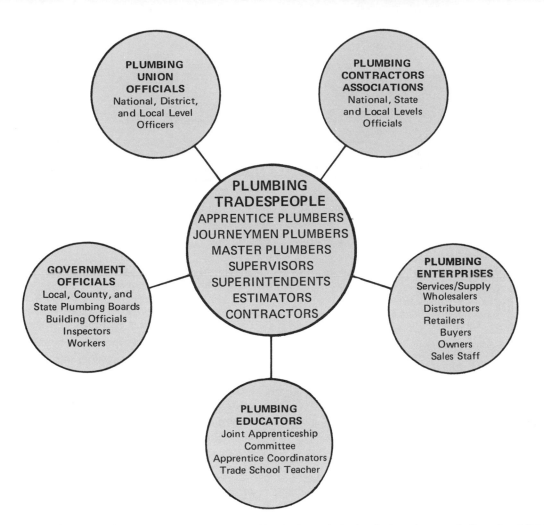

Fig. 21-1. Persons in the plumbing trades, represented by the inner circle also find employment in five related fields.

# Unit 21
# PLUMBING CAREER OPPORTUNITIES

### Objectives

This unit describes several different career areas for plumbers and the formal training provided through apprenticeship.

After reading this unit you will be able to:
- Identify five major areas which are a source of plumbing jobs.
- Explain the differences between the three levels of the plumbing apprenticeship program.
- Suggest three other advanced job classifications in plumbing.
- List qualifications for success in the plumbing trades.
- List basic educational requirements for entry into an apprenticeship in plumbing.

Plumbers contribute substantially to the general health and well-being of people. Without adequate provision for fresh drinking water and the sanitary removal of waste, waterborne disease would surely cause much illness, suffering and death. Because of the relationship between public health and the quality of plumbing systems, laws have been enacted which require the work of plumbers to be licensed and regulated by building codes.

### WHO EMPLOYS PLUMBERS?

In general, a plumber — or one with knowledge of plumbing materials and skills — will find employment in one of five different areas. These areas are shown in Fig. 21-1.

During the plumber's work career, he or she may change jobs several times and may move from one area into another.

Fig. 21-2. Plumbing businesses may employ one or more licensed plumbers and may specialize in installing/servicing whole systems. In larger companies, the owner may devote full time to management.

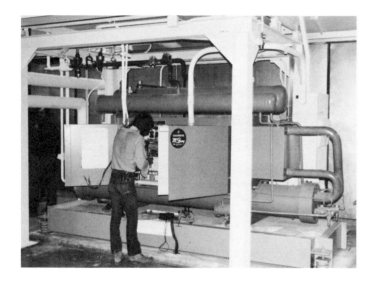

Fig. 21-3. Journeyman plumber makes final adjustments on the heating system in a large commercial building.

For example, a master plumber who has been working for a business which installs plumbing systems may take a job as an inspector. In this new position, the plumber, acting for the city, sees that new plumbing installations meet code requirements. The master plumber may also move on to become a contractor or start a business which sells plumbing materials. Regardless of the area of employment, those who would be successful will have considerable knowledge of plumbing practices, tools, materials and supplies.

## PLUMBING ENTERPRISES

Plumbing enterprises are the plumbing businesses which serve the needs of the community. They offer the greatest opportunity for the plumber and serve as the entry point for most beginners.

Sizes of these companies can vary greatly. Some are one-person operations working out of a small storefront.

Fig. 21-4. Plumbing supply dealers perform a valuable service by making all types of plumbing material readily available. They also take on the task of introducing new products to the plumbing tradespeople and to contractors.

Others employ many apprentices and journeymen who will be supervised by master plumbers.

The services they specialize in are varied. Some provide plumbing installation and repair, such as the company pictured in Fig. 21-2. They may concentrate on residential plumbing systems or they may contract for large systems such as are installed in commercial buildings or multifamily dwellings. Some may specialize in installing and maintaining piping systems for government agencies and public utilities. Still others may work only on plumbing for ships and aircraft. It is not uncommon for plumbers or plumbing firms to concentrate on maintenance work in industrial and commercial buildings, Fig. 21-3.

Another important plumbing specialization is in supplying the piping, fittings, fixtures and supplies plumbers and home owners need. Figs. 21-4 and 21-5 picture a company that supplies such materials for contractors and retail plumbing outlets.

Retailers — those who sell to the general public or do-it-yourself trade — make up another type of plumbing enterprise. Some of these businesses sell only plumbing but others may sell hardware, housewares, appliances and automotive parts and accessories. In any case, those who deal with the customers, must be plumbers or must have considerable knowledge of plumbing. They will wait on the customer, cut and thread pipe, advise on plumbing procedures or materials, maintain inventory and order new stock. Fig. 21-6 illustrates one aspect of such an employee's job.

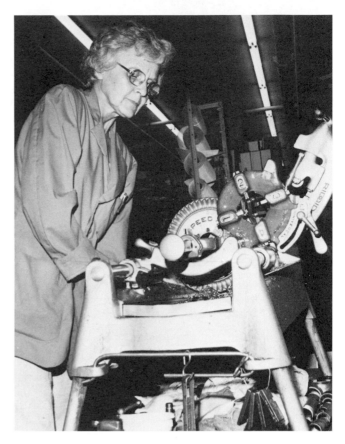

Fig. 21-6. Cutting and threading pipe are but two of the tasks performed by persons who work in plumbing sales. A thorough knowledge of piping, fittings and fixtures is a must. Occasionally they advise customers on installation procedures. (Ace Budget Centers)

Fig. 21-5. Dealers must stock a large variety of plumbing pipe, fittings and fixtures so that plumbers can have them as needed. (Westwater Supply Co.)

247

## GOVERNMENT AGENCIES

Government, at nearly every level, employs experienced plumbers to enforce the plumbing code and other laws regulating the installation of plumbing systems, Fig. 21-7. They are responsible for:

1. Licensing of plumbers wanting to enter the trade.
2. Reviewing the plumbing drawings in architectural plans.
3. Issuing plumbing permits for new construction or alterations of older plumbing systems.
4. Inspecting plumbing installations.

Fig. 21-8. A supervisor for the city water department oversees tapping of water main for a water service installation at a new building site.

Fig. 21-7. Government-employed plumbing experts are often required to study plans for new housing developments. They also must inspect and approve the plumbing installations in each building.

Many experienced plumbers are employed by water and sanitation departments of cities and villages. They become superintendents and supervisors who will oversee installation of water mains, sewers and water supply service, Fig. 21-8.

## PLUMBING EDUCATION

Plumbing educators generally are master plumbers with many years of experience. In addition to their knowledge of the craft, many will need to be experienced at management and administration. Their job is to:

1. Design training programs for beginning plumbers or for experienced plumbers who need updating.
2. Provide classroom or shop instruction, Fig. 21-9.
3. To coordinate the classroom instruction with the work experience of the apprentice.

In larger cities the educational program for plumbers is

Fig. 21-9. Instructing apprentices and journeymen on plumbing tasks is hard but interesting work. Instructors must have good communication skills and must be able to work with people.
(Mechanical Contractors Assoc. of Central Ohio)

directed by a joint apprenticeship committee. The functions of this committee include:

1. Selecting apprentices.
2. Conducting training programs.
3. Evaluating performance of apprentices.

Apprentice coordinators supervise on-the-job experiences which apprentices receive. In addition, they arrange for the related classwork where apprentices learn plumbing design, code requirements, plumbing math and many other types of technical information directly related to plumbing. See Fig. 21-10.

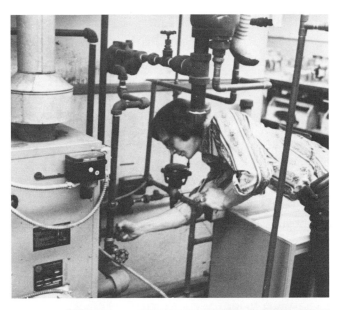

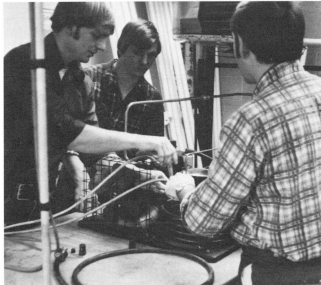

Fig. 21-10. Apprenticed plumbers learn to install heating and air conditioning equipment in their apprenticeship classes.

## PLUMBING ORGANIZATIONS

Organizations at the local, state and national level have been established for the benefit of plumbers and plumbing contractors. Plumbing unions provide benefits to the plumber and are the source of employees for union contractors. Plumbing contractors associations assist contractors with the problems of operating their business. Many of the people who work for these organizations must be qualified plumbers because of their need to thoroughly understand the plumbing field.

## ADVANCING IN THE PLUMBING TRADES

Some people have learned plumbing by working several years as helpers to experienced plumbers. However, those responsible for training plumbers generally recommend a formal apprenticeship. An outline of this program is shown in Useful Information, page 268. The trainee in the apprentice-

ship program moves up through several levels:
1. Apprentice.
2. Journeyman.
3. Another level, identified by some states, is the "master plumber."

Beyond these formal steps, the plumber can move on to other responsible positions such as supervisor, superintendent, contractor and estimator. However, these are opportunities which have nothing to do with the formal training of the plumber.

## APPRENTICES

Apprenticed plumbers are those learning the plumbing craft. The period of apprenticeship is generally five years. Part of the training is getting some practical experience, Fig. 21-11. In addition to working on the job under direction of a journeyman plumber, the apprentice must take 144 hours of classroom instruction every year. In these classes the apprentice learns to interpret the plumbing code and masters other technical information related to the craft. At the conclusion of the apprenticeship, a test is administered. If successfully completed, it permits the apprentice to be licensed as a journeyman.

Fig. 21-11. The apprenticed plumber works on the job part of the time performing normal plumbing tasks.

## JOURNEYMEN

Journeymen plumbers are full-fledged tradespeople who are licensed to practice their craft. Because of their training and experience, journeymen plumbers are able to do all types of plumbing work without the continuous surpervision given apprentices.

## MASTER PLUMBERS

Master plumbers are qualified by experience and knowledge to be plumbing contractors. They are generally required to work as journeymen for five years before they may take the examination for a master's license.

## SUPERVISORS AND SUPERINTENDENTS

Larger plumbing installations require supervisors and superintendents who oversee the work of crews made up of apprentices and journeymen. Supervisors are responsible for directing the work of a small group of workers. Superintendents oversee large plumbing jobs and generally have several supervisors working under their direction.

## ESTIMATORS

Estimators work for plumbing contractors. They make careful estimates of the materials and labor required to make plumbing installations. Based on these estimates, the contractor submits bids for jobs. Estimating requires complete understanding of both plumbing installation and the materials and supplies required. Only experienced plumbers who possess mathematical skill and the patience to prepare detailed, accurate estimates are employed to do this important work. Errors in the estimates can result in financial losses to the contractor. Therefore, this is a highly responsible position.

## CONTRACTORS

Plumbing contractors are master plumbers who have established a plumbing contracting business. Generally, they hire apprentices, journeymen and other master plumbers to work for them. Depending upon the size of the plumbing business, contractors may work with the crew or they may manage the business full time. In either case, the responsibilities of business management will be carried out by the plumbing contractor.

In small cities and rural communities the categories of plumbing tradespeople are frequently less well defined. However, the functions of each person described are carried out by those employed in the plumbing business. Sometimes in plumbing work, one person does everything, including managing the business, preparing estimates, obtaining supplies and installing the plumbing system.

## PLUMBING SUPPLY DEALERS

Plumbing supply dealers must have a thorough knowledge of the materials and supplies available, how they should be installed and what alternatives are available. As dealers, they find it necessary to keep up with new developments in plumbing materials, supplies, tools and equipment. Frequently, it is through them that new products are introduced to the plumbing craft. The most important function of the plumbing supply dealer is to deliver the needed materials to

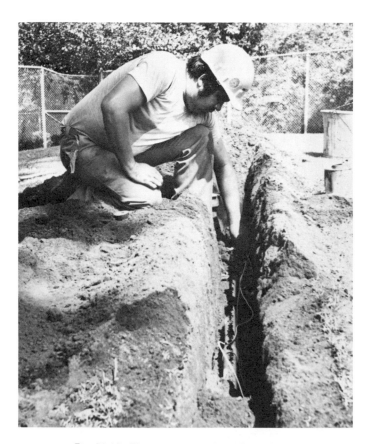

Fig. 21-12. These men work for a firm which specializes in sprinkler installations. Work is seasonal — usually from about May 1 through October — depending on climate.

the job site at the time they are needed. By combining knowledge of plumbing with business management skills the plumbing supply dealer makes a significant contribution to the plumbing industry.

## SPECIALIZATION

While many plumbing companies may do all types of plumbing — installation of new work, remodeling, maintenance and repair of residential structures as well as commercial and industrial buildings — others will specialize in one or two areas.

For example, some companies do nothing but plumbing maintenance and repair. Workers in Fig. 21-12 are employed by a landscaper who installs sprinkler systems in lawns and golf courses. Another type of specialization open to persons with plumbing experience is well drilling, Fig. 21-13.

Fig. 21-13. Well drilling is a specialty field which may attract some plumbers. It will take considerable experience to handle the huge drilling rig. In addition, the driller must understand the different rock formations and soil strata deep beneath the surface of the earth. (Mobile Drilling Co.)

## QUALIFICATIONS

Plumbers' work is sometimes strenuous and always active. This requires that anyone wishing to become a plumber be in good health. Plumbers spend most of their time indoors working on partially completed buildings which offer some protection from the weather. Plumbers must be able to work on ladders, scaffolds and in trenches.

It is important for plumbers to plan their work and be able to understand detailed instructions. One of the most important skills that a plumber must acquire is the ability to visualize completed piping systems before work on them is begun. Plumbers must enjoy working with their hands and be able to solve arithmetic problems accurately and rapidly.

## EDUCATION AND TRAINING

Those wishing to become apprenticed for plumbing should be high school graduates. While in high school, the future plumber should master the fundamentals of mathematics, including algebra. Courses in general science, physics, mechanical drawing and welding will make mastery of the trade easier.

Apprentices are paid at a lower rate than the journeyman wage scale. However, their pay increases periodically as they gain experience. Near the end of the apprenticeship period the pay scale for apprentices will be nearly equal to journeymen.

## EMPLOYMENT

Because of the need to continually build new homes, offices, schools and industrial buildings and the ever-increasing demand for repair and maintenance of existing plumbing systems, more qualified plumbers will be required. The opportunities for advancement to positions in management, trade associations, government and education offer additional incentives for people who possess the required skills.

## TEST YOUR KNOWLEDGE — UNIT 21

1. List the major employment areas in the plumbing field.
2. The most highly qualified plumber is known as a _____ plumber.
   a. Master.
   b. Apprentice.
   c. Journeyman.
3. People wishing to learn the plumbing trade begin their training as a (an) _____ plumber.
4. Only journeyman plumbers are licensed to be plumbing contractors. True or False?
5. Which of the following tasks is done by government officials responsible for enforcing the plumbing code:
   a. Inspect plans.
   b. Inspect work during construction.
   c. License plumbers.
   d. All of the above.
6. The person who supervises the on-the-job training of apprentices is known as the:
   a. Plumbing educator.
   b. Apprentice coordinator.
   c. Contractor.
   d. Supervisor.
7. Why is plumbing important to the public health and welfare?

## SUGGESTED ACTIVITIES

1. Interview plumbers regarding the career opportunities in plumbing. Report the findings of your interview to the class.
2. Obtain copies of the printed information describing the plumbing apprenticeship program in your community. Study these materials carefully and present a summary of your findings to the class.
3. Using publications such as the DICTIONARY OF OCCUPATION TITLES and the OCCUPATIONAL OUTLOOK HANDBOOK, summarize the opportunities in the plumbing field and report your findings to class.

# USEFUL INFORMATION

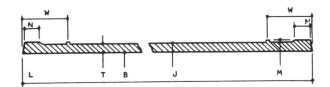

DIMENSIONS AND TOLERANCES (IN INCHES) OF SPIGOTS
AND BARRELS FOR "₵ NO-HUB®" PIPE AND FITTINGS

| SIZE | INSIDE DIAMETER BARREL | OUTSIDE DIAMETER BARREL | OUTSIDE DIAMETER SPIGOT | WIDTH SPIGOT BEAD | THICKNESS OF BARREL | | GASKET LUG |
|---|---|---|---|---|---|---|---|
| | B | J | M | N | T-NOM. | T-MIN. | W |
| 1 1/2 | 1.58 ± .06 | 1.90 ± .06 | 1.96 ± .06 | .25 ± .13 | .16 | .13 | 1.13 |
| 2 | 2.00 ± .06 | 2.31 ± .06 | 2.38 ± .06 | .25 ± .13 | .16 | .13 | 1.13 |
| 3 | 3.00 ± .06 | 3.31 ± .06 | 3.38 ± .06 | .25 ± .13 | .16 | .13 | 1.13 |
| 4 | 4.00 ± .06 | 4.38 ± .06 | 4.44 ± .06 | .31 ± .13 | .19 | .15 | 1.13 |
| 5 | 4.94 ± .09 | 5.30 ± .06 | 5.38 ± .09 | .31 ± .13 | .19 | .15 | 1.50 |
| 6 | 5.94 ± .09 | 6.30 ± .06 | 6.38 ± .09 | .31 ± .13 | .19 | .15 | 1.50 |

## LENGTH CONVERSIONS

| fractional inch | millimetres | decimal inch | millimetres | inches | centimetres | feet | metres |
|---|---|---|---|---|---|---|---|
| 1/32 | .7938 | 0.001 | .0254 | 1 | 2.54 | 1 | .3048 |
| 1/16 | 1.588 | 0.002 | .0508 | 1¼ | 3.175 | 1½ | .4572 |
| 3/32 | 2.381 | 0.003 | .0762 | 1½ | 3.81 | 2 | .6096 |
| 1/8 | 3.175 | 0.004 | .1016 | 1¾ | 4.445 | 2½ | .7620 |
| 5/32 | 3.969 | | | 2 | 5.08 | | |
| 3/16 | 4.763 | 0.005 | .1270 | 2¼ | 5.715 | 3 | .9144 |
| 7/32 | 5.556 | 0.006 | .1524 | 2½ | 6.35 | 3½ | 1.067 |
| 1/4 | 6.350 | 0.007 | .1778 | 2¾ | 6.985 | 4 | 1.219 |
| 9/32 | 7.144 | 0.008 | .2032 | 3 | 7.62 | 4½ | 1.372 |
| 5/16 | 7.938 | 0.009 | .2286 | 3¼ | 8.255 | 5 | 1.524 |
| 11/32 | 8.731 | 0.010 | 0.254 | 3½ | 8.89 | 5½ | 1.676 |
| 3/8 | 9.525 | 0.020 | 0.508 | 3¾ | 9.525 | 6 | 1.829 |
| 13/32 | 10.32 | 0.030 | 0.762 | 4 | 10.16 | 6½ | 1.981 |
| 7/16 | 11.11 | | | 4¼ | 10.80 | | |
| 15/32 | 11.91 | 0.040 | 1.016 | 4½ | 11.43 | 7 | 2.133 |
| 1/2 | 12.70 | 0.050 | 1.270 | 4¾ | 12.07 | 7½ | 2.286 |
| 17/32 | 13.49 | 0.060 | 1.524 | 5 | 12.70 | 8 | 2.438 |
| 9/16 | 14.29 | 0.070 | 1.778 | 5¼ | 13.34 | 8½ | 2.591 |
| 19/32 | 15.08 | | | 5½ | 13.97 | | |
| 5/8 | 15.88 | 0.080 | 2.032 | 5¾ | 14.61 | 9 | 2.743 |
| 21/32 | 16.67 | 0.090 | 2.286 | 6 | 15.24 | 9½ | 2.896 |
| 11/16 | 17.46 | 0.100 | 2.540 | 6½ | 16.51 | 10 | 3.048 |
| 23/32 | 18.26 | 0.200 | 5.080 | 7 | 17.78 | 10½ | 3.200 |
| 3/4 | 19.05 | 0.300 | 7.620 | 7½ | 19.05 | 11 | 3.353 |
| 25/32 | 19.84 | 0.400 | 10.16 | 8 | 20.32 | 11½ | 3.505 |
| 13/16 | 20.64 | 0.500 | 12.70 | 8½ | 21.59 | 12 | 3.658 |
| 27/32 | 21.43 | 0.600 | 15.24 | 9 | 22.86 | 15 | 4.572 |
| 7/8 | 22.23 | | | 9½ | 24.13 | | |
| 29/32 | 23.02 | 0.700 | 17.78 | 10 | 25.40 | 20 | 6.096 |
| 15/16 | 23.81 | 0.800 | 20.32 | 10½ | 26.67 | 25 | 7.620 |
| 31/32 | 24.61 | 0.900 | 22.86 | 11 | 27.94 | 50 | 15.24 |
| 1 | 25.40 | 1.000 | 25.40 | 11½ | 29.21 | 100 | 30.48 |

## CONVERSIONS: DECIMALS, FRACTIONS, MILLIMETRES

### MILLIMETERS TO DECIMALS

| mm | Decimal | mm | Decimal | mm | Decimal | mm | Decimal | mm | Decimal |
|---|---|---|---|---|---|---|---|---|---|
| 0.01 | .00039 | 0.41 | .01614 | 0.81 | .03189 | 21 | .82677 | 61 | 2.40157 |
| 0.02 | .00079 | 0.42 | .01654 | 0.82 | .03228 | 22 | .86614 | 62 | 2.44094 |
| 0.03 | .00118 | 0.43 | .01693 | 0.83 | .03268 | 23 | .90551 | 63 | 2.48031 |
| 0.04 | .00157 | 0.44 | .01732 | 0.84 | .03307 | 24 | .94488 | 64 | 2.51969 |
| 0.05 | .00197 | 0.45 | .01772 | 0.85 | .03346 | 25 | .98425 | 65 | 2.55906 |
| 0.06 | .00236 | 0.46 | .01811 | 0.86 | .03386 | 26 | 1.02362 | 66 | 2.59843 |
| 0.07 | .00276 | 0.47 | .01850 | 0.87 | .03425 | 27 | 1.06299 | 67 | 2.63780 |
| 0.08 | .00315 | 0.48 | .01890 | 0.88 | .03465 | 28 | 1.10236 | 68 | 2.67717 |
| 0.09 | .00354 | 0.49 | .01929 | 0.89 | .03504 | 29 | 1.14173 | 69 | 2.71654 |
| 0.10 | .00394 | 0.50 | .01969 | 0.90 | .03543 | 30 | 1.18110 | 70 | 2.75591 |
| 0.11 | .00433 | 0.51 | .02008 | 0.91 | .03583 | 31 | 1.22047 | 71 | 2.79528 |
| 0.12 | .00472 | 0.52 | .02047 | 0.92 | .03622 | 32 | 1.25984 | 72 | 2.83465 |
| 0.13 | .00512 | 0.53 | .02087 | 0.93 | .03661 | 33 | 1.29921 | 73 | 2.87402 |
| 0.14 | .00551 | 0.54 | .02126 | 0.94 | .03701 | 34 | 1.33858 | 74 | 2.91339 |
| 0.15 | .00591 | 0.55 | .02165 | 0.95 | .03740 | 35 | 1.37795 | 75 | 2.95276 |
| 0.16 | .00630 | 0.56 | .02205 | 0.96 | .03780 | 36 | 1.41732 | 76 | 2.99213 |
| 0.17 | .00669 | 0.57 | .02244 | 0.97 | .03819 | 37 | 1.45669 | 77 | 3.03150 |
| 0.18 | .00709 | 0.58 | .02283 | 0.98 | .03858 | 38 | 1.49606 | 78 | 3.07087 |
| 0.19 | .00748 | 0.59 | .02323 | 0.99 | .03898 | 39 | 1.53543 | 79 | 3.11024 |
| 0.20 | .00787 | 0.60 | .02362 | 1.00 | .03937 | 40 | 1.57480 | 80 | 3.14961 |
| 0.21 | .00827 | 0.61 | .02402 | 1 | .03937 | 41 | 1.61417 | 81 | 3.18898 |
| 0.22 | .00866 | 0.62 | .02441 | 2 | .07874 | 42 | 1.65354 | 82 | 3.22835 |
| 0.23 | .00906 | 0.63 | .02480 | 3 | .11811 | 43 | 1.69291 | 83 | 3.26772 |
| 0.24 | .00945 | 0.64 | .02520 | 4 | .15748 | 44 | 1.73228 | 84 | 3.30709 |
| 0.25 | .00984 | 0.65 | .02559 | 5 | .19685 | 45 | 1.77165 | 85 | 3.34646 |
| 0.26 | .01024 | 0.66 | .02598 | 6 | .23622 | 46 | 1.81102 | 86 | 3.38583 |
| 0.27 | .01063 | 0.67 | .02638 | 7 | .27559 | 47 | 1.85039 | 87 | 3.42520 |
| 0.28 | .01102 | 0.68 | .02677 | 8 | .31496 | 48 | 1.88976 | 88 | 3.46457 |
| 0.29 | .01142 | 0.69 | .02717 | 9 | .35433 | 49 | 1.92913 | 89 | 3.50394 |
| 0.30 | .01181 | 0.70 | .02756 | 10 | .39370 | 50 | 1.96850 | 90 | 3.54331 |
| 0.31 | .01220 | 0.71 | .02795 | 11 | .43307 | 51 | 2.00787 | 91 | 3.58268 |
| 0.32 | .01260 | 0.72 | .02835 | 12 | .47244 | 52 | 2.04724 | 92 | 3.62205 |
| 0.33 | .01299 | 0.73 | .02874 | 13 | .51181 | 53 | 2.08651 | 93 | 3.66142 |
| 0.34 | .01339 | 0.74 | .02913 | 14 | .55118 | 54 | 2.12598 | 94 | 3.70079 |
| 0.35 | .01378 | 0.75 | .02953 | 15 | .59055 | 55 | 2.16535 | 95 | 3.74016 |
| 0.36 | .01417 | 0.76 | .02992 | 16 | .62992 | 56 | 2.20472 | 96 | 3.77953 |
| 0.37 | .01457 | 0.77 | .03032 | 17 | .66929 | 57 | 2.24409 | 97 | 3.81890 |
| 0.38 | .01496 | 0.78 | .03071 | 18 | .70866 | 58 | 2.28346 | 98 | 3.85827 |
| 0.39 | .01535 | 0.79 | .03110 | 19 | .74803 | 59 | 2.32283 | 99 | 3.89764 |
| 0.40 | .01575 | 0.80 | .03150 | 20 | .78740 | 60 | 2.36220 | 100 | 3.93701 |

### FRACTIONS TO DECIMALS TO MILLIMETERS

| Fraction | Decimal | mm | Fraction | Decimal | mm |
|---|---|---|---|---|---|
| 1/64 | 0.0156 | 0.3969 | 33/64 | 0.5156 | 13.0969 |
| 1/32 | 0.0312 | 0.7938 | 17/32 | 0.5312 | 13.4938 |
| 3/64 | 0.0469 | 1.1906 | 35/64 | 0.5469 | 13.8906 |
| 1/16 | 0.0625 | 1.5875 | 9/16 | 0.5625 | 14.2875 |
| 5/64 | 0.0781 | 1.9844 | 37/64 | 0.5781 | 14.6844 |
| 3/32 | 0.0938 | 2.3812 | 19/32 | 0.5938 | 15.0812 |
| 7/64 | 0.1094 | 2.7781 | 39/64 | 0.6094 | 15.4781 |
| 1/8 | 0.1250 | 3.1750 | 5/8 | 0.6250 | 15.8750 |
| 9/64 | 0.1406 | 3.5719 | 41/64 | 0.6406 | 16.2719 |
| 5/32 | 0.1562 | 3.9688 | 21/32 | 0.6562 | 16.6688 |
| 11/64 | 0.1719 | 4.3656 | 43/64 | 0.6719 | 17.0656 |
| 3/16 | 0.1875 | 4.7625 | 11/16 | 0.6875 | 17.4625 |
| 13/64 | 0.2031 | 5.1594 | 45/64 | 0.7031 | 17.8594 |
| 7/32 | 0.2188 | 5.5562 | 23/32 | 0.7188 | 18.2562 |
| 15/64 | 0.2344 | 5.9531 | 47/64 | 0.7344 | 18.6531 |
| 1/4 | 0.2500 | 6.3500 | 3/4 | 0.7500 | 19.0500 |
| 17/64 | 0.2656 | 6.7469 | 49/64 | 0.7656 | 19.4469 |
| 9/32 | 0.2812 | 7.1438 | 25/32 | 0.7812 | 19.8438 |
| 19/64 | 0.2969 | 7.5406 | 51/64 | 0.7969 | 20.2406 |
| 5/16 | 0.3125 | 7.9375 | 13/16 | 0.8125 | 20.6375 |
| 21/64 | 0.3281 | 8.3344 | 53/64 | 0.8281 | 21.0344 |
| 11/32 | 0.3438 | 8.7312 | 27/32 | 0.8438 | 21.4312 |
| 23/64 | 0.3594 | 9.1281 | 55/64 | 0.8594 | 21.8281 |
| 3/8 | 0.3750 | 9.5250 | 7/8 | 0.8750 | 22.2250 |
| 25/64 | 0.3906 | 9.9219 | 57/64 | 0.8906 | 22.6219 |
| 13/32 | 0.4062 | 10.3188 | 29/32 | 0.9062 | 23.0188 |
| 27/64 | 0.4219 | 10.7156 | 59/64 | 0.9219 | 23.4156 |
| 7/16 | 0.4375 | 11.1125 | 15/16 | 0.9375 | 23.8125 |
| 29/64 | 0.4531 | 11.5094 | 61/64 | 0.9531 | 24.2094 |
| 15/32 | 0.4688 | 11.9062 | 31/32 | 0.9688 | 24.6062 |
| 31/64 | 0.4844 | 12.3031 | 63/64 | 0.9844 | 25.0031 |
| 1/2 | 0.5000 | 12.7000 | 1 | 1.0000 | 25.4000 |

### DECIMALS TO MILLIMETERS

| Decimal | mm | Decimal | mm |
|---|---|---|---|
| 0.001 | 0.0254 | 0.500 | 12.7000 |
| 0.002 | 0.0508 | 0.510 | 12.9540 |
| 0.003 | 0.0762 | 0.520 | 13.2080 |
| 0.004 | 0.1016 | 0.530 | 13.4620 |
| 0.005 | 0.1270 | 0.540 | 13.7160 |
| 0.006 | 0.1524 | 0.550 | 13.9700 |
| 0.007 | 0.1778 | 0.560 | 14.2240 |
| 0.008 | 0.2032 | 0.570 | 14.4780 |
| 0.009 | 0.2286 | 0.580 | 14.7320 |
| 0.010 | 0.2540 | 0.590 | 14.9860 |
| 0.020 | 0.5080 | 0.600 | 15.2400 |
| 0.030 | 0.7620 | 0.610 | 15.4940 |
| 0.040 | 1.0160 | 0.620 | 15.7480 |
| 0.050 | 1.2700 | 0.630 | 16.0020 |
| 0.060 | 1.5240 | 0.640 | 16.2560 |
| 0.070 | 1.7780 | 0.650 | 16.5100 |
| 0.080 | 2.0320 | 0.660 | 16.7640 |
| 0.090 | 2.2860 | 0.670 | 17.0180 |
| 0.100 | 2.5400 | 0.680 | 17.2720 |
| 0.110 | 2.7940 | 0.690 | 17.5260 |
| 0.120 | 3.0480 | 0.700 | 17.7800 |
| 0.130 | 3.3020 | 0.710 | 18.0340 |
| 0.140 | 3.5560 | 0.720 | 18.2880 |
| 0.150 | 3.8100 | 0.730 | 18.5420 |
| 0.160 | 4.0640 | 0.740 | 18.7960 |
| 0.170 | 4.3180 | 0.750 | 19.0500 |
| 0.180 | 4.5720 | 0.760 | 19.3040 |
| 0.190 | 4.8260 | 0.770 | 19.5580 |
| 0.200 | 5.0800 | 0.780 | 19.8120 |
| 0.210 | 5.3340 | 0.790 | 20.0660 |
| 0.220 | 5.5880 | 0.800 | 20.3200 |
| 0.230 | 5.8420 | 0.810 | 20.5740 |
| 0.240 | 6.0960 | 0.820 | 20.8280 |
| 0.250 | 6.3500 | 0.830 | 21.0820 |
| 0.260 | 6.6040 | 0.840 | 21.3360 |
| 0.270 | 6.8580 | 0.850 | 21.5900 |
| 0.280 | 7.1120 | 0.860 | 21.8440 |
| 0.290 | 7.3660 | 0.870 | 22.0980 |
| 0.300 | 7.6200 | 0.880 | 22.3520 |
| 0.310 | 7.8740 | 0.890 | 22.6060 |
| 0.320 | 8.1280 | 0.900 | 22.8600 |
| 0.330 | 8.3820 | 0.910 | 23.1140 |
| 0.340 | 8.6360 | 0.920 | 23.3680 |
| 0.350 | 8.8900 | 0.930 | 23.6220 |
| 0.360 | 9.1440 | 0.940 | 23.8760 |
| 0.370 | 9.3980 | 0.950 | 24.1300 |
| 0.380 | 9.6520 | 0.960 | 24.3840 |
| 0.390 | 9.9060 | 0.970 | 24.6380 |
| 0.400 | 10.1600 | 0.980 | 24.8920 |
| 0.410 | 10.4140 | 0.990 | 25.1460 |
| 0.420 | 10.6680 | 1.000 | 25.4000 |
| 0.430 | 10.9220 | | |
| 0.440 | 11.1760 | | |
| 0.450 | 11.4300 | | |
| 0.460 | 11.6840 | | |
| 0.470 | 11.9380 | | |
| 0.480 | 12.1920 | | |
| 0.490 | 12.4460 | | |

## NATURAL TRIGONOMETRIC FUNCTIONS

| Angle | sin | cos | tan | cot | sec | csc | Angle |
|---|---|---|---|---|---|---|---|
| 0° | .0000 | 1.0000 | .0000 | ...... | 1.0000 | ...... | 90° |
| 1 | .01745 | .99985 | .01745 | 57.2900 | 1.0001 | 57.2987 | 89 |
| 2 | .03490 | .99939 | .03492 | 28.6363 | 1.0006 | 28.6537 | 88 |
| 3 | .05234 | .99863 | .05241 | 19.0811 | 1.0014 | 19.1073 | 87 |
| 4 | .06976 | .99756 | .06993 | 14.3007 | 1.0024 | 14.3356 | 86 |
| 5 | .08715 | .99619 | .08749 | 11.4301 | 1.0038 | 11.4737 | 85 |
| 6 | .10453 | .99452 | .10510 | 9.5144 | 1.0055 | 9.5668 | 84 |
| 7 | .12187 | .99255 | .12278 | 8.1443 | 1.0075 | 8.2055 | 83 |
| 8 | .13917 | .99027 | .14054 | 7.1154 | 1.0098 | 7.1853 | 82 |
| 9 | .15643 | .98769 | .15838 | 6.3137 | 1.0125 | 6.3924 | 81 |
| 10 | .17365 | .98481 | .17633 | 5.6713 | 1.0154 | 5.7588 | 80 |
| 11 | .19081 | .98163 | .19438 | 5.1445 | 1.0187 | 5.2408 | 79 |
| 12 | .20791 | .97815 | .21256 | 4.7046 | 1.0223 | 4.8097 | 78 |
| 13 | .22495 | .97437 | .23087 | 4.3315 | 1.0263 | 4.4454 | 77 |
| 14 | .24192 | .97029 | .24933 | 4.0108 | 1.0306 | 4.1336 | 76 |
| 15 | .25882 | .96592 | .26795 | 3.7320 | 1.0353 | 3.8637 | 75 |
| 16 | .27564 | .96126 | .28674 | 3.4874 | 1.0403 | 3.6279 | 74 |
| 17 | .29237 | .95630 | .30573 | 3.2708 | 1.0457 | 3.4203 | 73 |
| 18 | .30902 | .95106 | .32492 | 3.0777 | 1.0515 | 3.2361 | 72 |
| 19 | .32557 | .94552 | .34433 | 2.9042 | 1.0576 | 3.0715 | 71 |
| 20 | .34202 | .93969 | .36397 | 2.7475 | 1.0642 | 2.9238 | 70 |
| 21 | .35837 | .93358 | .38386 | 2.6051 | 1.0711 | 2.7904 | 69 |
| 22 | .37461 | .92718 | .40403 | 2.4751 | 1.0785 | 2.6695 | 68 |
| 23 | .39073 | .92050 | .42447 | 2.3558 | 1.0864 | 2.5593 | 67 |
| 24 | .40674 | .91354 | .44523 | 2.2460 | 1.0946 | 2.4586 | 66 |
| 25 | .42262 | .90631 | .46631 | 2.1445 | 1.1034 | 2.3662 | 65 |
| 26 | .43837 | .89879 | .48773 | 2.0503 | 1.1126 | 2.2812 | 64 |
| 27 | .45399 | .89101 | .50952 | 1.9626 | 1.1223 | 2.2027 | 63 |
| 28 | .46947 | .88295 | .53171 | 1.8807 | 1.1326 | 2.1300 | 62 |
| 29 | .48481 | .87462 | .55431 | 1.8040 | 1.1433 | 2.0627 | 61 |
| 30 | .5000 | .86603 | .57735 | 1.7320 | 1.1547 | 2.0000 | 60 |
| 31 | .51504 | .85717 | .60086 | 1.6643 | 1.1666 | 1.9416 | 59 |
| 32 | .52992 | .84805 | .62487 | 1.6003 | 1.1792 | 1.8871 | 58 |
| 33 | .54464 | .83867 | .64941 | 1.5399 | 1.1922 | 1.8361 | 57 |
| 34 | .55919 | .82904 | .67451 | 1.4826 | 1.2062 | 1.7883 | 56 |
| 35 | .57358 | .81915 | .70021 | 1.4281 | 1.2208 | 1.7434 | 55 |
| 36 | .58778 | .80902 | .72654 | 1.3764 | 1.2361 | 1.7013 | 54 |
| 37 | .60181 | .79863 | .75355 | 1.3270 | 1.2521 | 1.6616 | 53 |
| 38 | .61566 | .78801 | .78128 | 1.2799 | 1.2690 | 1.6243 | 52 |
| 39 | .62932 | .77715 | .80978 | 1.2349 | 1.2868 | 1.5890 | 51 |
| 40 | .64279 | .76604 | .83910 | 1.1917 | 1.3054 | 1.5557 | 50 |
| 41 | .65606 | .75471 | .86929 | 1.1504 | 1.3250 | 1.5242 | 49 |
| 42 | .66913 | .74314 | .90040 | 1.1106 | 1.3456 | 1.4945 | 48 |
| 43 | .68200 | .73135 | .93251 | 1.0724 | 1.3673 | 1.4663 | 47 |
| 44 | .69466 | .71934 | .96569 | 1.0355 | 1.3902 | 1.4395 | 46 |
| 45 | .70711 | .70711 | 1.00000 | 1.0000 | 1.4142 | 1.4142 | 45 |
| Angle | cos | sin | cot | tan | csc | sec | Angle |

## TABLE OF NUMBERS 1 TO 100
### SQUARES, SQUARE ROOTS, CIRCUMFERENCES, AREAS

| No. or Dia. | Square | Square Root | Circum. | Area | No. or Dia. | Square | Square Root | Circum. | Area |
|---|---|---|---|---|---|---|---|---|---|
| 1 | 1 | 1.0000 | 3.142 | 0.7854 | 51 | 2,601 | 7.1414 | 160.22 | 2042.82 |
| 2 | 4 | 1.4142 | 6.283 | 3.1416 | 52 | 2,704 | 7.2111 | 163.36 | 2123.72 |
| 3 | 9 | 1.7321 | 9.425 | 7.0686 | 53 | 2,809 | 7.2801 | 166.50 | 2206.18 |
| 4 | 16 | 2.0000 | 12.566 | 12.5664 | 54 | 2,916 | 7.3485 | 169.65 | 2290.22 |
| 5 | 25 | 2.2361 | 15.708 | 19.6350 | 55 | 3,025 | 7.4162 | 172.79 | 2375.83 |
| 6 | 36 | 2.4495 | 18.850 | 28.2743 | 56 | 3,136 | 7.4833 | 175.93 | 2463.01 |
| 7 | 49 | 2.6458 | 21.991 | 38.4845 | 57 | 3,249 | 7.5598 | 179.07 | 2551.76 |
| 8 | 64 | 2.8284 | 25.133 | 50.2655 | 58 | 3,364 | 7.6158 | 182.21 | 2642.08 |
| 9 | 81 | 3.0000 | 28.274 | 63.6173 | 59 | 3,481 | 7.6811 | 185.35 | 2733.97 |
| 10 | 100 | 3.1623 | 31.416 | 78.5398 | 60 | 3,600 | 7.7460 | 188.50 | 2827.43 |
| 11 | 121 | 3.3166 | 34.558 | 95.0332 | 61 | 3,721 | 7.8102 | 191.64 | 2922.47 |
| 12 | 144 | 3.4641 | 37.699 | 113.097 | 62 | 3,844 | 7.8740 | 194.78 | 3019.07 |
| 13 | 169 | 3.6056 | 40.841 | 132.732 | 63 | 3,969 | 7.9373 | 197.92 | 3117.25 |
| 14 | 196 | 3.7417 | 43.982 | 153.938 | 64 | 4,096 | 8.0000 | 201.06 | 3216.99 |
| 15 | 225 | 3.8730 | 47.124 | 176.715 | 65 | 4,225 | 8.0623 | 204.20 | 3318.31 |
| 16 | 256 | 4.0000 | 50.265 | 201.062 | 66 | 4,356 | 8.1240 | 207.35 | 3421.19 |
| 17 | 289 | 4.1231 | 53.407 | 226.980 | 67 | 4,489 | 8.1854 | 210.49 | 3525.65 |
| 18 | 324 | 4.2426 | 56.549 | 254.469 | 68 | 4,624 | 8.2462 | 213.63 | 3631.68 |
| 19 | 361 | 4.3589 | 59.690 | 283.529 | 69 | 4,761 | 8.3066 | 216.77 | 3739.28 |
| 20 | 400 | 4.4721 | 62.832 | 314.159 | 70 | 4,900 | 8.3666 | 219.91 | 3848.45 |
| 21 | 441 | 4.5826 | 65.973 | 346.361 | 71 | 5,041 | 8.4261 | 223.05 | 3959.19 |
| 22 | 484 | 4.6904 | 69.119 | 380.133 | 72 | 5,184 | 8.4853 | 226.19 | 4071.50 |
| 23 | 529 | 4.7958 | 72.257 | 415.476 | 73 | 5,329 | 8.5440 | 229.34 | 4185.39 |
| 24 | 576 | 4.8990 | 75.398 | 452.389 | 74 | 5,476 | 8.6023 | 232.48 | 4300.84 |
| 25 | 625 | 5.0000 | 78.540 | 490.874 | 75 | 5,625 | 8.6603 | 235.62 | 4417.86 |
| 26 | 676 | 5.0990 | 81.681 | 530.929 | 76 | 5,776 | 8.7178 | 238.76 | 4536.46 |
| 27 | 729 | 5.1962 | 84.823 | 572.555 | 77 | 5,929 | 8.7750 | 241.90 | 4656.63 |
| 28 | 784 | 5.2915 | 87.965 | 615.752 | 78 | 6,084 | 8.8318 | 245.04 | 4778.36 |
| 29 | 841 | 5.3852 | 91.106 | 660.520 | 79 | 6,241 | 8.8882 | 248.19 | 4901.67 |
| 30 | 900 | 5.4772 | 94.248 | 706.858 | 80 | 6,400 | 8.9443 | 251.33 | 5026.55 |
| 31 | 961 | 5.5678 | 97.389 | 754.768 | 81 | 6,561 | 9.0000 | 254.47 | 5153.00 |
| 32 | 1,024 | 5.6569 | 100.531 | 804.248 | 82 | 6,724 | 9.0554 | 257.61 | 5281.02 |
| 33 | 1,089 | 5.7446 | 103.673 | 855.299 | 83 | 6,889 | 9.1104 | 260.75 | 5410.61 |
| 34 | 1,156 | 5.8310 | 106.814 | 907.920 | 84 | 7,056 | 9.1652 | 263.89 | 5541.77 |
| 35 | 1,225 | 5.9161 | 109.956 | 962.113 | 85 | 7,225 | 9.2195 | 267.04 | 5674.50 |
| 36 | 1,296 | 6.0000 | 113.097 | 1017.88 | 86 | 7,396 | 9.2736 | 270.18 | 5808.80 |
| 37 | 1,369 | 6.0828 | 116.239 | 1075.21 | 87 | 7,569 | 9.3274 | 273.32 | 5944.68 |
| 38 | 1,444 | 6.1644 | 119.381 | 1134.11 | 88 | 7,744 | 9.3808 | 276.46 | 6082.12 |
| 39 | 1,521 | 6.2450 | 122.522 | 1194.59 | 89 | 7,921 | 9.4340 | 279.60 | 6221.14 |
| 40 | 1,600 | 6.3246 | 125.66 | 1256.64 | 90 | 8,100 | 9.4868 | 282.74 | 6361.73 |
| 41 | 1,681 | 6.4031 | 128.81 | 1320.25 | 91 | 8,281 | 9.5394 | 285.88 | 6503.88 |
| 42 | 1,764 | 6.4807 | 131.95 | 1385.44 | 92 | 8,464 | 9.5917 | 289.03 | 6647.61 |
| 43 | 1,849 | 6.5574 | 135.09 | 1452.20 | 93 | 8,649 | 9.6437 | 292.17 | 6792.91 |
| 44 | 1,936 | 6.6332 | 138.23 | 1520.53 | 94 | 8,836 | 9.6954 | 295.31 | 6939.78 |
| 45 | 2,025 | 6.7082 | 141.37 | 1590.43 | 95 | 9,025 | 9.7468 | 298.45 | 7088.22 |
| 46 | 2,116 | 6.7823 | 144.51 | 1661.90 | 96 | 9,216 | 9.7980 | 301.59 | 7238.23 |
| 47 | 2,209 | 6.8557 | 147.65 | 1734.94 | 97 | 9,409 | 9.8489 | 304.73 | 7389.81 |
| 48 | 2,304 | 6.9282 | 150.80 | 1809.56 | 98 | 9,604 | 9.8995 | 307.88 | 7542.96 |
| 49 | 2,401 | 7.0000 | 153.94 | 1885.74 | 99 | 9,801 | 9.9499 | 311.02 | 7697.69 |
| 50 | 2,500 | 7.0711 | 157.08 | 1963.50 | 100 | 10,000 | 10.0000 | 314.16 | 7853.98 |

## Decimal Equivalents of 8ths, 16ths, 32nds, 64ths

| 8ths | 32nds | 64ths | 64ths |
|---|---|---|---|
| 1/8 = .125 | 1/32 = .03125 | 1/64 = .015625 | 33/64 = .515625 |
| 1/4 = .250 | 3/32 = .09375 | 3/64 = .046875 | 35/64 = .546875 |
| 3/8 = .375 | 5/32 = .15625 | 5/64 = .078125 | 37/64 = .578125 |
| 1/2 = .500 | 7/32 = .21875 | 7/64 = .109375 | 39/64 = .609375 |
| 5/8 = .625 | 9/32 = .28125 | 9/64 = .140625 | 41/64 = .640625 |
| 3/4 = .750 | 11/32 = .34375 | 11/64 = .171875 | 43/64 = .671875 |
| 7/8 = .875 | 13/32 = .40625 | 13/64 = .203125 | 45/64 = .703125 |
| **16ths** | 15/32 = .46875 | 15/64 = .234375 | 47/64 = .734375 |
| 1/16 = .0625 | 17/32 = .53125 | 17/64 = .265625 | 49/64 = .765625 |
| 3/16 = .1875 | 19/32 = .59375 | 19/64 = .296875 | 51/64 = .796875 |
| 5/16 = .3125 | 21/32 = .65625 | 21/64 = .328125 | 53/64 = .828125 |
| 7/16 = .4375 | 23/32 = .71875 | 23/64 = .359375 | 55/64 = .859375 |
| 9/16 = .5625 | 25/32 = .78125 | 25/64 = .390625 | 57/64 = .890625 |
| 11/16 = .6875 | 27/32 = .84375 | 27/64 = .421875 | 59/64 = .921875 |
| 13/16 = .8125 | 29/32 = .90625 | 29/64 = .453125 | 61/64 = .953125 |
| 15/16 = .9375 | 31/32 = .96875 | 31/64 = .484375 | 63/64 = .984375 |

## Decimal Equivalents of 7ths, 14ths, and 28ths

| 7th | 14th | 28th | Decimal | 7th | 14th | 28th | Decimal |
|---|---|---|---|---|---|---|---|
| | | 1 | .035714 | | | 15 | .535714 |
| | 1 | | .071429 | 4 | | | .571429 |
| | | 3 | .107143 | | | 17 | .607143 |
| 1 | | | .142857 | | 9 | | .642867 |
| | | 5 | .178571 | | | 19 | .678571 |
| | 3 | | .214286 | 5 | | | .714286 |
| | | 7 | .25 | | | 21 | .75 |
| 2 | | | .285714 | | 11 | | .785714 |
| | | 9 | .321429 | | | 23 | .821429 |
| | 5 | | .357143 | 6 | | | .857143 |
| | | 11 | .392857 | | | 25 | .892857 |
| 3 | | | .428571 | | 13 | | .928571 |
| | | 13 | .464286 | | | 27 | .964286 |
| | 7 | | .5 | | | | |

## Decimal Equivalents of 6ths, 12ths, and 24ths

| 6th | 12th | 24th | Decimal | 6th | 12th | 24th | Decimal |
|---|---|---|---|---|---|---|---|
| | | 1 | .041667 | 3 | | | .5 |
| | 1 | | .083333 | | | 13 | .541666 |
| | | 3 | .125 | | 7 | | .583333 |
| 1 | | | .166666 | | | 15 | .625 |
| | | 5 | .208333 | 4 | | | .666666 |
| | 3 | | .25 | | | 17 | .708333 |
| | | 7 | .291666 | | 9 | | .75 |
| 2 | | | .333333 | | | 19 | 791666 |
| | | 9 | .375 | 5 | | | .833333 |
| | 5 | | .416666 | | | 21 | .875 |
| | 11 | | .458333 | | 11 | | .916666 |
| | | | | | | 23 | .958333 |

## ENGLISH — METRIC CONVERSION FACTORS
### VOLUME AND MASS (WEIGHT)

| | |
|---|---|
| 1 cm³ = 0.06 CU. IN. | 1 CU. IN. = 16.4 cm³ |
| 1 m³ = 35.3 CU. FT. | 1 CU. FT. = 0.03 m³ |
| 1 m³ = 1.3 CU. YD. | 1 CU. YD. = 0.8 m³ |
| 1 L = 33.8 FL. OZ. | 1 FL. OZ. = 29.6 ml |
| 1 L = 4.2 CUPS | 1 CUP = 237 ml |
| 1 L = 2.1 PT. | 1 PT. = 0.47 L |
| 1 L = 1.06 QT. | 1 QT. = 0.95 L |
| 1 L = 0.26 GAL. | 1 GAL. = 3.79 L |
| 1 gram = 0.035 OZ. | 1 OZ. = 28.3 g |
| 1 kg = 2.2 LB. | 1 LB. = 0.45 kg |
| 1 t = 2205 LB. | 1 TON = 907.2 kg |

METRIC SYMBOLS USED:
cm³ = cubic centimetres    m³ = cubic metre
g = gram    ml = millilitre
kg = kilogram    t = tonne
L = litre

## DIMENSIONS AND STRENGTH OF CLAY PIPE

| NOMINAL SIZE (INCHES) | OUTSIDE DIAMETER BARREL (INCHES) | | NOMINAL BARREL THICKNESS (INCHES) | | MINIMUM CRUSHING STRENGTH LB./LINEAR FT. | |
|---|---|---|---|---|---|---|
| | MIN. | MAX. | STANDARD | X-STRENGTH | STANDARD | X-STRENGTH |
| 4 | 4 7/8 | 5 1/8 | 1/2 | 5/8 | 1200 | 2000 |
| 6 | 7 1/16 | 7 7/16 | 5/8 | 11/16 | 1200 | 2000 |
| 8 | 9 1/4 | 9 3/4 | 3/4 | 7/8 | 1400 | 2200 |
| 10 | 11 1/2 | 12 | 7/8 | 1 | 1600 | 2400 |

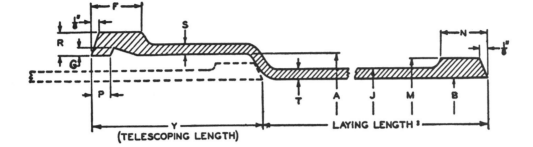

## DIMENSIONS OF HUBS, SPIGOTS, AND BARRELS FOR EXTRA-HEAVY CAST IRON SOIL PIPE AND FITTINGS

| SIZE | INSIDE DIAMETER OF HUB | OUTSIDE DIAMETER OF SPIGOT BEAD | OUTSIDE DIAMETER OF BARREL | TELESCOPING LENGTH | INSIDE DIAMETER OF BARREL | THICKNESS OF BARREL |
|---|---|---|---|---|---|---|
| | A | M | J | Y | B | T |
| INCHES | INCHES | INCHES | INCHES | INCHES | INCHES | INCHES |
| 2 | 3.06 | 2.75 | 2.38 | 2.50 | 2.00 | .19 |
| 3 | 4.19 | 3.88 | 3.50 | 2.75 | 3.00 | .25 |
| 4 | 5.19 | 4.88 | 4.50 | 3.00 | 4.00 | .25 |
| 5 | 6.19 | 5.88 | 5.50 | 3.00 | 5.00 | .25 |
| 6 | 7.19 | 6.88 | 6.50 | 3.00 | 6.00 | .25 |
| 8 | 9.50 | 9.00 | 8.62 | 3.50 | 8.00 | .31 |
| 10 | 11.62 | 11.13 | 10.75 | 3.50 | 10.00 | .37 |
| 12 | 13.75 | 13.13 | 12.75 | 4.25 | 12.00 | .37 |
| 15 | 17.00 | 16.25 | 15.88 | 4.25 | 15.00 | .44 |

| SIZE | THICKNESS OF HUB | | WIDTH OF HUB BEAD | WIDTH OF SPIGOT BEAD | DISTANCE FROM LEAD GROOVE TO END, PIPE AND FITTINGS | DEPTH OF LEAD GROOVE | |
|---|---|---|---|---|---|---|---|
| | HUB BODY | OVER BEAD | | | | | |
| | S (MIN.) | R (MIN.) | F | N | P | G (MIN.) | G (MAX.) |
| INCHES | INCHES | INCHES | INCHES | INCHES | INCHES | INCHES | INCHES |
| 2 | 0.18 | 0.37 | 0.75 | 0.69 | 0.28 | 0.10 | 0.13 |
| 3 | .25 | .43 | .81 | .75 | .28 | .10 | .13 |
| 4 | .25 | .43 | .88 | .81 | .28 | .10 | .13 |
| 5 | .25 | .43 | .88 | .81 | .28 | .10 | .13 |
| 6 | .25 | .43 | .88 | .81 | .28 | .10 | .13 |
| 8 | .34 | .59 | 1.19 | 1.12 | .38 | .15 | .19 |
| 10 | .40 | .65 | 1.19 | 1.12 | .38 | .15 | .19 |
| 12 | .40 | .65 | 1.44 | 1.38 | .47 | .15 | .19 |
| 15 | .46 | .71 | 1.44 | 1.38 | .47 | .15 | .19 |

(Cast Iron Soil Pipe Institute)

## DIMENSIONS OF HUBS, SPIGOTS, AND BARRELS FOR SERVICE CAST IRON SOIL PIPE AND FITTINGS

| SIZE | INSIDE DIAMETER OF HUB | OUTSIDE DIA. OF SPIGOT BEAD | OUTSIDE DIAMETER OF BARREL | TELESCOPING LENGTH | INSIDE DIAMETER OF BARREL | THICKNESS OF BARREL |
|---|---|---|---|---|---|---|
| | A | M | J | Y | B | T |
| INCHES | INCHES | INCHES | INCHES | INCHES | INCHES | INCHES |
| 2 | 2.94 | 2.62 | 2.30 | 2.50 | 1.96 | 0.17 |
| 3 | 3.94 | 3.62 | 3.30 | 2.75 | 2.96 | .17 |
| 4 | 4.94 | 4.62 | 4.30 | 3.00 | 3.94 | .18 |
| 5 | 5.94 | 5.62 | 5.30 | 3.00 | 4.94 | .18 |
| 6 | 6.94 | 6.62 | 6.30 | 3.00 | 5.94 | .18 |
| 8 | 9.25 | 8.75 | 8.38 | 3.50 | 7.94 | .23 |
| 10 | 11.38 | 10.88 | 10.50 | 3.50 | 9.94 | .28 |
| 12 | 13.50 | 12.88 | 12.50 | 4.25 | 11.94 | .28 |
| 15 | 16.75 | 16.00 | 15.62 | 4.25 | 15.00 | .31 |

(Cast Iron Soil Pipe Institute)                    Refer to illustration on page 257.

| SIZE | THICKNESS OF HUB | | WIDTH OF HUB BEAD | WIDTH OF SPIGOT BEAD | DISTANCE FROM LEAD GROOVE TO END, PIPE AND FITTINGS | DEPTH OF LEAD GROOVE | |
|---|---|---|---|---|---|---|---|
| | HUB BODY | OVER BEAD | | | | | |
| | S (MIN.) | R (MIN.) | F | N | P | G (MIN.) | G (MAX.) |
| INCHES | INCHES | INCHES | INCHES | INCHES | INCHES | INCHES | INCHES |
| 2 | 0.13 | 0.34 | 0.75 | 0.69 | 0.28 | 0.10 | 0.13 |
| 3 | .16 | .37 | .81 | .75 | .28 | .10 | .13 |
| 4 | .16 | .37 | .88 | .81 | .28 | .10 | .13 |
| 5 | .16 | .37 | .88 | .81 | .28 | .10 | .13 |
| 6 | .18 | .37 | .88 | .81 | .28 | .10 | .13 |
| 8 | .19 | .44 | 1.19 | 1.12 | .38 | .15 | .19 |
| 10 | .27 | .53 | 1.19 | 1.12 | .38 | .15 | .19 |
| 12 | .27 | .53 | 1.44 | 1.38 | .47 | .15 | .19 |
| 15 | .30 | .58 | 1.44 | 1.38 | .47 | .15 | .19 |

(Cast Iron Soil Pipe Institute)                    Refer to illustration on page 257.

## DIMENSIONS OF COPPER PIPE AND TUBE

| NOMINAL SIZE (INCHES) | OD ALL SIZES | TYPE K | | | TYPE L | | | TYPE M | | | DWV | | |
|---|---|---|---|---|---|---|---|---|---|---|---|---|---|
| | | WALL THK. | ID | LB./FT. | WALL THK. | ID | LB./FT. | WALL THK. | ID | LB./FT. | WALL THK. | ID | LB./FT. |
| 1/4 | .375 | .035 | .305 | .145 | .030 | .315 | .126 | N.A. | N.A. | N.A. | NOT | | |
| 3/8 | .500 | .049 | .402 | .269 | .035 | .430 | .198 | .025 | .450 | .145 | AVAILABLE | | |
| 1/2 | .625 | .049 | .527 | .344 | .040 | .545 | .285 | .028 | .569 | .204 | IN | | |
| 5/8 | .750 | .049 | .652 | .418 | .042 | .666 | .362 | N.A. | N.A. | N.A. | THESE | | |
| 3/4 | .875 | .065 | .745 | .641 | .045 | .785 | .455 | .032 | .811 | .328 | SIZES | | |
| 1 | 1.125 | .065 | .995 | .839 | .050 | 1.025 | .655 | .035 | 1.055 | .465 | | | |
| 1 1/4 | 1.375 | .065 | 1.245 | 1.04 | .055 | 1.265 | .884 | .042 | 1.291 | .682 | .040 | 1.295 | .650 |
| 1 1/2 | 1.625 | .072 | 1.481 | 1.36 | .060 | 1.505 | 1.14 | .049 | 1.527 | .940 | .042 | 1.541 | .809 |
| 2 | 2.125 | .083 | 1.959 | 2.06 | .070 | 1.985 | 1.75 | .058 | 2.009 | 1.46 | .042 | 2.041 | 1.07 |
| 2 1/2 | 2.625 | .095 | 2.435 | 2.93 | .080 | 2.465 | 2.48 | .065 | 2.495 | 2.03 | N.A. | N.A. | N.A. |
| 3 | 3.125 | .109 | 2.907 | 4.00 | .090 | 2.945 | 3.33 | .072 | 2.981 | 2.68 | .045 | 3.035 | 1.69 |
| 3 1/2 | 3.625 | .120 | 3.385 | 5.12 | .100 | 3.425 | 4.29 | .083 | 3.459 | 3.58 | N.A. | N.A. | N.A. |
| 4 | 4.125 | .134 | 3.857 | 6.51 | .110 | 3.905 | 5.38 | .095 | 3.935 | 4.66 | .058 | 4.009 | 2.87 |

## BUILDING MATERIAL SYMBOLS

| | PLAN | ELEVATION | SECTION |
|---|---|---|---|
| WOOD | FLOOR AREAS LEFT BLANK | SIDING PANEL | FRAMING / FINISH |
| BRICK | FACE, COMMON | FACE OR COMMON | SAME AS PLAN VIEW |
| STONE | CUT, RUBBLE | CUT RUBBLE | CUT RUBBLE |
| CONCRETE | | | SAME AS PLAN VIEW |
| CONCRETE BLOCK | | | SAME AS PLAN VIEW |
| EARTH | NONE | NONE | |
| GLASS | | | LARGE SCALE / SMALL SCALE |
| INSULATION | SAME AS SECTION | INSULATION | LOOSE FILL OR BATT / BOARD |
| PLASTER | SAME AS SECTION | PLASTER | STUD / LATH AND PLASTER |
| STRUCTURAL STEEL | INDICATE BY NOTE | INDICATE BY NOTE | |
| SHEET METAL FLASHING | INDICATE BY NOTE | | SHOW CONTOUR |
| TILE | FLOOR | WALL | |

## ELECTRICAL SYMBOLS

- CEILING OUTLETS FOR FIXTURES
- LIGHTING PANEL
- POWER PANEL
- WALL FIXTURE OUTLET
- SINGLE-POLE SWITCH
- CEILING OUTLET WITH PULL SWITCH
- DOUBLE-POLE SWITCH
- WALL OUTLET WITH PULL SWITCH
- THREE-WAY SWITCH
- DUPLEX CONVENIENCE OUTLET
- FOUR-WAY SWITCH
- WATERPROOF CONVENIENCE OUTLET
- SWITCH WITH PILOT LIGHT
- CONVENIENCE OUTLET  1 = SINGLE  3 = TRIPLE
- PUSH BUTTON
- RANGE OUTLET
- BELL
- CONVENIENCE OUTLET WITH SWITCH
- OUTSIDE TELEPHONE CONNECTION
- SPECIAL PURPOSE (SEE SPECS.)
- TELEVISION CONNECTION
- SWITCH WIRING
- FLOOR OUTLET
- EXTERIOR CEILING FIXTURE
- CEILING LIGHT FIXTURE
- FLUORESCENT CEILING FIXTURE
- PULL CHAIN LIGHT FIXTURE
- FLUORESCENT WALL FIXTURE
- EXTERIOR LIGHT FIXTURE

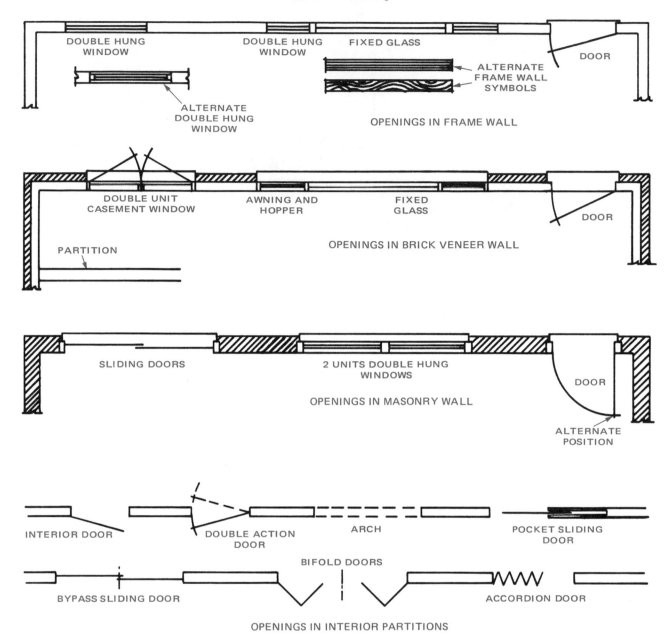

DOUBLE HUNG WINDOW

DOUBLE HUNG WINDOW

FIXED GLASS

DOOR

ALTERNATE FRAME WALL SYMBOLS

ALTERNATE DOUBLE HUNG WINDOW

OPENINGS IN FRAME WALL

DOUBLE UNIT CASEMENT WINDOW

AWNING AND HOPPER

FIXED GLASS

DOOR

PARTITION

OPENINGS IN BRICK VENEER WALL

SLIDING DOORS

2 UNITS DOUBLE HUNG WINDOWS

DOOR

OPENINGS IN MASONRY WALL

ALTERNATE POSITION

INTERIOR DOOR

DOUBLE ACTION DOOR

ARCH

POCKET SLIDING DOOR

BYPASS SLIDING DOOR

BIFOLD DOORS

ACCORDION DOOR

OPENINGS IN INTERIOR PARTITIONS

## APPROXIMATE FRICTION LOSS IN THERMOSPLASTIC PIPE FITTINGS IN EQUIVALENT FEET OF PIPE*

| NOMINAL PIPE SIZE, IN | 3/8 | 1/2 | 3/4 | 1 | 1-1/4 | 1-1/2 | 2 | 2-1/2 | 3 | 3-1/2 | 4 | 5 |
|---|---|---|---|---|---|---|---|---|---|---|---|---|
| TEE, SIDE OUTLET | 3 | 4 | 5 | 6 | 7 | 8 | 12 | 15 | 16 | 20 | 22 | 28 |
| 90 DEG. L | 1-1/2 | 1-1/2 | 2 | 2-3/4 | 4 | 4 | 6 | 8 | 8 | 10 | 12 | 14 |
| 45 DEG. L | 3/4 | 3/4 | 1 | 1-3/8 | 1-3/4 | 2 | 2-1/2 | 3 | 4 | 4-1/2 | 5 | 6 |
| INSERT COUPLING | — | 1/2 | 3/4 | 1 | 1-1/4 | 1-1/2 | 2 | 3 | 3 | — | 4 | — |
| MALE-FEMALE INSERT ADAPTERS | — | 1 | 1-1/2 | 2 | 2-3/4 | 3-1/2 | 4-1/2 | — | 6-1/2 | — | 9 | — |

(Plastics Pipe Institute)

## PRESSURE LOSS AND VELOCITY RELATIONSHIPS FOR WATER FLOWING IN COPPER TUBE

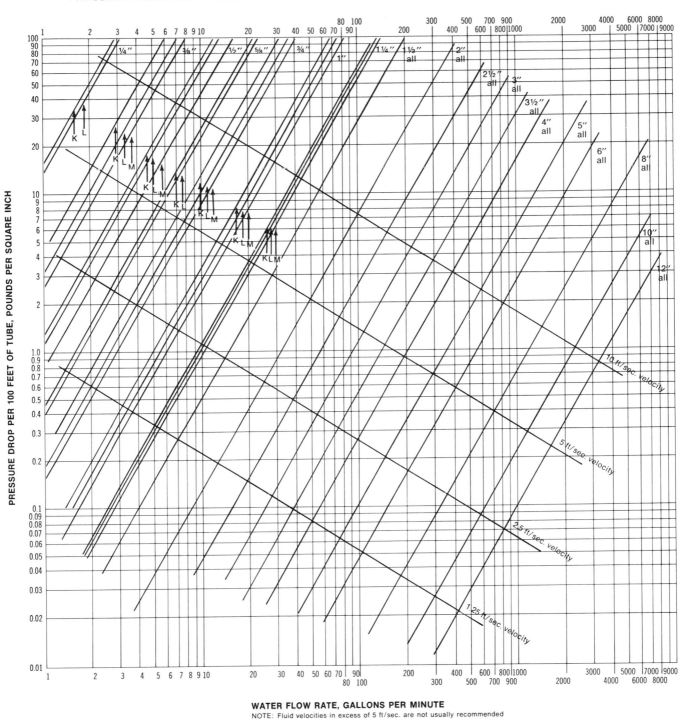

**WATER FLOW RATE, GALLONS PER MINUTE**

NOTE: Fluid velocities in excess of 5 ft/sec. are not usually recommended

(Copper Development Assoc. Inc.)

# CARRYING CAPACITY AND FRICTION LOSS FOR SCHEDULE 40 THERMOPLASTIC PIPE

(Independent variables: Gallons per minute and nominal pipe size O.D.)

Dependent variables: Velocity, friction head and pressure drop per 100 feet of pipe, interior smooth.)

For each pipe size the three sub‑columns are: **V** = Velocity (feet per second), **H** = Friction Head (feet), **P** = Friction Loss (pounds per square inch), per 100 feet of pipe.

## Nominal sizes ½ in. – 3 in.

| Gallons per minute | ½ in. V | ½ in. H | ½ in. P | ¾ in. V | ¾ in. H | ¾ in. P | 1 in. V | 1 in. H | 1 in. P | 1¼ in. V | 1¼ in. H | 1¼ in. P | 1½ in. V | 1½ in. H | 1½ in. P | 2 in. V | 2 in. H | 2 in. P | 2½ in. V | 2½ in. H | 2½ in. P | 3 in. V | 3 in. H | 3 in. P |
|---|---|---|---|---|---|---|---|---|---|---|---|---|---|---|---|---|---|---|---|---|---|---|---|---|
| 1 | 1.13 | 2.08 | 0.90 | 0.63 | 0.51 | 0.22 | | | | | | | | | | | | | | | | | | |
| 2 | 2.26 | 4.16 | 1.80 | 1.26 | 1.02 | 0.44 | 0.77 | 0.55 | 0.24 | 0.44 | 0.14 | 0.06 | 0.33 | 0.07 | 0.03 | | | | | | | | | |
| 5 | 5.64 | 23.44 | 10.15 | 3.16 | 5.73 | 2.48 | 1.93 | 1.72 | 0.75 | 1.11 | 0.44 | 0.19 | 0.81 | 0.22 | 0.09 | 0.49 | 0.066 | 0.029 | 0.30 | 0.038 | 0.016 | 0.22 | 0.015 | 0.007 |
| 7 | 7.90 | 43.06 | 18.64 | 4.43 | 10.52 | 4.56 | 2.72 | 3.17 | 1.37 | 1.55 | 0.81 | 0.35 | 1.13 | 0.38 | 0.17 | 0.69 | 0.11 | 0.048 | 0.49 | 0.051 | 0.023 | 0.31 | 0.021 | 0.009 |
| 10 | 11.28 | 82.02 | 35.51 | 6.32 | 20.04 | 8.68 | 3.86 | 6.02 | 2.61 | 2.21 | 1.55 | 0.67 | 1.62 | 0.72 | 0.31 | 0.98 | 0.21 | 0.091 | 0.68 | 0.09 | 0.039 | 0.44 | 0.03 | 0.013 |
| 15 | | | | 9.48 | 42.46 | 18.39 | 5.79 | 12.77 | 5.53 | 3.31 | 3.28 | 1.42 | 2.42 | 1.53 | 0.66 | 1.46 | 0.45 | 0.19 | 1.03 | 0.19 | 0.082 | 0.66 | 0.07 | 0.030 |
| 20 | | | | 12.65 | 72.34 | 31.32 | 7.72 | 21.75 | 9.42 | 4.42 | 5.59 | 2.42 | 3.23 | 2.61 | 1.13 | 1.95 | 0.76 | 0.33 | 1.37 | 0.32 | 0.14 | 0.88 | 0.11 | 0.048 |
| 25 | | | | | | | 9.65 | 32.88 | 14.22 | 5.52 | 8.45 | 3.66 | 4.04 | 3.95 | 1.71 | 2.44 | 1.15 | 0.50 | 1.71 | 0.49 | 0.21 | 1.10 | 0.17 | 0.074 |
| 30 | | | | | | | 11.58 | 46.08 | 19.95 | 6.63 | 11.85 | 5.13 | 4.85 | 5.53 | 2.39 | 2.93 | 1.62 | 0.70 | 2.05 | 0.68 | 0.29 | 1.33 | 0.23 | 0.10 |
| 35 | | | | | | | | | | 7.73 | 15.76 | 6.82 | 5.66 | 7.36 | 3.19 | 3.41 | 2.15 | 0.93 | 2.39 | 0.91 | 0.39 | 1.55 | 0.31 | 0.13 |
| 40 | | | | | | | | | | 8.84 | 20.18 | 8.74 | 6.47 | 9.43 | 4.08 | 3.90 | 2.75 | 1.19 | 2.73 | 1.16 | 0.50 | 1.77 | 0.40 | 0.17 |
| 45 | | | | | | | | | | 9.94 | 25.10 | 10.87 | 7.27 | 11.73 | 5.08 | 4.39 | 3.43 | 1.49 | 3.08 | 1.44 | 0.62 | 1.99 | 0.50 | 0.22 |
| 50 | | | | | | | | | | 11.05 | 30.51 | 13.21 | 8.08 | 14.25 | 6.17 | 4.88 | 4.16 | 1.80 | 3.42 | 1.75 | 0.76 | 2.21 | 0.60 | 0.26 |
| 60 | | | | | | | | | | | | | 9.70 | 19.98 | 8.65 | 5.85 | 5.84 | 2.53 | 4.10 | 2.46 | 1.07 | 2.65 | 0.85 | 0.37 |
| 70 | | | | | | | | | | | | | | | | 6.83 | 7.76 | 3.36 | 4.79 | 3.27 | 1.42 | 3.09 | 1.13 | 0.49 |
| 75 | | | | | | | | | | | | | | | | 7.32 | 8.82 | 3.82 | 5.13 | 3.71 | 1.61 | 3.31 | 1.28 | 0.55 |
| 80 | | | | | | | | | | | | | | | | 7.80 | 9.94 | 4.30 | 5.47 | 4.19 | 1.81 | 3.53 | 1.44 | 0.62 |
| 90 | | | | | | | | | | | | | | | | 8.78 | 12.37 | 5.36 | 6.15 | 5.21 | 2.26 | 3.98 | 1.80 | 0.78 |
| 100 | | | | | | | | | | | | | | | | 9.75 | 15.03 | 6.51 | 6.84 | 6.33 | 2.74 | 4.42 | 2.18 | 0.94 |
| 125 | | | | | | | | | | | | | | | | | | | 8.55 | 9.58 | 4.15 | 5.52 | 3.31 | 1.43 |
| 150 | | | | | | | | | | | | | | | | | | | 10.26 | 13.41 | 5.81 | 6.63 | 4.63 | 2.00 |
| 175 | | | | | | | | | | | | | | | | | | | | | | 7.73 | 6.16 | 2.67 |
| 200 | | | | | | | | | | | | | | | | | | | | | | 8.83 | 7.88 | 3.41 |
| 250 | | | | | | | | | | | | | | | | | | | | | | 11.04 | 11.93 | 5.17 |

## Nominal sizes 4 in. – 12 in.

| Gallons per minute | 4 in. V | 4 in. H | 4 in. P | 5 in. V | 5 in. H | 5 in. P | 6 in. V | 6 in. H | 6 in. P | 8 in. V | 8 in. H | 8 in. P | 10 in. V | 10 in. H | 10 in. P | 12 in. V | 12 in. H | 12 in. P |
|---|---|---|---|---|---|---|---|---|---|---|---|---|---|---|---|---|---|---|
| 20 | 0.51 | 0.03 | 0.013 | | | | | | | | | | | | | | | |
| 25 | 0.64 | 0.04 | 0.017 | | | | | | | | | | | | | | | |
| 30 | 0.77 | 0.06 | 0.026 | 0.49 | 0.02 | 0.009 | | | | | | | | | | | | |
| 35 | 0.89 | 0.08 | 0.035 | 0.57 | 0.03 | 0.013 | | | | | | | | | | | | |
| 40 | 1.02 | 0.11 | 0.048 | 0.65 | 0.03 | 0.013 | | | | | | | | | | | | |
| 45 | 1.15 | 0.13 | 0.056 | 0.73 | 0.04 | 0.017 | | | | | | | | | | | | |
| 50 | 1.28 | 0.16 | 0.069 | 0.81 | 0.05 | 0.022 | 0.56 | 0.02 | 0.009 | | | | | | | | | |
| 60 | 1.53 | 0.22 | 0.095 | 0.97 | 0.07 | 0.030 | 0.67 | 0.03 | 0.013 | | | | | | | | | |
| 70 | 1.79 | 0.30 | 0.13 | 1.14 | 0.10 | 0.043 | 0.79 | 0.04 | 0.017 | | | | | | | | | |
| 75 | 1.92 | 0.34 | 0.15 | 1.22 | 0.11 | 0.048 | 0.84 | 0.05 | 0.022 | | | | | | | | | |
| 80 | 2.05 | 0.38 | 0.16 | 1.30 | 0.13 | 0.056 | 0.90 | 0.05 | 0.022 | | | | | | | | | |
| 90 | 2.30 | 0.47 | 0.20 | 1.46 | 0.16 | 0.069 | 1.01 | 0.06 | 0.026 | | | | | | | | | |
| 100 | 2.56 | 0.58 | 0.25 | 1.62 | 0.19 | 0.082 | 1.12 | 0.08 | 0.035 | 0.65 | 0.03 | 0.012 | | | | | | |
| 125 | 3.20 | 0.88 | 0.38 | 2.03 | 0.29 | 0.125 | 1.41 | 0.12 | 0.052 | 0.81 | 0.035 | 0.015 | | | | | | |
| 150 | 3.84 | 1.22 | 0.53 | 2.44 | 0.40 | 0.17 | 1.69 | 0.16 | 0.069 | 0.97 | 0.04 | 0.017 | | | | | | |
| 175 | 4.48 | 1.63 | 0.71 | 2.84 | 0.54 | 0.235 | 1.97 | 0.22 | 0.096 | 1.14 | 0.055 | 0.024 | | | | | | |
| 200 | 5.11 | 2.08 | 0.90 | 3.25 | 0.69 | 0.30 | 2.25 | 0.28 | 0.12 | 1.30 | 0.07 | 0.030 | 0.82 | 0.027 | 0.012 | | | |
| 250 | 6.40 | 3.15 | 1.36 | 4.06 | 1.05 | 0.45 | 2.81 | 0.43 | 0.19 | 1.63 | 0.11 | 0.048 | 1.03 | 0.035 | 0.015 | | | |
| 300 | 7.67 | 4.41 | 1.91 | 4.87 | 1.46 | 0.63 | 3.37 | 0.60 | 0.26 | 1.94 | 0.16 | 0.069 | 1.23 | 0.05 | 0.022 | | | |
| 350 | 8.95 | 5.87 | 2.55 | 5.69 | 1.95 | 0.85 | 3.94 | 0.79 | 0.34 | 2.27 | 0.21 | 0.091 | 1.44 | 0.065 | 0.028 | 1.01 | 0.027 | 0.012 |
| 400 | 10.23 | 7.52 | 3.26 | 6.50 | 2.49 | 1.08 | 4.49 | 1.01 | 0.44 | 2.59 | 0.27 | 0.12 | 1.64 | 0.09 | 0.039 | 1.16 | 0.04 | 0.017 |
| 450 | | | | 7.31 | 3.09 | 1.34 | 5.06 | 1.26 | 0.55 | 2.92 | 0.33 | 0.14 | 1.85 | 0.11 | 0.048 | 1.30 | 0.05 | 0.022 |
| 500 | | | | 8.12 | 3.76 | 1.63 | 5.62 | 1.53 | 0.66 | 3.24 | 0.40 | 0.17 | 2.05 | 0.16 | 0.069 | 1.45 | 0.06 | 0.026 |
| 750 | | | | | | | 8.43 | 3.25 | 1.41 | 4.86 | 0.85 | 0.37 | 3.08 | 0.28 | 0.12 | 2.17 | 0.12 | 0.052 |
| 1000 | | | | | | | 11.24 | 5.54 | 2.40 | 6.48 | 1.45 | 0.63 | 4.11 | 0.48 | 0.21 | 2.89 | 0.20 | 0.087 |
| 1250 | | | | | | | | | | 8.11 | 2.20 | 0.95 | 5.14 | 0.73 | 0.32 | 3.62 | 0.31 | 0.13 |
| 1500 | | | | | | | | | | 9.72 | 3.07 | 1.33 | 6.16 | 1.01 | 0.44 | 4.34 | 0.43 | 0.19 |
| 2000 | | | | | | | | | | | | | 8.21 | 1.72 | 0.74 | 5.78 | 0.73 | 0.32 |
| 2500 | | | | | | | | | | | | | 10.27 | 2.61 | 1.13 | 7.23 | 1.11 | 0.49 |
| 3000 | | | | | | | | | | | | | | | | 8.68 | 1.55 | 0.67 |
| 3500 | | | | | | | | | | | | | | | | 10.12 | 2.07 | 0.90 |
| 4000 | | | | | | | | | | | | | | | | 11.07 | 2.66 | 1.15 |

(Plastics Pipe Institute)

## CARRYING CAPACITY AND FRICTION LOSS FOR SCHEDULE 80 THERMOPLASTIC PIPE

(Independent variables: Gallons per minute and nominal pipe size O.D.

Dependent variables: Velocity, friction head and pressure drop per 100 feet of pipe, interior smooth.)

For each pipe size the three reported quantities are: **V** = Velocity (feet per second), **Hf** = Friction Head (feet), **P** = Friction Loss (pounds per square inch).

| GPM | ½ V | ½ Hf | ½ P | ¾ V | ¾ Hf | ¾ P | 1 V | 1 Hf | 1 P | 1¼ V | 1¼ Hf | 1¼ P | 1½ V | 1½ Hf | 1½ P | 2 V | 2 Hf | 2 P | 2½ V | 2½ Hf | 2½ P | 3 V | 3 Hf | 3 P | 4 V | 4 Hf | 4 P | 5 V | 5 Hf | 5 P | 6 V | 6 Hf | 6 P | 8 V | 8 Hf | 8 P | 10 V | 10 Hf | 10 P | 12 V | 12 Hf | 12 P |
|---|---|---|---|---|---|---|---|---|---|---|---|---|---|---|---|---|---|---|---|---|---|---|---|---|---|---|---|---|---|---|---|---|---|---|---|---|---|---|---|---|---|---|
| 1 | 1.48 | 4.02 | 1.74 | 0.74 | 0.86 | 0.37 | | | | | | | | | | | | | | | | | | | | | | | | | | | | | | | | | | | | |
| 2 | 2.95 | 8.03 | 3.48 | 1.57 | 1.72 | 0.74 | 0.94 | 0.88 | 0.38 | 0.52 | 0.21 | 0.09 | 0.38 | 0.10 | 0.041 | | | | | | | | | | | | | | | | | | | | | | | | | | | |
| 5 | 7.39 | 45.23 | 19.59 | 3.92 | 9.67 | 4.19 | 2.34 | 2.75 | 1.19 | 1.30 | 0.66 | 0.29 | 0.94 | 0.30 | 0.126 | 0.56 | 0.10 | 0.040 | 0.39 | 0.05 | 0.022 | 0.25 | 0.02 | 0.009 | | | | | | | | | | | | | | | | | | |
| 7 | 10.34 | 83.07 | 35.97 | 5.49 | 17.76 | 7.69 | 3.28 | 5.04 | 2.19 | 1.82 | 1.21 | 0.53 | 1.32 | 0.55 | 0.24 | 0.78 | 0.15 | 0.065 | 0.54 | 0.07 | 0.032 | 0.35 | 0.028 | 0.012 | | | | | | | | | | | | | | | | | | |
| 10 | | | | 7.84 | 33.84 | 14.65 | 4.68 | 9.61 | 4.16 | 2.60 | 2.30 | 1.00 | 1.88 | 1.04 | 0.45 | 1.12 | 0.29 | 0.13 | 0.78 | 0.12 | 0.052 | 0.50 | 0.04 | 0.017 | | | | | | | | | | | | | | | | | | |
| 15 | | | | 11.76 | 71.70 | 31.05 | 7.01 | 20.36 | 8.82 | 3.90 | 4.87 | 2.11 | 2.81 | 2.20 | 0.95 | 1.68 | 0.62 | 0.27 | 1.17 | 0.26 | 0.11 | 0.75 | 0.09 | 0.039 | | | | | | | | | | | | | | | | | | |
| 20 | | | | | | | 9.35 | 34.68 | 15.02 | 5.20 | 8.30 | 3.59 | 3.75 | 3.75 | 1.62 | 2.23 | 1.06 | 0.45 | 1.56 | 0.44 | 0.19 | 1.00 | 0.15 | 0.065 | 0.57 | 0.04 | 0.017 | | | | | | | | | | | | | | | |
| 25 | | | | | | | 11.69 | 52.43 | 22.70 | 6.50 | 12.55 | 5.43 | 4.69 | 5.67 | 2.46 | 2.79 | 1.60 | 0.69 | 1.95 | 0.67 | 0.29 | 1.25 | 0.22 | 0.095 | 0.72 | 0.06 | 0.026 | | | | | | | | | | | | | | | |
| 30 | | | | | | | 14.03 | 73.48 | 31.82 | 7.80 | 17.59 | 7.62 | 5.63 | 7.95 | 3.44 | 3.35 | 2.25 | 0.97 | 2.34 | 0.94 | 0.41 | 1.49 | 0.31 | 0.13 | 0.86 | 0.08 | 0.035 | 0.54 | 0.03 | 0.013 | | | | | | | | | | | | |
| 35 | | | | | | | | | | 9.10 | 23.40 | 10.13 | 6.57 | 10.58 | 4.58 | 3.91 | 2.99 | 1.29 | 2.73 | 1.25 | 0.54 | 1.74 | 0.42 | 0.18 | 1.00 | 0.11 | 0.048 | 0.63 | 0.04 | 0.017 | | | | | | | | | | | | |
| 40 | | | | | | | | | | 10.40 | 29.97 | 12.98 | 7.50 | 13.55 | 5.87 | 4.47 | 3.83 | 1.65 | 3.12 | 1.60 | 0.69 | 1.99 | 0.54 | 0.23 | 1.15 | 0.14 | 0.061 | 0.72 | 0.06 | 0.026 | | | | | | | | | | | | |
| 45 | | | | | | | | | | 11.70 | 37.27 | 16.14 | 8.44 | 16.85 | 7.30 | 5.03 | 4.76 | 2.07 | 3.51 | 1.99 | 0.86 | 2.24 | 0.67 | 0.29 | 1.29 | 0.17 | 0.074 | 0.81 | 0.07 | 0.030 | | | | | | | | | | | | |
| 50 | | | | | | | | | | 13.00 | 45.30 | 19.61 | 9.38 | 20.48 | 8.87 | 5.58 | 5.79 | 2.51 | 3.90 | 2.42 | 1.05 | 2.49 | 0.81 | 0.35 | 1.43 | 0.21 | 0.091 | 0.90 | 0.10 | 0.043 | 0.63 | 0.03 | 0.013 | | | | | | | | | |
| 60 | | | | | | | | | | | | | 11.26 | 28.70 | 12.43 | 6.70 | 8.12 | 3.52 | 4.68 | 3.39 | 1.47 | 2.99 | 1.14 | 0.49 | 1.72 | 0.30 | 0.13 | 1.08 | 0.13 | 0.056 | 0.75 | 0.04 | 0.017 | | | | | | | | | |
| 70 | | | | | | | | | | | | | | | | 7.82 | 10.80 | 4.68 | 5.46 | 4.51 | 1.95 | 3.49 | 1.51 | 0.65 | 2.01 | 0.39 | 0.17 | 1.26 | 0.14 | 0.061 | 0.88 | 0.05 | 0.022 | | | | | | | | | |
| 75 | | | | | | | | | | | | | | | | 8.38 | 12.27 | 5.31 | 5.85 | 5.12 | 2.22 | 3.74 | 1.72 | 0.74 | 2.15 | 0.45 | 0.19 | 1.35 | 0.16 | 0.069 | 0.94 | 0.06 | 0.026 | | | | | | | | | |
| 80 | | | | | | | | | | | | | | | | 8.93 | 13.83 | 5.99 | 6.24 | 5.77 | 2.50 | 3.99 | 1.94 | 0.84 | 2.29 | 0.50 | 0.22 | 1.44 | 0.20 | 0.087 | 1.00 | 0.07 | 0.030 | | | | | | | | | |
| 90 | | | | | | | | | | | | | | | | 10.05 | 17.20 | 7.45 | 7.02 | 7.18 | 3.11 | 4.48 | 2.41 | 1.04 | 2.58 | 0.63 | 0.27 | 1.62 | 0.24 | 0.10 | 1.13 | 0.08 | 0.035 | | | | | | | | | |
| 100 | | | | | | | | | | | | | | | | 11.17 | 20.90 | 9.05 | 7.80 | 8.72 | 3.78 | 4.98 | 2.93 | 1.27 | 2.87 | 0.76 | 0.33 | 1.80 | 0.37 | 0.16 | 1.25 | 0.10 | 0.043 | | | | | | | | | |
| 125 | | | | | | | | | | | | | | | | | | | 9.75 | 13.21 | 5.72 | 6.23 | 4.43 | 1.92 | 3.59 | 1.16 | 0.50 | 2.25 | 0.52 | 0.23 | 1.57 | 0.16 | 0.068 | 0.90 | 0.045 | 0.019 | | | | | | |
| 150 | | | | | | | | | | | | | | | | | | | 11.70 | 18.48 | 8.00 | 7.47 | 6.20 | 2.68 | 4.30 | 1.61 | 0.70 | 2.70 | 0.69 | 0.30 | 1.88 | 0.22 | 0.095 | 1.07 | 0.06 | 0.026 | | | | | | |
| 175 | | | | | | | | | | | | | | | | | | | | | | 8.72 | 8.26 | 3.58 | 5.02 | 2.15 | 0.93 | 3.15 | 0.88 | 0.38 | 2.20 | 0.29 | 0.12 | 1.25 | 0.075 | 0.033 | | | | | | |
| 200 | | | | | | | | | | | | | | | | | | | | | | 9.97 | 10.57 | 4.58 | 5.73 | 2.75 | 1.19 | 3.60 | 1.34 | 0.58 | 2.51 | 0.37 | 0.16 | 1.43 | 0.095 | 0.041 | 0.90 | 0.036 | 0.015 | | | |
| 250 | | | | | | | | | | | | | | | | | | | | | | 12.46 | 16.00 | 6.93 | 7.16 | 4.16 | 1.81 | 4.50 | 1.87 | 0.81 | 3.14 | 0.56 | 0.24 | 1.79 | 0.14 | 0.061 | 1.14 | 0.045 | 0.02 | | | |
| 300 | | | | | | | | | | | | | | | | | | | | | | | | | 8.60 | 5.83 | 2.52 | 5.40 | 2.49 | 1.08 | 3.76 | 0.78 | 0.34 | 2.14 | 0.20 | 0.087 | 1.36 | 0.07 | 0.03 | | | |
| 350 | | | | | | | | | | | | | | | | | | | | | | | | | 10.03 | 7.76 | 3.36 | 6.30 | 3.19 | 1.38 | 4.39 | 1.04 | 0.45 | 2.50 | 0.27 | 0.12 | 1.59 | 0.085 | 0.037 | 1.12 | 0.037 | 0.016 |
| 400 | | | | | | | | | | | | | | | | | | | | | | | | | 11.47 | 9.93 | 4.30 | 7.19 | 3.97 | 1.72 | 5.02 | 1.33 | 0.58 | 2.86 | 0.34 | 0.15 | 1.81 | 0.11 | 0.048 | 1.28 | 0.05 | 0.022 |
| 450 | | | | | | | | | | | | | | | | | | | | | | | | | | | | 8.09 | 4.82 | 2.09 | 5.64 | 1.65 | 0.71 | 3.21 | 0.42 | 0.18 | 2.04 | 0.14 | 0.061 | 1.44 | 0.06 | 0.026 |
| 500 | | | | | | | | | | | | | | | | | | | | | | | | | | | | 8.99 | | | 6.27 | 2.00 | 0.87 | 3.57 | 0.51 | 0.22 | 2.27 | 0.17 | 0.074 | 1.60 | 0.07 | 0.030 |
| 750 | | | | | | | | | | | | | | | | | | | | | | | | | | | | | | | 9.40 | 4.16 | 1.80 | 5.36 | 1.08 | 0.47 | 3.40 | 0.36 | 0.16 | 2.40 | 0.15 | 0.065 |
| 1000 | | | | | | | | | | | | | | | | | | | | | | | | | | | | | | | 12.54 | 7.23 | 3.13 | 7.14 | 1.84 | 0.80 | 4.54 | 0.61 | 0.26 | 3.20 | 0.26 | 0.11 |
| 1250 | | | | | | | | | | | | | | | | | | | | | | | | | | | | | | | | | | 8.93 | 2.78 | 1.20 | 5.67 | 0.92 | 0.40 | 4.01 | 0.40 | 0.17 |
| 1500 | | | | | | | | | | | | | | | | | | | | | | | | | | | | | | | | | | 10.71 | 3.89 | 1.68 | 6.80 | 1.29 | 0.56 | 4.81 | 0.55 | 0.24 |
| 2000 | | | | | | | | | | | | | | | | | | | | | | | | | | | | | | | | | | | | | 9.07 | 2.19 | 0.95 | 6.41 | 0.94 | 0.41 |
| 2500 | | | | | | | | | | | | | | | | | | | | | | | | | | | | | | | | | | | | | 11.34 | 3.33 | 1.44 | 8.01 | 1.42 | 0.62 |
| 3000 | | | | | | | | | | | | | | | | | | | | | | | | | | | | | | | | | | | | | | | | 9.61 | 1.99 | 0.86 |
| 3500 | | | | | | | | | | | | | | | | | | | | | | | | | | | | | | | | | | | | | | | | 11.21 | 2.65 | 1.15 |
| 4000 | | | | | | | | | | | | | | | | | | | | | | | | | | | | | | | | | | | | | | | | 12.82 | 3.41 | 1.48 |

(Plastics Pipe Institute)

## FLOW OF WATER IN COPPER WATER TUBE

### TYPE K

| Flow, Gallons per min. | TYPE K COPPER WATER TUBE — Pressure Loss due to friction, in lb. per sq. in. per 100 ft. | | | | | | | | | | |
|---|---|---|---|---|---|---|---|---|---|---|---|
| | ¼" | ⅜" | ½" | ⅝" | ¾" | 1" | 1¼" | 1½" | 2" | 2½" | 3" |
| 1 | | 4.66 | 1.29 | .467 | .248 | .063 | | | | | |
| 2 | | 15.7 | 4.34 | 1.58 | .836 | .211 | .073 | | | | |
| 3 | | 32.0 | 8.83 | 3.21 | 1.70 | .430 | .148 | .065 | | | |
| 4 | | 53.0 | 14.6 | 5.32 | 2.82 | .713 | .246 | .108 | | | |
| 5 | | 78.4 | 21.6 | 7.87 | 4.17 | 1.05 | .363 | .159 | .042 | | |
| 6 | | 108. | 29.9 | 10.8 | 5.75 | 1.45 | .500 | .219 | .058 | | |
| 7 | | 141. | 39.1 | 14.2 | 7.53 | 1.90 | .655 | .287 | .076 | | |
| 8 | | 179. | 49.4 | 17.9 | 9.53 | 2.41 | .828 | .363 | .096 | | |
| 9 | | 220. | 60.7 | 22.1 | 11.7 | 2.96 | 1.02 | .446 | .118 | | |
| 10 | | 264. | 73.0 | 26.5 | 14.1 | 3.56 | 1.23 | .537 | .142 | .050 | .022 |
| 12 | | | 101. | 36.5 | 19.4 | 4.90 | 1.69 | .739 | .196 | .070 | .030 |
| 15 | | | 149. | 54.1 | 28.7 | 7.24 | 2.50 | 1.09 | .289 | .103 | .044 |
| 20 | | | | 87.5 | 47.5 | 12.0 | 4.13 | 1.81 | .479 | .170 | .073 |
| 25 | | | | 132. | 70.3 | 17.8 | 6.12 | 2.68 | .709 | .252 | .109 |
| 30 | | | | | 96.7 | 24.5 | 8.42 | 3.69 | .976 | .347 | .149 |
| 35 | | | | | 127. | 32.0 | 11.0 | 4.83 | 1.28 | .455 | .196 |
| 40 | | | | | | 40.5 | 14.0 | 6.11 | 1.62 | .575 | .248 |
| 45 | | | | | | 49.8 | 17.2 | 7.51 | 1.99 | .707 | .304 |
| 50 | | | | | | 59.9 | 20.6 | 9.04 | 2.39 | .850 | .366 |
| 60 | | | | | | | 28.4 | 12.4 | 3.29 | 1.17 | .504 |
| 70 | | | | | | | 37.2 | 16.3 | 4.31 | 1.53 | .661 |
| 80 | | | | | | | 47.1 | 20.6 | 5.45 | 1.94 | .835 |
| 90 | | | | | | | | 25.4 | 6.71 | 2.38 | 1.03 |
| 100 | | | | | | | | 30.5 | 8.07 | 2.87 | 1.24 |
| 125 | | | | | | | | | 11.9 | 4.24 | 1.83 |
| 150 | | | | | | | | | 16.4 | 5.84 | 2.52 |
| 175 | | | | | | | | | 21.5 | 7.66 | 3.30 |
| 200 | | | | | | | | | 27.2 | 9.68 | 4.17 |

### TYPE L

| Flow, Gallons per min. | TYPE L COPPER WATER TUBE — Pressure Loss due to friction, in lb. per sq. in. per 100 ft. | | | | | | | | | | |
|---|---|---|---|---|---|---|---|---|---|---|---|
| | ¼" | ⅜" | ½" | ⅝" | ¾" | 1" | 1¼" | 1½" | 2" | 2½" | 3" |
| 1 | 14.8 | 3.38 | 1.10 | .422 | .193 | .054 | | | | | |
| 2 | 50.1 | 11.5 | 3.70 | 1.42 | .652 | .183 | .068 | | | | |
| 3 | 102. | 23.2 | 7.53 | 2.90 | 1.33 | .374 | .137 | .060 | | | |
| 4 | 169. | 38.5 | 12.5 | 4.81 | 2.20 | 619 | .228 | .100 | | | |
| 5 | 250. | 56.9 | 18.4 | 7.11 | 3.25 | .915 | .337 | .147 | .040 | | |
| 6 | | 78.4 | 25.4 | 9.79 | 4.48 | 1.26 | .464 | .203 | .054 | | |
| 7 | | 103. | 33.3 | 12.8 | 5.87 | 1.65 | .608 | .266 | .071 | | |
| 8 | | 130. | 42.1 | 16.2 | 7.42 | 2.09 | .768 | .336 | .090 | | |
| 9 | | 160. | 51.7 | 19.9 | 9.13 | 2.57 | .944 | .413 | .111 | | |
| 10 | | 192. | 62.2 | 24.0 | 11.0 | 3.09 | 1.14 | .497 | .133 | .048 | .020 |
| 12 | | | 85.7 | 33.0 | 15.1 | 4.25 | 1.56 | .685 | .184 | .066 | .028 |
| 15 | | | 127. | 48.9 | 22.4 | 6.29 | 2.31 | 1.01 | .272 | .097 | .042 |
| 20 | | | | 80.9 | 37.0 | 10.4 | 3.83 | 1.68 | .450 | .161 | .069 |
| 25 | | | | 120. | 54.8 | 15.4 | 5.67 | 2.48 | .666 | .238 | .102 |
| 30 | | | | | 75.4 | 21.2 | 7.81 | 3.42 | .917 | .327 | .140 |
| 35 | | | | | 98.9 | 27.8 | 10.2 | 4.48 | 1.20 | .429 | .184 |
| 40 | | | | | | 35.2 | 12.9 | 5.66 | 1.52 | .542 | .233 |
| 45 | | | | | | 43.2 | 15.9 | 6.96 | 1.87 | .667 | .286 |
| 50 | | | | | | 52.0 | 19.1 | 8.37 | 2.25 | .802 | .344 |
| 60 | | | | | | | 26.3 | 11.5 | 3.09 | 1.10 | .474 |
| 70 | | | | | | | 34.5 | 15.1 | 4.05 | 1.45 | .621 |
| 80 | | | | | | | 43.6 | 19.1 | 5.12 | 1.83 | .785 |
| 90 | | | | | | | | 23.5 | 6.30 | 2.25 | .965 |
| 100 | | | | | | | | 28.3 | 7.58 | 2.71 | 1.16 |
| 125 | | | | | | | | | 11.2 | 4.00 | 1.72 |
| 150 | | | | | | | | | 15.4 | 5.51 | 2.37 |
| 175 | | | | | | | | | 20.2 | 7.22 | 3.10 |
| 200 | | | | | | | | | 25.6 | 9.13 | 3.92 |

### TYPE M

| Flow Gallons per min. | TYPE M — Pressure Loss due to friction, in lb. per sq. in. per 100 ft. | | | | | | | | |
|---|---|---|---|---|---|---|---|---|---|
| | ⅜" | ½" | ¾" | 1" | 1¼" | 1½" | 2" | 2½" | 3" |
| 1 | 2.69 | .87 | .153 | | | | | | |
| 2 | 9.70 | 3.12 | .555 | .155 | | | | | |
| 3 | 20.6 | 6.62 | 1.18 | .328 | | | | | |
| 4 | | 11.3 | 1.99 | .558 | | | | | |
| 5 | | 17.0 | 3.00 | .840 | .314 | | | | |
| 6 | | 23.8 | 4.22 | 1.18 | .442 | | | | |
| 8 | | 40.5 | 7.18 | 2.02 | .753 | .333 | | | |
| 10 | | | 10.9 | 3.04 | 1.14 | .503 | .133 | | |
| 12 | | | 15.3 | 4.26 | 1.59 | .701 | .185 | | |
| 16 | | | 25.9 | 7.28 | 2.71 | 1.21 | .317 | | |
| 20 | | | | 10.9 | 4.08 | 1.80 | .477 | .165 | |
| 25 | | | | 16.6 | 6.20 | 2.72 | .719 | .249 | |
| 30 | | | | 23.1 | 8.62 | 3.81 | 1.01 | .349 | |
| 35 | | | | 30.8 | 11.5 | 5.22 | 1.34 | .463 | |
| 40 | | | | | 14.7 | 6.50 | 1.72 | .594 | .251 |
| 50 | | | | | 22.3 | 9.83 | 2.59 | .897 | .381 |
| 60 | | | | | 31.2 | 13.8 | 3.64 | 1.26 | .533 |
| 70 | | | | | 41.6 | 18.4 | 4.85 | 1.68 | .711 |
| 80 | | | | | | 23.5 | 6.20 | 2.15 | .906 |
| 90 | | | | | | 29.2 | 7.70 | 2.67 | 1.13 |
| 100 | | | | | | 35.5 | 9.35 | 3.24 | 1.37 |
| 120 | | | | | | | 13.2 | 4.55 | 1.92 |
| 140 | | | | | | | 17.5 | 6.03 | 2.55 |
| 170 | | | | | | | 25.0 | 8.67 | 3.66 |
| 200 | | | | | | | 33.8 | 11.7 | 4.94 |

*Suitable allowances should be made for fittings, etc.
*To convert pressure loss figures in Table to feet, multiply by 2.31.
(Copper Development Assoc. Inc.)

## HEAD LOSS FOR STANDARD GALVANIZED PIPE

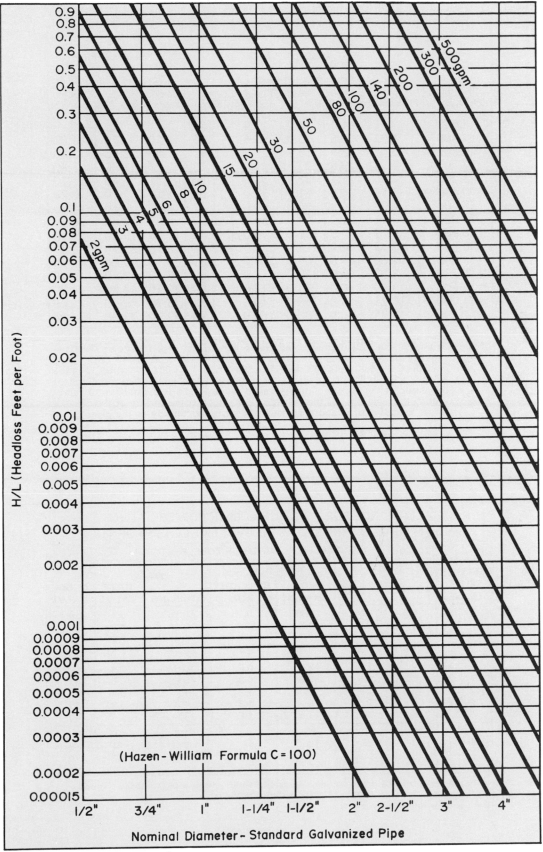

(Hazen - William Formula C = 100)

H/L (Headloss Feet per Foot)

Nominal Diameter - Standard Galvanized Pipe

(Environmental Protection Agency)

## FLOW CAPACITY OF CAST-IRON SOIL PIPE FOR SEWERS UNDER PRESSURE, BASED ON THE WILLIAMS-HAZEN FORMULA
## (2 AND 3 INCH DIAMETER PIPE)

| | | Loss of Head per 1000 feet | | | | | | |
|---|---|---|---|---|---|---|---|---|
| | | 2″ Diameter Pipe | | | | 3″ Diameter Pipe | | |
| Gallons Flow in 24 hr. dy. | Velocity in Feet per Second | Loss of Head in Feet | | | Velocity in Feet per Second | Loss of Head in Feet | | |
| | | Smooth c = 120 | Aver. c = 100 | Rough c = 80 | | Smooth c = 120 | Aver. c = 100 | Rough c = 80 |
| 8,640 | .61 | 1.4 | 2.0 | 2.9 | | | | |
| 14,400 | 1.02 | 3.6 | 5.0 | 7.6 | .45 | .50 | .70 | 1.00 |
| 28,800 | 2.04 | 12.9 | 18.2 | 27.5 | .91 | 1.80 | 2.50 | 3.80 |
| 36,000 | 2.55 | 19.6 | 27.3 | 41.6 | 1.13 | 2.71 | 3.80 | 5.80 |
| 43,200 | 3.06 | 27.3 | 38.4 | 58.0 | 1.36 | 3.81 | 5.40 | 8.10 |
| 50,400 | 3.57 | 36.6 | 51.0 | 78.0 | 1.59 | 5.10 | 7.10 | 10.70 |
| 57,600 | 4.08 | 46.8 | 66.0 | 99.0 | 1.82 | 6.50 | 9.10 | 13.80 |
| 72,000 | 5.11 | 71.0 | 99.0 | 150.0 | 2.27 | 9.80 | 13.80 | 20.80 |
| 86,400 | 6.13 | 99.0 | 139.0 | 210.0 | 2.72 | 13.70 | 19.20 | 29.10 |
| 100,800 | 7.15 | 132.0 | 184.0 | 280.0 | 3.18 | 18.30 | 25.70 | 38.80 |
| 108,000 | 7.66 | 149.0 | 209.0 | 318.0 | 3.41 | 20.70 | 29.00 | 43.80 |
| 115,200 | 8.17 | 169.0 | 237.0 | 358.0 | 3.63 | 23.40 | 32.80 | 49.60 |
| 129,600 | 9.19 | 210.0 | 294.0 | 447.0 | 4.09 | 29.10 | 40.80 | 62.00 |
| 144,000 | 10.21 | 256.0 | 358.0 | 540.0 | 4.54 | 35.20 | 49.60 | 75.00 |
| 172,800 | 12.25 | 360.0 | 500.0 | 760.0 | 5.45 | 49.70 | 70.00 | 106.00 |

Note: From 1905 to 1920 Williams and Hazen carried out experiments on the flow-capacity of pipes ranging in diameter from 1 inch to about 15 feet. They developed the following formula: $V = Cr^{0.63} s^{0.54} 0.001^{-0.04}$ where V is the velocity of the fluid, C is the factor proportional to the surface condition of the inside of the pipe, r is the hydraulic radius, and s is the quantity h representing the head lost per foot of pipe. The smoother the surface of the pipe the larger the value of C and the greater the carrying capacity of the pipe.

(Cast Iron Soil Pipe Institute)

## ALLOWANCE FOR FRICTION LOSS IN COPPER VALVES AND FITTINGS EXPRESSED AS EQUIVALENT LENGTH OF TUBE

| | EQUIVALENT LENGTH OF TUBE, FEET | | | | | | |
|---|---|---|---|---|---|---|---|
| FITTING SIZE, INCHES | STANDARD Ls | | 90 DEG. T | | | | |
| | 90 DEG. | 45 DEG. | SIDE BRANCH | STRAIGHT RUN | COUPLING | GATE VALVE | GLOBE VALVE |
| 3/8 | 0.5 | 0.3 | 0.75 | 0.15 | 0.15 | 0.1 | 4 |
| 1/2 | 1 | 0.6 | 1.5 | 0.3 | 0.3 | 0.2 | 7.5 |
| 3/4 | 1.25 | 0.75 | 2 | 0.4 | 0.4 | 0.25 | 10 |
| 1 | 1.5 | 1.0 | 2.5 | 0.45 | 0.45 | 0.3 | 12.5 |
| 1 1/4 | 2 | 1.2 | 3 | 0.6 | 0.6 | 0.4 | 18 |
| 1 1/2 | 2.5 | 1.5 | 3.5 | 0.8 | 0.8 | 0.5 | 23 |
| 2 | 3.5 | 2 | 5 | 1 | 1 | 0.7 | 28 |
| 2 1/2 | 4 | 2.5 | 6 | 1.3 | 1.3 | 0.8 | 33 |
| 3 | 5 | 3 | 7.5 | 1.5 | 1.5 | 1 | 40 |
| 3 1/2 | 6 | 3.5 | 9 | 1.8 | 1.8 | 1.2 | 50 |
| 4 | 7 | 4 | 10.5 | 2 | 2 | 1.4 | 63 |
| 5 | 9 | 5 | 13 | 2.5 | 2.5 | 1.7 | 70 |
| 6 | 10 | 6 | 15 | 3 | 3 | 2 | 84 |

NOTE: Allowances are for streamlined soldered fittings and recessed threaded fittings. For threaded fittings, double the allowances shown in the table.
(Copper Development Assoc. Inc.)

# Useful Information

## FLOW CAPACITY OF CAST-IRON SOIL PIPE FOR SEWERS UNDER PRESSURE, BASED ON THE WILLIAMS-HAZEN FORMULA
### (4 AND 6 INCH DIAMETER PIPE)

| Gallons Flow in 24 Hr. Day | Loss of Head per 1000 feet | | | | | | | |
|---|---|---|---|---|---|---|---|---|
| | | 4" Diameter Pipe | | | | 6" Diameter Pipe | | |
| | Velocity in feet per sec. | Loss of Head in Ft. | | | Velocity in feet per sec. | Loss of Head in Ft. | | |
| | | Smooth c=140 | Aver. c=130 | Rough c=100 | | Smooth c=140 | Aver. c=130 | Rough c=100 |
| 20,000 | .36 | .17 | .19 | .32 | | | | |
| 30,000 | .53 | .36 | .41 | .67 | | | | |
| 40,000 | .71 | .61 | .70 | 1.13 | | | | |
| 50,000 | .89 | .92 | 1.05 | 1.71 | .39 | .13 | .15 | .24 |
| 60,000 | 1.07 | 1.29 | 1.47 | 2.40 | .47 | .18 | .20 | .33 |
| 70,000 | 1.24 | 1.72 | 1.96 | 3.20 | .55 | .24 | .27 | .44 |
| 80,000 | 1.42 | 2.20 | 2.52 | 4.10 | .63 | .30 | .35 | .57 |
| 90,000 | 1.60 | 2.72 | 3.12 | 5.04 | .71 | .38 | .43 | .71 |
| 100,000 | 1.78 | 3.30 | 3.81 | 6.19 | .79 | .46 | .53 | .86 |
| 110,000 | 1.95 | 3.95 | 4.55 | 7.40 | .87 | .55 | .63 | 1.03 |
| 120,000 | 2.13 | 4.63 | 5.17 | 8.65 | .95 | .65 | .74 | 1.21 |
| 140,000 | 2.49 | 6.20 | 7.10 | 11.60 | 1.10 | .87 | .99 | 1.62 |
| 160,000 | 2.84 | 7.90 | 9.10 | 14.70 | 1.26 | 1.10 | 1.26 | 2.06 |
| 180,000 | 3.19 | 9.80 | 11.30 | 18.30 | 1.42 | 1.37 | 1.57 | 2.56 |
| 200,000 | 3.56 | 12.00 | 13.80 | 22.20 | 1.58 | 1.67 | 1.91 | 3.10 |
| 220,000 | 3.91 | 14.20 | 16.40 | 26.70 | 1.73 | 1.99 | 2.29 | 3.71 |
| 240,000 | 4.27 | 16.70 | 19.30 | 31.20 | 1.89 | 2.33 | 2.69 | 4.35 |
| 260,000 | 4.63 | 19.4 | 22.40 | 36.10 | 2.05 | 2.71 | 3.10 | 5.00 |
| 280,000 | 4.99 | 25.3 | 29.10 | 47.10 | 2.21 | 3.11 | 3.58 | 5.80 |
| 300,000 | 7.12 | 43.2 | 49.50 | | 2.36 | 3.54 | 4.06 | 6.60 |
| 400,000 | | | | | 3.15 | 6.00 | 6.90 | 11.30 |
| 500,000 | | | | | 3.94 | 9.10 | 10.40 | 16.90 |
| 600,000 | | | | | 4.73 | 12.80 | 14.60 | 23.80 |
| 700,000 | | | | | 5.52 | 17.00 | 19.50 | 31.60 |
| 800,000 | | | | | 6.30 | 21.60 | 24.90 | 40.40 |
| 900,000 | | | | | 7.09 | 26.90 | 30.90 | 50.00 |
| 1,000,000 | | | | | 7.88 | 32.90 | 37.80 | 61.00 |

(Cast Iron Soil Pipe Institute)

## ALLOWANCE IN EQUIVALENT LENGTH OF GALVANIZED PIPE FOR FRICTION LOSS IN VALVES AND THREADED FITTINGS

| DIAMETER OF FITTING | 90 DEG. STD. L | 45 DEG. STD. L | 90 DEG. SIDE T | COUPLING OR STRAIGHT RUN | GATE VALVE | GLOBE VALVE | ANGLE VALVE |
|---|---|---|---|---|---|---|---|
| INCHES | FEET | FEET | FEET | FEET | FEET | FEET | FEET |
| 3/8 | 1 | 0.6 | 1.5 | 0.3 | 0.2 | 8 | 4 |
| 1/2 | 2 | 1.2 | 3 | 0.6 | 0.4 | 15 | 8 |
| 3/4 | 2.5 | 1.5 | 4 | 0.8 | 0.5 | 20 | 12 |
| 1 | 3 | 1.8 | 5 | 0.9 | 0.6 | 25 | 15 |
| 1 1/4 | 4 | 2.4 | 6 | 1.2 | 0.8 | 35 | 18 |
| 1 1/2 | 5 | 3 | 7 | 1.5 | 1.0 | 45 | 22 |
| 2 | 7 | 4 | 10 | 2 | 1.3 | 55 | 28 |
| 2 1/2 | 8 | 5 | 12 | 2.5 | 1.6 | 65 | 34 |
| 3 | 10 | 6 | 15 | 3 | 2 | 80 | 40 |
| 3 1/2 | 12 | 7 | 18 | 3.6 | 2.4 | 100 | 50 |
| 4 | 14 | 8 | 21 | 4 | 2.7 | 125 | 55 |
| 5 | 17 | 10 | 25 | 5 | 3.3 | 140 | 70 |
| 6 | 20 | 12 | 30 | 6 | 4 | 165 | 80 |

(Environmental Protection Agency)

## TYPICAL FIVE-YEAR TRAINING PROGRAM FOR PLUMBING APPRENTICES

1. INTRODUCTION
   A. History of plumbing.
   B. Job ethics and responsibility.
   C. Job safety.
   D. Plumbing terms.

2. HAND AND POWER TOOLS
   A. Types of tools.
   B. Use and care.
   C. Safety.
   D. Special tools.

3. MATHEMATICS
   A. Whole numbers.
   B. Fractions.
   C. Decimals.
   D. Measurement.
      1. Linear.
      2. Square.
      3. Cubic.
      4. Weight.
   E. Computation.
      1. Area.
      2. Volume.
   F. Practical estimating.
      1. Material required.
      2. Labor factor.
   G. Applied geometry.
      1. Construction.
      2. Triangles.
      3. Circles.

4. LEVELING INSTRUMENTS
   A. Builders' level and transit.
      1. Setting up.
      2. Vernier scale.
   B. Leveling rods and targets.
   C. Signals.
   D. Leveling practice.
      1. Determining elevations.
      2. Laying out pipe lines.

5. RIGGING AND HOISTING
   A. Fiber rope.
      1. Use and care.
      2. Knots, bends and hitches.
      3. Splicing.
   B. Wire rope.
      1. Use and care.
      2. Attachments.
      3. Safety.
   C. Hoisting.
      1. Chains and hooks.
      2. Slings.
      3. Safe working loads.

6. PLUMBING SYSTEMS
   A. Water supply.
      1. Hot and cold piping.
      2. Auxiliary equipment.
   B. Sanitary disposal.
      1. Vents.
      2. Traps.
      3. Ejectors.
   C. Pipe and fittings.
      1. Type.
      2. Function.
   D. Code requirements.
      1. City and county.
      2. State and federal.

7. COMMERCIAL PLUMBING SYSTEMS
   A. Water supply.
      1. Laundries.
      2. Car washes.
      3. Hospitals.
      4. Swiming pools.
      5. Fire protection equipment.
   B. Sewage disposal.
      1. Hotels and hospitals.
      2. Schools and stadiums.
   C. Auxiliary pumps.

8. DRAFTING AND BLUEPRINT READING
   A. Introduction.
      1. Scale reading.
      2. Symbols and abbreviations.
      3. Line representation.
   B. Dimension.
      1. Size.
      2. Location.
   C. Isometric sketching.
      1. Shop description.
      2. Pipe intersections.
   D. Views.
      1. Three-view.
      2. Sections.
   E. Floor plans.
      1. Schedules.
      2. Special symbols.
   F. Elevations.
      1. Exterior.
      2. Interior.
   G. Specifications.
   H. Interpretation of prints.
      1. Residential.
      2. Commercial.
      3. Industrial.

9. HYDRAULICS AND PNEUMATICS
   A. Static water pressure.
      1. Measurement of pressure.
      2. Bursting pressure.
   B. Measuring rate of flow.
      1. Gravity flow.
      2. Pressure flow.
      3. Hot water flow.
      4. Hydraulic gradient.
   C. Air and gas leaks.
   D. Pressure losses.
      1. Fittings.
      2. Appurtenances.
   E. Water hammer effects.
   F. Characteristics of plumbing traps.
   G. Stacks.
      1. Flow.
      2. Solids.
      3. Horizontal runs.

10. PRIVATE WATER SUPPLIES
    A. Well construction.
    B. Well equipment.
    C. Surface sources.

11. MATERIAL, FITTINGS AND
    A. Pipe.
       1. Cast iron.
       2. Wrought iron.
       3. Steel.
       4. Copper tube.
       5. Lead.
       6. Cement and concrete.
       7. Fiber and plastic.
    B. Fittings.
       1. Cast Iron.
          a. Bell and spigot.
          b. Flanged.
          c. Threaded.
       2. Malleable iron.
       3. Soil and drainage.
       4. Brass and copper.
       5. Cement and concrete.

12. VALVES, FAUCETS AND METERS
    A. Faucets.
       1. Bathtub and shower.
       2. Lavatory.
       3. Kitchen sink.
       4. Special.
    B. Valves.
       1. Shower.
       2. Gate.

3. Globe.
4. Check.
5. Plug.
6. Backwater.
7. Balanced.
8. Float controlled.
9. Relief or safety.
10. Pressure regulators.
11. Foot.
12. Butterfly.
13. Needle.
C. Meters.
   1. Displacement.
   2. Velocity.

13. BUILDING CODES
A. Ventilation.
B. Lighting.
C. Size.
D. Floors.
E. Walls.
F. Bathroom fixtures.
   1. Types.
   2. Standards.
G. Protecting the water supply.

14. WELDING
A. Safety.
   1. Protective clothing.
   2. Ventilation.
   3. Electrical.
   4. Radiation.
B. Oxyacetylene.
   1. Equipment and accessories.
   2. Setup and operating procedures.
   3. Welding positions.
   4. Flame cutting.
C. Electric.

1. Basic electric theory.
2. Welding machines.
3. Polarity.
4. Equipment and accessories.
5. Procedure.
6. Welding positions.
D. Practical application.
   1. Safety.
   2. Setup procedure.
   3. Beads and weaves.
   4. Joints.
   5. Position welds.
      a. Pipe welding.
      b. Tank welding.
      c. Hanger welding.
   6. Flame cutting.

15. SOLDERING, BRAZING AND LEAD WIPING
A. Principles.
   1. Alloys.
   2. Fluxes.
B. Procedure.
   1. Soldering, copper method.
   2. Soldering, torch method.
C. Safety.
D. Joints.
E. Positions and branches.
F. Practical application.
   1. Safety.
   2. Setup procedure.
   3. Sweating and flaring, copper tube and fittings.
   4. Caulking joints on cast-iron soil pipe.
   5. Fabricating roof flashings.
   6. Soldering sheet lead seams.

16. SMALL PUMPS
A. Reciprocating pumps.
   1. Operating principles.
   2. Maintenance.
B. Jet wall pumps.
   1. Operating principles.
   2. Maintenance.
C. Centrifugal pumps.
   1. Operating principles.
      a. Volute pump.
      b. Turbine pump.
   2. Maintenance.
   3. Installation.

17. INSTALLATION AND SERVICE OF PIPE AND FITTINGS
A. Installation.
   1. Threaded pipe.
   2. Cast-iron pipe.
   3. Copper tubing.
   4. Silicon-iron pipe.
   5. Nonmetallic sewers and mains.
   6. Glass drain line.
   7. Plastic pipe.
   8. Selected pipes, fittings and valves.
B. Maintenance and repair.
   1. Faucets.
   2. Flush valves, ball cocks, vacuum breakers and types of water closets.
   3. Nonscald shower valves.
   4. Control valves, pressure reducing valves and gauges.
   5. Temperature and pressure relief valve.
   6. Water heater controls.

# DICTIONARY OF PLUMBING TERMS

ABRASIVE: Any material which erodes another material by rubbing. The commonly used abrasive materials include aluminum oxide, silicon carbide and garnet.

ABS (ACRYLONITRILE-BUTADIENE-STYRENE): Plastic material used in manufacturing drainage pipe and fittings.

ABSOLUTE PRESSURE: The pressure measured by a gauge plus a correction for the effect of air pressure on the gauge (14.7 psi at sea level).

ADAPTER: Fitting which joins pipes of different materials or different sizes.

AERATOR: Device which adds air to water; fills flowing water with bubbles.

AGA: American Gas Association.

AIR CONDITIONER: A device used to control temperature, humidity, circulation and freshness of air in a building.

AIR GAP: The vertical distance from the top of the flood rim (highest point water can reach within a fixture) to the faucet or spout which supplies fresh water to the fixture. Designed to prevent backflow.

AIR LOCK: Air trapped within a pipe which restricts or blocks the flow of liquid through the pipe.

ANAEROBIC BACTERIA: Bacteria which live and work in the absence of free oxygen (air).

ANCHOR: A specially designed fastener used to attach pipes, fixtures and other parts to the building.

ANGLE VALVE: A valve designed such that the inlet and outlet are at a 90 deg. angle to each other.

ANTIHAMMER DEVICE: An air chamber such as a closed length of pipe or a coil which is designed to absorb the shock caused by a rapidly closed valve.

AREA DRAIN: Any drain installed in a low area to collect rain water and channel it to the normal storm water drainage.

AREA OF A CIRCLE: Determined by using the formula: $\pi r^2 = 3.14 \times$ (radius squared).

ASBESTOS JOINT RUNNER: A tool made of an asbestos rope and a clamp. Used to hold molten lead in the bell of a cast pipe until it has cooled.

ASHRAE: American Society of Heating, Refrigeration and Air Conditioning Engineers.

ASTM: American Society for Testing and Materials.

AWWA: American Water Works Association.

BACKFILL: Material — usually earth, sand, or gravel — used to fill an excavated trench.

BACKFLOW: The flow of liquids (possibly contaminated) into the water supply piping. This is most likely to happen when water pressure drops in the water supply piping.

BACKFLOW CONNECTION: Any arrangement of pipe, fixtures or accessories which can cause backflow to occur.

BACK PRESSURE: In plumbing systems, (compressing of trapped air) which resists the flow of waste through the DWV piping.

BACK SIPHONAGE: See BACKFLOW.

BACK VENT: A branch vent connected to the main vent stack and extending to a location near a fixture trap. Its purpose is to prevent the trap from siphoning.

BALLCOCK: A valve or faucet controlled by a change in the water level; generally consists of a device floating near the surface of the water to operate the valve.

BAR HANGER: Supporting bracket for a sink hung on a wall.

BASEMENT: Usable space beneath the ground floor of a building.

BASIN WRENCH: Tool with a swiveling jaw used to install or remove nuts in hard to reach places.

BASKET: A recessed strainer fitted into the drain opening of a sink.

BATTER BOARDS: A simply framed structure of two horizontal boards meeting at right angles and supported by stakes; used to lay out the foundation of a building. One set of batter boards is located at each corner.

BEARING PARTITION: An interior wall of a building which carries the load of the structure above in addition to its own weight.

B & S: Abbreviation for bell and spigot.

BELL OR HUB: The enlarged end of some types of pipe which fit over the next pipe section.

BELL TILE: Clay pipe sections with one end enlarged to join with the next pipe section.

BENCHMARK: A fixed location of known elevation from which land measurements can be made.

BEND: A change of direction in piping.

BENDING PIN: A tool used to straighten or stretch lead pipe.

BIB: Another name for faucet.

BIDET: A plumbing fixture designed to facilitate washing the perineal area of the body.

BLIND VENT: An illegal vent which stops in a wall thus giving the appearance of a vent but not actually functioning as a vent.

BLOWOFF: The controlled discharge of excess pressure.

BLUEPRINT: A method of copying drawings which produces

white line copy on a blue background.

BOCA: Building Officials Conference of America.

BONNET: The upper portion of the gate valve body into which the disc of a gate valve rises when it is opened.

BOX NAIL: A nail with a head and shaft similar to a common nail except that it is thinner and less likely to split a board.

BRACKET HANGER: Hanger supporting a wall-hung sink.

BRANCH: Any additions to the main pipe in a piping system, connecting to a fixture.

BRANCH INTERVAL: The vertical distance between the connection of branch pipes to the main DWV stack. Generally an 8 in. minimum is required.

BRANCH VENT: A vent which connects a branch of the drainage piping to the main vent stack.

BRAZE: Means of joining metal with an alloy having a melting point higher than common solder but lower than the metal being brazed.

BRIDGING: Pieces of wood or metal installed on a diagonal between floor joists. Their purpose is to distribute the load on the floor to more than one joist.

BRITISH THERMAL UNIT (Btu): A unit for measuring heat.

BUILDING CODE: A set of rules governing the quality of construction in a community. The purpose of these rules is to protect the public health and safety.

BUILDING DRAIN: Part of DWV piping system which extends from the base of the stack to the sanitary sewer.

BUILDING DRAINAGE SYSTEM: The complete system of pipes installed for the purpose of carrying away waste water and sewage.

BUILDING LINES: The outside of the building foundation.

BUILDING SEWER: See BUILDING DRAIN.

BUILDING MAIN: Water supply piping beginning at the source of supply and ending at the first branch inside the building.

BUILDING STORM DRAIN: Drainage piping which connects the storm sewer to a drainage system which collects rain water, ground water and surface runoff.

BUILDING TRAP: A trap placed in the building drain to prevent entry of sewer gases from the sewer main.

BURR: A sharp, roughened, in-turned edge on a piece of pipe which has been cut but not reamed.

BUSHING: A pipe fitting with both male and female threads. Used in a fitting to reduce the size. It is used to connect pipes of different sizes.

C: Symbol for Celsius, the SI metric temperature scale.

CAP: A female pipe fitting which is closed at one end; used to close off the end of a piece of pipe.

CAPILLARY ATTRACTION: Movement of liquid upward through cellular structure of fibrous strands or through structure of other solids.

CAST IRON (CI): Used in manufacture of soil pipe.

CAST-IRON PIPE: Any pipe made from cast iron. The rotational casting process is used.

CAULK: Material used to seal joints in cast-iron drainage pipe.

CAULKING: A method of making a bell and spigot pipe joint watertight by packing it with oakum, lead and/or other material.

CAULKING RECESS: Space between the inside of the hub (bell) of one piece of pipe and the spigot of the pipe length to which it is joined. This space is filled with caulking to seal the joint.

CEILING JOISTS: Horizontal framing members which rest on top of the double plate and support the ceiling in a typical wood framed structure.

CEMENT: Material which bonds other materials together.

C-to-C: Abbreviation for center-to-center.

CESSPOOL (Dry well): Deep pit which receives liquid waste and permits the excess liquid to be absorbed into the ground. Different from a septic tank because the rate of liquid intake and outflow are not the same.

CFM: Abbreviation for cubic feet per minute.

CHALK LINE: Marking tool consisting of a string coated with chalk. Usually the string is stored in a chalk box. The metal or plastic container is equipped with a reel to wind up the chalk line and provides a means of coating the line with chalk each time it is unwound.

CHAIN WRENCH: Adjustable tool for holding and turning large pipe up to 4 in. diameter. A flexible chain replaces usual jaws.

CHASE: Specifically, a Pipe Chase. A space or recess in the walls of a building where pipes are run.

CHECK VALVE: A device preventing backflow in pipes. Water can flow readily in one direction but any reversal of the flow causes the check valve to close.

CI: Abbreviation for cast iron.

CISP: Abbreviation for cast-iron soil pipe.

CISPI: Cast Iron Soil Pipe Institute.

CIRCUMFERENCE OF A CIRCLE: Distance around the perimeter of a circle. Circumference of a pipe is the distance around a pipe. Can be found by multiplying the diameter by pi (3.14).

CLARIFIED SEWAGE: Sewage from which part or all of the suspended matter has been removed.

CLEANOUT: Removable drainage fitting which permits access to the inside of drainage piping for the purpose of removing obstructions.

CLOSE NIPPLE: The shortest length of a given size pipe which can be threaded externally from both ends; used to closely connect two internally threaded pipe fittings.

CLOSED-END NUT: A type of cap nut.

CLOSET BEND: An elbow drainage fitting connecting water closet to branch drain.

CLOSET BOLT: A bolt used to attach a water closet securely to the closet flange.

CLOSET SPUD: Connector between base of ballcock assembly in a water closet tank and water supply pipe.

CO: Abbreviation for clean out.

COCK: See FAUCET.

CODE: A set of regulations which have been adopted by a governmental unit for the purpose of protecting the public health and safety. In plumbing, these codes regulate the quality of materials, the design and installation of plumbing systems.

COLD CHISEL: A hand tool used with a hammer to cut metal.

COLLAR BEAM: Horizontal roof framing member which is attached to the rafters at a point some distance above the ceiling joists. The purpose of a collar beam is to transfer part of the load on the rafters on one side of the building to the rafters on the other side. Collar beams should not be cut by the plumber in the process of installing vent stacks.

COMBINATION FIXTURE: A fixture designed to be both a kitchen sink and a wash basin.

COMBINATION SQUARE: A layout tool with a metal blade calibrated in inches or centimetres. The head of the tool is movable and has fixed surfaces at angles of 45 and 90 deg. to the blade. Some combination square heads contain a small level and a scribe.

COMMON NAIL: The standard type nail with a large flat head and relatively heavy body; used in framing and other applications where an exposed head is not undesirable.

COMMON RAFTER: The sloping member of the roof frame which extends from the ridge board (highest point of the roof) to the double plate on top of the walls.

COMMON VENT: A vent serving two or more drainage stacks.

COMPASS: A layout tool used to draw arcs and circles.

COMPOUND: See PIPE JOINT COMPOUND.

COMPRESSION: Stress resulting from forces which attempt to shorten a piece of material. Also: Term used to indicate an increase in pressure on a fluid (air or water) as a result of the action of some mechanical device (pump).

COMPRESSION FAUCET OR VALVE: A faucet or valve designed to stop the flow of water by the action of a flat disc (washer) closing against a seat.

CONCRETE: A mixture of portland cement, fine aggregate (sand), coarse aggregate (gravel) and water.

CONCRETE NAIL: A nail made from hard steel and specifically designed to be driven into concrete, concrete block and other similar materials.

CONDUCTOR: Vertical pipe which connects the roof drain to the storm drain.

CONTINUOUS VENT: Upward continuation of the waste piping to produce a vent.

CONTINUOUS WASTE: Two or more fixtures connected to the same trap.

CONTROL STOP: Device installed in supply piping to regulate or shut off flow of water entering a flush valve.

COPPER PIPE STRAPS: Straps made from copper which are used to secure copper pipe to the structure of the building.

CORPORATION STOP: A valve installed in the building water service line at the water main; also called corporation cock.

COUNTERFLASHING: A flashing usually used on chimneys at the roof line to cover shingle flashing and to prevent moisture entry.

COUPLING: A pipe fitting containing female threads on both ends. Couplings are used to join two or more lengths of pipe in a straight run or to join a pipe and a fixture.

CPVC: Abbreviation for chlorinated polyvinyl chloride, a type of plastic used to make pipe that will carry hot water and chemicals.

CROSS: A pipe fitting with four female openings at right angles to one another.

CROSS CONNECTION: Any link between contaminated water and potable water in the water supply system.

CROSSOVER: Connection between two piping runs in the same piping system or the connection of two different piping systems which contain potable water.

CROWN OF A TRAP: The point in a trap where the direction of flow changes from upward to downward.

CROWN VENT: A vent which is connected to a trap at the crown.

CROWN WEIR: The point in the curve of the trap directly below the crown. This is the point at which the water level will normally remain when the fixture is not discharging through the trap.

CURB BOX: A cylindrical casting placed in the ground over the corporation stop. It extends to ground level and permits a special key to be inserted to turn off the corporation cock. Also called a "buffalo box."

CURB COCK: A control valve installed in house building service between the corporation stop and the structure. Also called a curb stop.

CRAWL SPACE: The space between the floor framing and the ground in a building which has no basement.

CRUDE SEWAGE: Untreated sewage.

CS: Abbreviation for Commercial Standard; a voluntary standard that establishes quality, methods of testing, certification, rating and labeling of manufactured items.

CU. FT.: Abbreviation for cubic foot or feet.

CU. IN.: Abbreviation for cubic inch or inches.

CW: Abbreviation for cold water.

D: Abbreviation for the word "penny," (always used as small letter d). Used to designate the length of nails.

DEAD END: A branch of a drainage piping system which ends in a closed fitting. A dead end is not used to admit water or air into the piping system.

DEEP SEAL TRAP: A trap generally located in the building drain for the purpose of resisting abnormal back pressure of sewer gas; also prevents loss of trap seal over long periods of nonuse.

DEG.: Degree (of a circle).

DEHUMIDIFIER: Device used to remove moisture from air in an enclosed space.

DEVELOPED LENGTH: Length of pipe and fittings measured along the centerline.

DF: Abbreviation for drinking fountain.

DIAMETER: Distance across a circle or cylinder.

DIE: A tool used to cut external threads by hand or machine.

DIE STOCK: Tool used to turn dies when cutting external threads.

DISPOSAL FIELD: See LEACH BED.

DOMESTIC SEWAGE: Sewage primarily from residential buildings; same as sanitary sewage.

DOPE: See PIPE JOINT COMPOUND.

DOUBLE HUB: Cast-iron sewer pipe having a bell on both ends.

DOUBLE JOISTS: Floor or ceiling framing members which have been doubled to provide added strength under partition walls and around openings.

DOWNSPOUT: Vertical pipe usually made from sheet metal or plastic which carries water from the gutters to the ground or to a storm drain.

DRAIN: Any pipe in the drainage piping system which carries waste water.

DRAIN, BUILDING: The nearly horizontal piping which connects the building drainage piping to the sanitary sewer or private sewage treatment system.

DRAINAGE FITTING: Any pipe fitting which is designed specifically to be used in drainage piping. The distinctive feature of this type fitting is the lip or shoulder on the inside of the fitting. When the pipe is installed the shoulder produces a smooth, unbroken interior which permits solid matter to flow through the pipe more easily.

DRAINAGE PIPING: All or any portion of the drainage piping system.

DRAINAGE SYSTEM: The complete set of pipe and fittings which carry waste water from the fixtures to the building drain.

DRAIN, COMBINED: The part of the drainage piping within a building which carries both sanitary sewage and storm water.

DRAIN, SUBSOIL: Part of the storm drain system which conveys ground water to the storm sewers. An example of a subsoil drain is the piping which is placed around the foundation to carry away ground water which might otherwise seep through the foundation wall.

DRAINS, STORM: Piping systems which carry subsoil water and rainwater from a building to the storm sewer.

DRIFT PLUG: A plug which is driven through a soft metal pipe to remove dents.

DROP L: An L pipe fitting which has "ears" permitting it to be attached directly to the building frame.

DROP T: A T pipe fitting with "ears" that permits it to be attached with screws directly to the building frame.

DRUM TRAP: A trap installed in the drainage piping. A vertically oriented cylinder, it has an inlet near the bottom and an outlet near the top. Because the cylinder holds water, it serves to prevent sewer gas from entering the building.

DRY VENT: Any vent which does not carry waste water.

DRY WELL: A well, frequently composed of a hole filled with aggregate, which is designed to permit water to seep into the ground. It is used to receive rain water and, sometimes, the effluent from a septic tank.

DUCTILITY: The property of a material which allows it to be formed into thin sections without breaking. Copper is more ductile than steel.

DWG: Abbreviation for drawing.

DWV: Abbreviation for drainage, waste and vent system.

EASEMENT: The right of a person, governmental agency, company or corporation to use land owned by another for some specific purpose (e.g., the right of public utility to install service through a person's property).

ECCENTRIC FITTING: A pipe fitting in which the centerline of the openings is offset.

EFFECTIVE OPENING: The cross-sectional area of the opening where water is discharged from a water supply pipe.

EFFLUENT: The outflow from sewage treatment equipment.

EIGHTH-BEND: A pipe fitting which causes the run of pipe to make a 45 deg. turn.

ELASTIC LIMIT: The greatest stress a piece of material can withstand and still return to its original shape when the stress is removed.

ELBOW: A pipe fitting having two openings which causes a run of pipe to change directions.

EROSION: The gradual wearing away of material as a result of abrasive action.

EVAPORATION: Loss of water (especially in a drainage trap) to the atmosphere.

EWC: Abbreviation for electric water cooler.

EXCAVATION LINES: Lines laid out on the job site (usually with lime) to indicate where digging for foundation and piping is to be done.

EXISTING WORK: That part of the plumbing system which is in place when an addition or alteration is begun.

EXPANSION JOINT: A joint which permits pipe to move as a result of expansion without breaking or damaging fixtures and fittings.

EXTRA HEAVY: A term used to designate the heaviest and strongest grades of cast-iron and steel pipe.

F: Abbreviation for Fahrenheit.

FALL: In pipe installation, the amount of slope given to horizontal runs of pipe.

FAUCET: A valve whose purpose is to permit controlled amounts of water to be obtained from the water pipe as needed; generally used at a fixture.

FEMALE THREAD: Any internal thread.

FERRULE: A cast-iron pipe fitting which, when installed in the bell of a cast-iron pipe, permits a threaded cleanout to close the opening.

FG: Abbreviaton for finish grade.

FIELD TILE: Short lengths of clay pipe which are installed as subsurface drains. Water enters through a gap at the joints and is carried away through the pipe.

FILL: Sand, gravel or other loose earth used to raise the ground level around a structure.

FINISHING: The third major stage of the plumbing process. It includes the installation of fixtures and other components which are exposed.

FINISHING NAIL: A nail with a small head designed to be countersunk and covered with putty so it is not exposed in the completed project.

FITTING GAIN: Amount of space inside a fitting required by a pipe.

FITTINGS: The parts of the piping system, (but not the valves) which serve to join lengths of pipe.

FIXTURE: A device such as a sink, lavatory, bathtub, water closet or shower stall. Attached to plumbing systems, it receives water from the water supply piping and provides a means for waste to enter the drainage piping.

FIXTURES, BATTERY OF: Any two or more similar fixtures which are served by the same horizontal run of drainage piping (for example, a group of lavatories in a public restroom connected to the same horizontal branch of the drainage piping system).

FIXTURES, COMBINATION: A fixture, such as a kitchen

sink/laundry basin, which is specifically designed to perform two or more functions.

FIXTURE BRANCH: The water supply piping which connects a fixture to the water supply piping.

FIXTURE DRAIN: Drainage piping including a trap which connects a fixture and a branch waste pipe.

FIXTURE SUPPLY PIPE: The water supply pipe which connects the fixture to the stub-out. This pipe is generally exposed within the finished structure and is often chrome plated soft copper so it can be bent easily.

FIXTURE VENT: A part of the DWV piping system. It connects with the drainage piping near the point where the fixture trap is installed and extends to a point above the roof of the structure. The purpose of the fixture vent is to permit air to enter the drainage piping.

FIXTURE UNIT: A means of rating the amount of discharge from a given fixture so that drainage piping is large enough to carry the required amount of waste. Also: A flow of 1 cfm.

FLASHING: Rust-resistant materials such as copper or stainless steel which are installed at joints between roofs and walls, roofs and chimneys to prevent water from entering.

FLANGE: A rim or collar attached to one end of a pipe to give support or a finished appearance.

FLANGE NUT: Connects flared copper pipe to a threaded flare fitting.

FLANGE UNIT: Union secured with nuts and bolts.

FLOAT BALL: Metal or plastic ball used to control inlet valve in water closet tanks.

FLOAT ARM: Thin rod threaded at each end which connects the float ball to the inlet valve of ball cock assembly.

FLOOD LEVEL: The point in a fixture above which water overflows.

FLOOR DRAIN: A fitting which is located in the floor (generally concrete) to carry waste water into the drainage piping.

FLOOR FLANGE: A fitting attached at floor level to the end of a closet bend so that the water closet can be bolted to the drainage piping.

FLUSH BALL: In a water closet tank assembly, the rubber ball-shaped closure which controls flow of water into the bowl.

FLUSH BUSHING: A pipe fitting used to reduce the diameter of a female threaded pipe fitting. A flush bushing has no shoulder and, therefore, is flush with the face of the fitting when it is installed.

FLUSH VALVE: A valve installed in the bottom of flush tanks of water closets and similar fixtures to control the flushing of the fixture. It releases water from the tank into the bowl. Also: A pressure-controlled valve which releases water from a supply pipe into a water closet bowl.

FLUSH VALVE SEAT: Opening between tank and bowl in a water closet, against which the flush ball is fitted.

FLUSHOMETER: A valve which permits a preestablished amount of water to enter a fixture, such as a water closet or urinal, for the purpose of flushing it. When a flushometer is installed, a flush tank is not needed.

FLUX: A chemical substance which prevents oxides from forming on the surface of metals as they are heated for soldering, brazing and welding.

FOLLOWER: The sleeve on a pipe die which aligns the die with the pipe.

FOOTING: The part of the foundation of a building which rests directly on the ground. The footing distributes the weight of the building over a sufficiently large amount of ground so that the building will not settle excessively.

FOOTING PADS: Separate sections of footing not under the foundation walls. Used to support columns in the structure.

FORCE CUP: A rubber cup attached to a wooden handle. Used for unclogging water closets and drains. Also called a plunger or "plumbers' friend."

FORCED AIR: Air blown by a fan from a furnace or air conditioner.

FORCED AIR FURNACE: Any furnace which uses a fan to circulate heated air.

FOUNDATION: That part of a building which is below the first framed floor. Includes the foundation wall and footing.

FOUNDATION DRAIN: Piping around base of foundation to collect ground water and convey it into a sump.

FPT: Abbreviaton for female pipe thread.

FREEZELESS WATER FAUCET: A water faucet designed to be installed through an exterior wall. It is made so that the valve seat is approximately 12 in. inside the wall.

FROSTLINE: The depth of frost penetration in soil. This depth varies in different parts of the country depending on the normal temperature range. Water supply and drainage pipes should be installed below this depth to prevent freezing.

FS: Abbreviation for federal specification.

FTG: Abbreviation for fitting.

FU: Abbreviation for fixture unit.

FURNACE: A central heating device exclusive of any duct work or piping.

GA (ga): Abbreviation for gage or gauge.

GAL: Abbreviation for gallon (231 cu. in.).

GASKET: Any semihard material placed between two surfaces to make a wateright seal when the surfaces are drawn together by bolts or other fasteners.

GALV: Abbreviation for galvanized.

GALVANIZED IRON: Iron which has been coated with zinc to prevent rust.

GATE VALVE: A valve which utilizes a disc moving at a right angle to the flow of water to regulate the rate of flow. When a gate valve is fully opened there is no obstruction to the flow of water.

GIRDER: A large beam of steel or wood which supports other framing members in the structure.

GLOBE VALVE: A spherically shaped valve body which controls the flow of water with a compression disc. The disc, opened and closed by means of a stem, mates with a ground seat to stop water flow.

GPD: Abbreviation for gallons per day.

GPM (or gal. per min.): Abbreviation for gallons per minute.

GRAB BAR: A metal bar installed near a shower stall, bathtub, or other fixture. Provides extra support for personal safety.

# Dictionary of Plumbing Terms

GRADE: Slope of a horizontal run of pipe. Also: Elevation of land after some phase of earthmoving.

GREASE TRAP: A drum type trap installed in the drainage piping from any fixture likely to receive large quantities of grease. Purpose is to separate grease from water by allowing the grease to float to the top.

GROUND KEY VALVE: A valve which controls the flow of water or other fluid with a tapered cylindrical plug rotating in the valve seat.

GROUND WATER: Water in the subsoil.

GV: Abbreviation for gate valve.

HACKSAW: A metal-cutting saw with a replaceable blade.

HANDLE PULLER: Tool for removing handles from faucets and valves.

HANGER: Support for pipe.

HB: Abbreviation for hose bib.

HD: (hd): Abbreviation for head.

HEADER: A water supply pipe to which two or more branch pipes are connected to service fixtures. Also: A framing member in the wall, floor, ceiling or roof framing which supports joists that have been cut off.

HEADROOM: Space between the floor and the lowest pipe, duct or part of the framing.

HORIZONTAL BRANCH: Any horizontal pipe in the DWV piping system which extends from a stack to the fixture trap. Also: Called a lateral.

HORIZONTAL PIPE: Any pipe which is installed so that it makes an angle of less than 45 deg. from level.

HOSE BIB: A water faucet made with a threaded outlet for the attachment of a hose.

HOUSE DRAIN: The horizontal part of the drainage piping which connects the DWV piping system within the structure to the sanitary sewer or private sewage treatment equipment.

HR. (hr.): Abbreviation for hour.

HUB: The enlarged end of a hub and spigot cast-iron pipe.

HUBLESS PIPE: See NO-HUB PIPE.

HUMIDIFIER: A device for adding moisture to the air in closed spaces.

HW: Abbreviation for hot water.

HYDRANT: Water supply outlet with a valve located below ground. Designed for obtaining relatively large quantities of water from the water supply piping. Hydrants supply water for fire fighting and sprinkling.

HYDRATED LIME: Slaked lime, an ingredient of mortar. Also: Called quick lime.

HYDRAULICS: Engineering science pertaining to the pressure and flow of liquids.

HYDRONICS: Practice of heating and/or cooling with water.

IAPMO: International Association of Plumbing and Mechanical Officials.

ICBOA: International Conference of Building Officials Association.

ID: Abbreviation for inside diameter.

IN. (in.): Abbreviation for inch.

INCREASER: A fitting installed in a vent stack before the stack goes through the roof. Its purpose is to enlarge the stack. This reduces the possibility of water vapor condensing and freezing to the point of closing the opening.

IPS: Abbreviation for iron pipe size.

INTERCEPTOR: Any device installed in the drainage piping to prevent the passage of grease or solid materials such as sand.

INTERCONNECTION: Any pipe which joins two or more water supply systems.

JOINT RUNNER: A tool composed of asbestos rope and a clamp used to aid the plumber in leading joints in horizontal runs of bell and spigot cast-iron pipe.

JOISTS: Horizontal framing member in the floor or ceiling which carry the load to the supporting walls.

KEEL: Colored marking crayon used for marking pipe.

K GRADE COPPER PIPE: Copper pipe suitable for installation underground.

KW (kw): Abbreviation for kilowatt.

L (l): Abbreviation for length.

LAV.: Abbreviation for lavatory.

LATRINE: A multiseat toilet which has a single water trough under several seats.

LAY OUT: The act of measuring and marking location of something.

LAYOUT: The arrangement of a house, room or part of a job.

LAVATORY: A fixture designed for washing hands and face. Generally installed in a bathroom.

LB. (lb.): Abbreviation for pound.

LEACH BED: System of underground piping which permits absorption of liquid waste into the earth. Also called a disposal field or leach field.

LEACHING WELL: See CESSPOOL.

L COPPER PIPE: A type of copper pipe which may be used to convey water above ground.

LEVEL: A tool used to determine if something is horizontal or vertical.

LINE LEVEL: A small, lightweight level designed to be hung from a string line to determine if the line is horizontal.

LITRE: A unit of volume measure in the SI metric system; equal to 61.02 cu. in.

LONG QUARTER BEND: A 90 deg. fitting with one section much longer.

LONG-SWEEP FITTING: Any drainage fitting which has a long radius curve at bends.

LOT LINE: The line(s) forming the legal boundry of a piece of property.

LPG: Liquefied petroleum gas; used for home heating.

MAIN WATER LINE: The large water supply pipe to which branches are connected.

MAIN SEWER: The large sewer to which the building drains of several houses are connected.

MALE THREAD: Threads on the outside of a pipe, fitting, or valve.

MALLEABLE IRON: Cast iron which has been heat treated to reduce its brittleness. Iron fittings are made from malleable iron.

MALLET: A soft-face hammer (rawhide or plastic) used to drive parts without damaging them.

MANHOLE: An opening in the sanitary or storm sewer system to permit access.

MANIFOLD: A pipe which has many outlets close together.

MASTER PLUMBER: A plumber licensed to install and to assume responsibility for contractual agreements pertaining to plumbing.

MASONRY BIT: A bit designed to drill holes in mortar, concrete and masonry.

MAX.: Abbreviation for maximum.

MCA: Mechanical Contractors Association.

METER STOP: A valve used on a water main between the street and a water meter; permits installation or removal of the meter.

MGD: Abbreviation for million gallons per day.

MI: Abbreviation for malleable iron.

MIN. (min.): Abbreviation for minute or minimum.

MITER BOX: Hardwood or metal saw guide in "U" shape. Sides are slotted to guide hand saw for 45 and 90 deg. cuts.

MIXING FAUCET: Separate faucets having a common spout permitting control of water temperature.

MOISTURE BARRIER: A material such as polyethylene plastic that retards the passage of vapor or moisture into walls or through concrete floors.

MORTAR: A mixture of portland cement, hydrated lime, sand and water used to bond joints in clay pipe or in bonding masonry.

MORTARBOARD: A board or table-like stand used by a mason at the job site. It holds mortar being used to bond masonry units.

MPT: Abbreviation for male pipe thread.

MS: Mild steel.

M TYPE: Lightest type of rigid copper pipe.

NAPHCC: National Association of Plumbing Heating and Cooling Contractors.

NBFU: National Board of Fire Underwriters.

NBS: National Bureau of Standards.

NEEDLE VALVE: Similar to a globe valve. It has a needle which seats into a small opening to control the flow of fluid.

NEOPRENE: A synthetic rubber with superior resistance to oils; often used as gasket and washer material.

NFPA: National Fire Protection Association.

NIPPLES: Short lengths of pipe (usually less than 12 in.) with male threads on both ends; used to join fittings.

NO-HUB PIPE: Soil pipe having smooth ends, without spigot or hub.

NOMINAL SIZE: The approximate dimension(s) of standard material. For example, 1/2 in. galvanized pipe is not actually 1/2 in. in diameter on either the inside or outside. However, because the inside dimension is near 1/2 in. it is referred to as 1/2 in.

NONBEARING WALL: A wall within a structure which supports no load other than its own weight.

NONRISING STEM VALVE: A type of gate valve in which the stem does not rise when the valve is opened.

NOZZLE: A fitting attached to the outlet of a pipe or hose which varies the volume of water and causes the shape of the stream of water to be changed to a spray of varying diameter.

NPS: Abbreviation for nominal pipe size.

OAKUM: Loosely woven hemp rope which has been treated with oil or other water-proofing agent. Used to caulk joints in bell and spigot pipe and fittings.

OC: Abbreviation for on center.

OD: Abbreviation for outside diameter.

OFFSET, FITTING: Fitting with two bends, one offsetting the other. Connects two parallel pipes.

O-RING: A rubber seal used around stems of some valves to prevent water from leaking past.

OUTSIDE WALL: Any wall of a structure exposed to the weather on one side.

OXIDIZED SEWAGE: Sewage which has been exposed to oxygen which causes organic substances to become stable.

OZ.: Abbreviation for ounce.

OVERFLOW TUBE: Vertical tube in water closet tank preventing overfilling of tank.

PACKING: A loosely packed waterproof material installed in the packing box of valves to prevent leaking around the stem.

PACKING NUT: Special nut holding the stem in a faucet or valve while compressing the packing.

PARTITION OR PARTITION WALL: An interior wall that divides spaces within a building. Generally, it does not support the structure above it.

PENNY: Measure of the length of nails. Abbreviated as "d."

PERIMETER HEATING: A method of installing central heating systems so that registers are placed on outside walls of the building under windows.

PETCOCK: Small ground key type valve used with soft copper tube.

PILOT LIGHT: A relatively small flame which burns constantly. Its purpose is to ignite the main supply of gas when a gas-fired heating or cooking unit is turned on.

PIPE DIE: A tool for cutting external pipe threads.

PIPE JOINT COMPOUND: Putty-like material used for sealing threaded pipe joints.

PIPE STRAP: A metal strap used for supporting or holding pipe in place.

PIPE, SOIL: A pipe for conveying waste which contains fecal matter (human waste).

PIPE, VERTICAL: Any pipe or part thereof which is installed in a vertical position.

PIPE, WASTE: A pipe which conveys only liquid and other waste, not fecal matter.

PIPE, WATER RISER: A water supply pipe which rises vertically from a horizontal pipe.

PIPE WRENCH: A wrench with adjustable, slightly curved, toothed jaws; designed to grip pipe firmly as pressure is applied to the handle. Also called a Stillson wrench.

PIPE, WATER DISTRIBUTION: Pipes which carry water from the service pipe to fixtures in the building.

PIPES, WATER SERVICE: That portion of the water piping which extends from the main to the meter.

PIPING: A generic term used to refer to all the pipes in a building.

PITCH: Degree of slope or grade given a horizontal run of pipe.

PLENUM: Chamber attached directly to a furnace which receives heated air. From this relatively large chamber, ducts carry the air to each of the registers.

PLUG: A pipe fitting with external threads and head that is used for closing the opening in another fitting.

PLUMB: Exactly perpendicular (vertical); at a right angle to the horizontal.

PLUMB BOB: A tool consisting of a weight suspended by a string. When allowed to hang freely, the string line will assume a position which is exactly vertical.

PLUMBER: A person trained and experienced in the skill of plumbing.

PLUMBERS' FRIEND: Plunger or force cup, a tool consisting of a rubber cup and handle. It is used under water to force blockage through sewer lines.

PLUMBERS' FURNACE: A heating source used to melt lead, heat soldering irons or melt solder. May be fueled by propane, gasoline or other petroleum based fuels.

PLUMBERS' SOIL: A mixture of glue and lampblack. Used in lead work to prevent lead from sticking to selected parts of the lead pipe and fittings.

PLUMBING: A general term which includes the methods, materials, fixtures and tools used in the installation, maintenance and alteration of piping, fixtures and appliances in: 1—Sanitary sewers. 2—Storm sewers. 3—DWV piping systems. 4—Water supply piping systems.

PLUMBING FIXTURES: Devices which receive water and discharge it and/or waterborne wastes into the DWV system.

PLUMBING INSPECTOR: A person authorized to inspect plumbing and drainage for compliance with the code for the municipality.

PLUMBING SYSTEM: All of the water supply and distribution pipes, plumbing fixtures and drainage pipe, building drains and building sewer which are part of a building.

PNEUMATICS: Study of compressible gases, their properties and reactions in containment.

POLYETHYLENE: Plastic used to make pipe and fittings primarily for gas piping. Also: Plastic sheet material used in the building trade as a vapor barrier and to protect building materials from the weather during construction.

POP-OFF VALVE: A safety valve which opens automatically when pressure exceeds a predetermined limit.

PORCELAIN: White ceramic material made of koalin (fine clay), quartz and feldspar; used for bathroom fixtures. When used as a finish for metal fixtures, it is called vitreous enamel.

POTABLE WATER: Water which is satisfactory for drinking and domestic purposes.

PPM (ppm): Abbreviation for parts per million.

PRECIPITATION: The total measurable amount of water received in the form of snow, rain, hail and sleet. It is usually expressed in inches per day, month or year.

PRESSURE HEAD: Amount of force or pressure created by a depth of one ft. of water.

PRESSURE REGULATOR: A valve which reduces water pressure in the supply piping.

PRIVATE SEWER: See SEWER, PRIVATE.

PROPANE: Hydrocarbon derived from crude petroleum and natural gas. Used as a fuel for plumbers' furnaces or torches.

PSI (psi): Abbreviation for pounds per square inch.

PSIG (psig): Abbreviation for pounds per square inch, gauge.

PUBLIC SEWER: A sewer that is publicly owned.

PUNCH LIST: A list made by the home builder or owner near the end of construction, indicating what must be done before the house is completely finished and ready for occupancy.

PUTTY: A soft, prepared mixture used to seal sink rims, water closet bases and other places where a sealant is needed.

PVC (polyvinyl chloride): A type of plastic used to make plumbing pipe and fittings for water distribution, irrigation and natural gas distribution.

PYTHAGOREAN THEOREM: A theorem is a mathematical truth or statement on which a formula is based. The Pythagorean theorem states that "the square of the hypotenuse (third side) of a right-angle triangle is equal to the sum of the squares of the other two sides." The formula: (side 1)$^2$ + (side 2)$^2$ = (side 3)$^2$ can be used to find the length of a diagonal pipe which must connect two parallel pipes at different levels.

QUARTER BEND: A drainage pipe fitting which makes a 90 deg. angle.

QT. (qt.): Abbreviation for quart.

RAD.: Abbreviation for radiator.

RADIANT HEATING: A method of heating which depends primarily upon heat being transferred by radiation. An example is an electric heating system installed in the ceiling plaster.

RADIATOR, HOT WATER OR STEAM: The room heating element which is connected to a hot water or steam boiler.

RAFTERS: Sloping framing members making up part of the roof. Rafters are designed to support both the live and dead loads of the roof.

RD: Abbreviation of roof drain.

REAMING: Removing the burr from the inside of a pipe which has been cut with a pipe cutter.

REAMER: A tool used in reaming.

RECOVERY RATE: Speed at which a water heater will heat cold water to desired temperature.

RED.: Abbreviation for reducer.

REDUCER: A pipe fitting having one opening smaller than the other. Reducers are used to change from a relatively large diameter pipe to a smaller one.

REFILL TUBE: Copper or rubber tube from ball cock to overflow tube in water closet assembly.

REINFORCING ROD (rerod): Embossed steel rods placed in concrete slabs, beams or columns to increase their strength.

REINFORCEMENT WIRE: Heavy woven wire placed in concrete to give added strength.

RELIEF VALVE: A pressure safety valve placed on water heaters, hot water and boiler tanks to relieve pressure when it exceeds a preset level.

RELIEF VENT (revent): A branch from the main vent which provides air to a trap that is some distance from the main stack.

RIDGE: The horizontal line formed by the intersection of two sloping roof surfaces.

RIGID COPPER TUBING: Hard copper pipe used when installing water lines, particularly where they can be seen.

RISE: In construction, the vertical distance from the top of the double plate to the top of the ridge.

RISER: A vertical water supply pipe extending from a horizontal water supply pipe to a fixture. Also: The vertical boards in a staircase which close the openings between the treads.

RISING STEM: A type of valve stem which moves up and down as the valve is opened and closed.

ROD: To agitate or tamp freshly poured concrete for purpose of removing air pockets and increasing density.

ROOF DRAIN: A drain installed in a flat or nearly flat roof to receive water and conduct it into a leader, downspout or conductor.

ROTATING BALL FAUCET: A single-handle faucet which controls water flow and temperature with a channeled rotating plastic ball. Holes in the ball are aligned with orifices for hot and cold water.

ROUGH-IN: Earliest stage of plumbing installation sometimes divided into two stages: 1—First rough brings water and sewer lines inside the building foundation. 2—Second rough is the installation of all piping which will be enclosed in the walls of the finished building.

ROUGH-IN MEASUREMENTS: Measurements which indicate where water supply and DWV piping must terminate to serve the fixtures which will be installed later.

RUN: One or more lengths of pipe which continue in a straight line. Also: In a roof, the horizontal distance from the centerline of the ridge to the outside of the wall framing.

S: Abbreviation for hydraulic slope (in inches per ft.).

SADDLE FITTING: A fitting used to install a branch from an existing run of pipe. First, a hole is made in the pipe. Then the saddle fitting is clamped to it so that the opening is inside the fitting.

SAFETY VALVE: A combination temperature and pressure relief valve generally installed in a hot water tank to prevent an explosion from overheating or excessive pressure inside the tank.

SAN.: Abbreviation for sanitary.

SANITARY SEWER: The piping system which carries away wastes.

SAND TRAP or INTERCEPTOR: A device designed to allow sand and other heavy particles to settle out before the water enters the water supply piping.

S & W: Abbreviation for soil and waste.

SANITARY SEWAGE: Water and waterborne waste containing human excrement as well as other liquid household wastes.

SANITARY SEWER: A sewer especially designed to carry sewage.

SANITARY T BRANCH: A drainage fitting having three openings and formed in the shape of a T.

SANITARY Y BRANCH: A drainage fitting shaped like a Y.

SBCC: Southern Building Code Congress.

SCAFFOLD: Any platform erected temporarily to support workers and materials while work is being done.

SCALE DRAWING: A drawing of any object which has been carefully reduced to a fraction of real size so that all parts are in the correct proportion.

SCUTTLE: A small opening in a ceiling providing access to an attic or roof.

SEAL OF A TRAP: The depth of water held in a trap under normal operating conditions.

SEC. (sec.): Abbreviation for second.

SECONDARY BRANCH: Any branch off the primary branch of a building drain.

SELF-SIPHONAGE: An unsafe condition in which water is drained from a trap causing the seal to be broken. The water normally in the trap is drawn out of the trap by a partial vacuum in the stack. Condition is corrected by installing a proper vent.

SEPARATOR: See INTERCEPTOR.

SEPTIC TANK: A watertight tank in a private waste disposal system which receives household sewage. Within the septic tank, solid matter is separated from the water before the water is discharged.

SERVICE BOX: See CURB BOX.

SERVICE L (street L): A 45 or 90 deg. elbow with external threads on one end and internal threads on the other.

SERVICE PIPE: The water supply pipe from the main in the street or other source of supply to the building served.

SERVICE T: A T with external threads on one end and internal threads on the other end and on the branch.

SEWAGE: All water and waterborne waste discharged through the fixture.

SEWER: A piping system designed to convey sewage.

SEWER, BUILDING (house sewer): Horizontal sewage piping which extends from the building to the sewer main.

SEWER, BUILDING STORM: The piping from the building storm drain to the public storm sewer.

SEWER, PRIVATE: A sewer owned and maintained privately. It may convey sewage from building(s) to a public sewer or to a privately owned sewage disposal system.

SEWER, STORM: A sewer used to carry rainwater, surface water or similar water wastes which do not include sanitary sewage.

SHUT-OFF VALVE: A valve installed in a waterline whenever a cut off is required.

SIDE OUTLET: An opening at the side of a fitting. A T or Y fitting having one side opening.

SIDE VENT: A vent connected to a drain at an angle of 45 deg. or less.

SILL COCK: A faucet used on the outside of a building to which the garden hose can be attached.

SINGLE LEVER FAUCET: Any of several types of washerless faucets using a single control and springs, balls or cartridges to control flow and temperature.

SINK: A fixture commonly used in a kitchen or in connection with the preparation of food. It holds a small amount of water for a variety of cleaning tasks.

SIPHONAGE: A partial vacuum created by the flow of liquids in pipes.

SIZE OF PIPE: Approximately equal to the inside diameter of the pipe. The nominal dimension by which the pipe is designated.

SLAB: A large, flat, concrete section such as a basement

floor, driveway or patio.

**SLIP COUPLING:** A pipe coupling which has no stop to prevent it from slipping over a pipe. Used to make watertight joints in plastic and copper pipe during a repair or alteration of the original piping.

**SLIP JOINT:** A connection in which one pipe slides inside another. The purpose of a slip joint is to permit pipes to expand and contract without breaking or to make assembly easier.

**SLIP NUT:** A nut used on P traps and similar connections. A gasket is compressed around the joint by the slip nut to form a watertight seal.

**SLIP-ON FLANGE:** A flange that slips onto the end of a pipe without threads and is welded or soldered in place.

**SLOP SINK:** A deeper fixture than an ordinary sink. Frequently installed in custodians' rooms.

**SOIL PIPE:** Any pipe which carries sanitary sewage. Also: Cast-iron drainage pipe with bell and spigot.

**SOIL STACK:** A general term for the vertical main of a DWV system.

**SOLDER:** Metal alloy composed of tin and lead used to join copper pipe and fittings.

**SOLDERING IRON:** A tool composed of copper which is heated in a furnace and used to melt solder when joining pieces of metal.

**SOLDER JOINT:** The means of joining copper pipe to slip-on fittings using solder.

**SPAN:** The horizontal distance between vertical supports of a beam, joist or arch.

**SPEC.:** Abbreviation for specification.

**SPECIFICATIONS:** A document which describes the quality of materials and work quality required for a given building. Specifications are the source of information about the quality of pipe, fixtures, etc., to be included in the plumbing system.

**SPIGOT:** The plain end of a cast-iron pipe. The spigot is inserted into the bell end of the next pipe to make a watertight joint.

**SPLASH GUARD:** A specially formed block which is placed under the outlet of a downspout to prevent erosion of the soil.

**SPLINE:** Projections on a shaft that are mated to a handle or wheel so that both will rotate as one.

**SPOUT:** End of a faucet which serves as a passageway for water.

**SPUD:** See CLOSET SPUD.

**SQ. (sq.):** Abbreviation for square.

**SQ. FT. (sq. ft.):** Abbreviation for square foot or feet.

**SQ. IN. (sq. in.):** Abbreviation for square inch or inches.

**SS:** Abbreviation for service sink.

**STD. (std.):** Abbreviation for standard.

**STACK:** A general term used for any vertical run of the DWV system.

**STACK VENT:** The vertical extension through the roof, including all of the DWV piping above the highest horizontal drain connected to the stack.

**STAR DRILL:** A tool made from steel which has a star shaped chisel on one end and a face which is hit with a hammer on the other end. This tool is used to make holes in concrete and masonry.

**STEAM HEATING:** Heating system in which steam from a boiler is piped to radiators in the rooms.

**STEM:** Shaft of a faucet which holds the washer and to which the handle is attached.

**STILLSON WRENCH:** See PIPE WRENCH.

**STOCK or DIE STOCK:** A tool which is used to turn a die when cutting threads.

**STOPCOCK:** A small ground key valve.

**STOP and WASTE COCK:** A valve which can be used to stop the flow of water in a pipe and permit the water downstream from the valve to be drained from the piping.

**STOPPER:** A plug which controls waste water drainage from a lavatory or bathtub. Usually controlled remotely by a handle on the fixture. Sometimes called a pop-up plug.

**STORM DRAIN:** A drain which conveys rainwater, subsurface water or other waste which does not need to be treated in a private or public sewage treatment facility.

**STORM SEWER:** A sewer used for carrying away water collected by storm drains. Generally conveys the water to a stream or lake.

**STORY:** That part of a building between any floor and the floor or roof immediately above.

**STRAP WRENCH:** Tool for gripping pipe. Strap is made of nylon web treated with latex.

**STREET L:** An elbow fitting with one male end and one female end. Same as a service L.

**STREET T:** A T with one female and one male threaded opening plus an outlet opening with female threads.

**STORM WATER:** The excess rainfall which runs off during or after a rain.

**STUD:** One of a series of vertical wood or metal structural members in walls and partitions.

**SUBFLOOR:** Rough floor consisting of boards or plywood panels applied directly over the floor joist.

**SUBSOIL DRAIN:** A drain which receives only subsurface water and conveys it to a storm drain.

**SUMP:** A tank or pit installed in the basement of a building to collect subsurface water so it can be pumped to a storm drain.

**SUMP PUMP:** Rotary type pump which lifts water from sump into drain pipe.

**SURVEY:** A description of a piece of property including the measurements and marking of land.

**SV.:** Abbreviation for service.

**SWAGE:** To increase or decrease the diameter of a pipe by using a special tool which is forced into or around the pipe.

**SWEAT SOLDERING:** Method of soldering in which the parts to be joined are first coated with a thin layer of solder, then joined while exposed to a flame.

**T:** Abbreviation for temperature.

**TAMP:** To firmly compact earth during backfilling.

**TAP:** A tool which is rotated by hand or machine to produce internal threads.

**TAPERED REAMER:** Tool for deburring and cleaning inside ends of pipes.

**TAPPED T:** A cast-iron T with at least one branch tapped to receive a threaded pipe or fitting.

T or T FITTING: A fitting shaped like the letter T. Each leg of the T can be joined to a pipe or another fitting.

THERMOSTAT: An automatic device consisting of a temperature sensing unit which turns an energy source on and off; used in heating and cooling.

TRAP: A drainage fitting which produces a water seal to prevent sewer gas from entering the building.

TRUNK LINE: The main piping from which building drains or water supply piping branch.

TUBING: Any thin-walled pipe which can be bent easily.

TUBING CUTTER: A tool used to cut tubing.

U or URN.: Abbreviation for urinal.

UNION: A fitting used to join two lengths of pipe. Permits disconnecting the two pieces of pipe without cutting.

UNIT VENT: One vent pipe which serves two or more traps.

V: Symbol for volume.

V (v): Symbol for valve.

VACUUM: Air pressure below atmospheric pressure.

VACUUM BREAKER: A device which prevents the formation of a vacuum in a water supply pipe. Installed to prevent backflow.

VALVE: A device which will control the flow of liquid within or from a pipe.

VALVE BODY: The main part of a valve into which the stem and other parts are installed.

VAPOR BARRIER: A material which prevents moisture from penetrating a wall, ceiling or floor. Roofing felt and polyethylene plastic sheets are commonly used for this purpose.

VENT: That part of the drain, waste, vent piping which permits air to circulate and protects the seals in traps from siphonage and back pressure.

VENT, CIRCUIT: Vent installed where two similar fixtures discharge into horizontal waste branch.

VENT, COMMON: A vent which serves two or more fixture traps.

VENT, LOOPED: Vent which drops below flood rim of fixture before being connected to main vent.

VENT, RELIEF: Vent installed at point where waste piping changes direction.

VENT STACK: The vertical portion of the vent piping which extends through the roof of the building.

VENT, WET: A pipe which serves as both a vent and a drain.

VENTING, INDIVIDUAL: Venting of each trap.

VIBRATOR: A tool used to remove air pockets from con-

crete as walls are poured.

VITRIFIED CLAY PIPE: Pipe made of clay and fired; generally used for sewers.

VTR: Abbreviation for vent through roof.

W: Abbreviation for waste.

WARM AIR HEATING: Any heating system which depends upon the circulation of warm air.

WASTE: Liquid discharged from a fixture. The liquid contains no fecal matter.

WASTE PIPE: A pipe which conveys liquid waste that does not contain fecal matter.

WATER CLOSET: A water flush plumbing fixture designed to receive human excrement and discharge it into the DWV piping. Sometimes called a toilet.

WATER CONDITIONER: A device used to remove dissolved minerals from water. Removal of the minerals frequently improves the taste of the water and reduces the likelihood of mineral deposits building up in the plumbing. An additional advantage is the fact that "soft" water requires less soap in laundering and generally cleans better.

WATER HAMMER: A banging sound in water supply pipes caused by sudden stopping of water flow.

WATER MAIN: Large water supply pipe, generally located near the street, which serves a large number of buildings.

WATER SUPPLY SYSTEM: All the piping and valves from the source of water to the point of use.

WC: Abbreviation for water closet.

WETTING ACTION: Reducing the tendency of a solid to repel a liquid flowing over its surface. Also: The act of reducing surface tension of a liquid to make it flow more readily.

WH: Abbreviation for wall hydrant.

WHITEPRINT: A drawing reproduced on white background with colored lines.

WORKING DRAWINGS: Drawings showing exactly how a building should be constructed.

XH: Abbreviation for extra heavy.

XXH: Abbreviation for double extra heavy.

Y or WYE BRANCH: A section of pipe which joins the main run of pipe at an angle. The fitting which makes the joint is in the shape of the letter Y.

YARNING IRON: A tool used to pack oakum into the bell and spigot pipe joints before they are leaded.

ZONING: Building restrictions which regulate the size, location and type of structures to be built within a specific geographic area.

# INDEX

# Index

# Index